Springer Transactions in Civil and Environmental Engineering

Springer Transactions in Civil and Environmental Engineering (STICEE) publishes the latest developments in Civil and Environmental Engineering. The intent is to cover all the main branches of Civil and Environmental Engineering, both theoretical and applied, including, but not limited to: Structural Mechanics, Steel Structures, Concrete Structures, Reinforced Cement Concrete, Civil Engineering Materials, Soil Mechanics, Ground Improvement, Geotechnical Engineering, Foundation Engineering, Earthquake Engineering, Structural Health and Monitoring, Water Resources Engineering, Engineering Hydrology, Solid Waste Engineering, Environmental Engineering, Wastewater Management, Transportation Engineering, Sustainable Civil Infrastructure, Fluid Mechanics, Pavement Engineering, Soil Dynamics, Rock Mechanics, Timber Engineering, Hazardous Waste Disposal Instrumentation and Monitoring, Construction Management, Civil Engineering Construction, Surveying and GIS Strength of Materials (Mechanics of Materials), Environmental Geotechnics, Concrete Engineering, Timber Structures.

Within the scopes of the series are monographs, professional books, graduate and undergraduate textbooks, edited volumes and handbooks devoted to the above subject areas.

The proposal for each volume is reviewed by the main editor and/or the advisory board. The chapters in each volume are individually reviewed single blind by expert reviewers (at least four reviews per chapter) and the main editor.

Bal Krishna Rastogi · Girish Chandra Kothyari ·
Khayingshing Luirei
Editors

Natural Hazards and Risk Mitigation

Natural Hazards in Himalaya and Risk Mitigation

Springer

Editors
Bal Krishna Rastogi
Indian Society of Earthquake Science
Hyderabad, Telangana, India

Girish Chandra Kothyari
University of Petroleum Engineering
and Energy Studies
Dehradun, Uttarakhand, India

Khayingshing Luirei
Wadia Institute of Himalayan Geology
Dehradun, Uttarakhand, India

ISSN 2363-7633 ISSN 2363-7641 (electronic)
Springer Transactions in Civil and Environmental Engineering
ISBN 978-981-97-7657-3 ISBN 978-981-97-7658-0 (eBook)
https://doi.org/10.1007/978-981-97-7658-0

This Springer imprint is published by the registered company Springer Nature Singapore Pte Ltd.
The registered company address is: 152 Beach Road, #21-01/04 Gateway East, Singapore 189721, Singapore

Preface

The Himalaya mountain chain faces frequent natural hazards like earthquakes and landslides, because of its young age, high tectonic activity due to continent–continent collision of Indian and Eurasian plates, uplifting mountains at 1 cm/y, soft rocks in large part and unstable setting; while floods/cloud bursts, snow avalanches and forest fires are due to high elevation, high rainfall and annual cycle of cold and warm climate causing high rate of weathering. Risk due to hazards is enhanced by unplanned human intervention. It is important to assess hazards in the Himalayas and learn methods of coping with them for sustainable development of the region. Different disasters in India on an average annually kill 5000 people and affect 60 million (World Disasters Report). Half of this loss is in the Himalayas. Floods in India alone annually cause a loss of 1500 human lives and 1 million houses despite of preparedness and advance warning measures (the Ministry of Agriculture, Government of India). Landslides are the perennial problem in the entire Himalayan belt and in a limited way in west as well as east coast regions. The causes and favourable conditions for landslides are discussed in some chapters. The country is divided into five zones on the basis of seismicity. Most of the Himalayan states, particularly the north-eastern states and some of the western part of the country lie in seismic zone 5, the highest degree of vulnerability for earthquakes. Four earthquakes exceeding magnitude 8 on the Richter scale occurred in this region since 1897. The book deals with four types of hazards, viz. earthquakes, landslides, floods and glaciers. Prediction of such hazards is not possible, except to some extent for floods, hence, preparedness measures need to be undertaken for earthquake, landslide and flood hazards.

The book is outcome of the presentations in the 2023 Symposium on '**Earthquake, Landslide and Glacial Hazards**' organized by Indian Society of Earthquake Science (ISES) and University of Kashmir at Srinagar. Chapter 1 by Rastogi gives salient points about geology and seismotectonics of India. Chapter 2 summarizes the locations and causes of earthquakes, landslides and floods in Himalaya. Tectonics and seismicity part covers Peninsular India also giving list of significant earthquakes. Important landslides and floods in Himalaya are listed. Chapter 3 by Sridevi Jade reviews the work on geodetic deformation and GPS measurements in Himalaya. About the landslides—Chap. 4 by Halder et al. describes geology and landslides in Himalaya and suggests that presence of chlorite rich clay inhibits landslides. Chib et al. (Chap. 5) give results of investigations of Toldi Landslide in Udhampur district of J&K. Neeraj et al. (Chap. 6) describe Stability Assessment of Rockfall-Prone Slides in Garhwal. Iftikhar et al. (Chap. 7) give a Statistical Model for Landslide Susceptibility Assessment in Kashmir. In Chap. 8, Manish Mehta assess Existing and Potential Changes in Himalayan Glaciers in Climate Change Prospective. Afroz et al. identify the factors which convert earthquake into disaster (Chap. 9). Nazia et al. (Chap. 10) describe seismic microzonation of an area in Gorakhpur district. Anurag et al. describe liquefaction potential of a bridge site in Bihar in Chap. 11. Wani et al. in Chap. 12 describe Time–Frequency Analysis of Strong Ground Motions from the 1994 Northridge Earthquake, while, in Chap. 13, they describe modelling of the Effect of Soil Structure Interaction on Response. Chapter 14 by Arjun Pandey and Jayangondaperumal estimate rupture lengths and slip of pre-historic, historic and contemporary large earthquakes in the Kumaon–Garhwal Central seismic gap of Himalaya providing geological constraints through the review of past descriptions as well as of paleoseismological investigations.

My co-editors are Dr. Girish Chandra Kothyari and Dr. Ashing Luirei. Dr. A. Vasanthi has done editing and formatting of several chapters. Reviewers, Prof. Sharad Purohit, Dr. S. P. Sati, Dr. Rakesh Kumar Dumka and Dr. Raghuveer Singh Negi, have done a thorough job. I am thankful to co-editors, reviewers and authors.

Hyderabad, India Bal Krishna Rastogi

Contents

About the Editors

Dr. Bal Krishna Rastogi formerly Head, Seismology at NGRI, Hyderabad and Director General Institute of Seismological Research (2006–2015) which he established, is presently President of Indian Society of Earthquake Science. Rastogi is M.Sc. and Ph.D. (Geophys.) from ISM, Dhanbad. He has worked in/visited for more than five years most prestigious institutes of the world, including University of South Carolina, University of Colorado, Boulder, IISEE, Tokyo, University of Tokyo, US Geological Survey, Menlo Park, University of Hamburg, Germany, Grenoble University IPSN, Paris and Institute of Physics Earth, Moscow. His significant contributions include (i) identification of Bombay High, the oil-bearing structure in 1968, (ii) development of seismic characteristics to differentiate reservoir-induced earthquakes from ordinary earthquakes, (iii) assessment of seismic hazard on regional basis and in specific areas, and (iv) characterization of intraplate earthquakes and their foreshock-aftershock patterns. Professor Rastogi is Editor-in-Chief of the Journal of Ind. Soc. Earthq. Sc., and he has edited three special volumes. He has a US patent for Earthquake Prediction method. Professor Rastogi has guided eight Ph.D. students and 100+ M.Sc. dissertations. He is author of the book *Dams and Earthquakes*, which is translated into Russian and Chinese and followed by engineers worldwide. He is Fellow of Ind. Geophysical Un. and Guj. Sc. Academy, and is recipient of 2014 Bharat Jyoti Award of In. Intl. Friendship Soc., New Delhi, 2014 SKOCH Order of Merit Award, 2013–2014 National e-Governance award, 2011 Indian Geophysical Un Decennial medal, ISM Distinguished Alumni Award, European Council Madam Curie Bursary, JICA Fellowship, and NSF Fellowship.

Dr. Girish Chandra Kothyari works as Senior Associate Professor, University of Petroleum and Energy Studies Dehradun from 2022. Earlier, he worked at the Institute of Seismological Research, Gandhinagar for 13 years. He has guided 16 M.Sc. dissertations, three Ph.D. students and is currently guiding four Ph.D. students. He is recipient of National Geoscience Award (NGA) 2017 Ministry of Mines Government of India in the field of applied geosciences and several other awards. He has authored or co-authored some 100 SCI research papers. His Google scholar citations are 1789, with h-index of 25.

Dr. Khayingshing Luirei is Senior Scientist at the Wadia Institute of Himalayan Geology, Dehradun. He has completed Ph.D. under Prof. P. D. Pant at Kumaun University on the topic "Active faults and neotectonics of Kumaun Himalaya including landslide hazard zonation." He is a recipient of Wadia Institute of Himalayan Geology best paper award 2010. Dr. Luirei has authored some 50 SCI research papers on active faults, neotectonics, slope stability and landslides. He has guided two Ph.D. students.

Chapter 1
Structure and Seismotectonics of the Himalaya: Some New Insights

B. K. Rastogi

Abstract Some new insights about the structure and seismotectonics of the Himalaya are given in this chapter including thick or doubling of the crust and the role of detachment surface, Low-Velocity Zones (LVZs), transverse faults and subducting ridges of the Indian plate in Himalayan earthquakes. Strain rates inferred from GPS are described for different sectors of the Himalaya–Myanmar–Andaman belt and estimates of return periods of large earthquakes are highlighted. Seismic gap areas in the Himalaya inferred from past large earthquakes are described.

1.1 Geology of the Himalaya

The world's highest and youngest Himalaya-mountain belt started forming some 50 million years ago, in the Eocene time of the Tertiary period (Valdiya 1973, 2010). The 2500 km long convex Himalaya from Kashmir to NE India has a width of about 320 km from the Indus–Tsangpo Suture (ITS) in southern Tibet to the Main Frontal Thrust (MFT) (contact between the mountains and plains). The Indian plate underthrusts northwards beneath Eurasia and the Himalayan wedge moves southwards generating a thick crust due to folding and thrusting along three E–W trending groups of mega-thrusts, viz. Main Central Thrust (MCT), Main Boundary Thrust (MBT) and Main Frontal Thrust (MFT) (Figs. 1.1 and 1.2). The mountains rise at a rate of ~1 cm/y.

ITS represents the initial boundary of the Indian and Eurasian plates as it is having an ophiolite chain. The thrust sheets are progressively younger, southwards. The Main Central Thrust (MCT) formed ~25–18 My ago demarcates the High Himalayas in the North. The Main Boundary Thrust (MBT) formed 10 My ago delimits the metamorphic rocks of the Lesser Himalayas to the south and the Main Frontal Thrust (MFT) which is surface expression of MHT. MCT and MBT were formed over the Main Himalayan Thrust (MHT). MHT is the upper surface of the subducting Indian plate over which the Himalayan rocks move south. It is also called a detachment

B. K. Rastogi (✉)
President Ind. Society of Earthquake Science, 1-2-63/3 Habsiguda, Hyderabad 500 007, India
e-mail: bkrastogi12@gmail.com

B. K. Rastogi et al. (eds.), *Natural Hazards and Risk Mitigation*, Springer Transactions in Civil and Environmental Engineering, https://doi.org/10.1007/978-981-97-7658-0_1

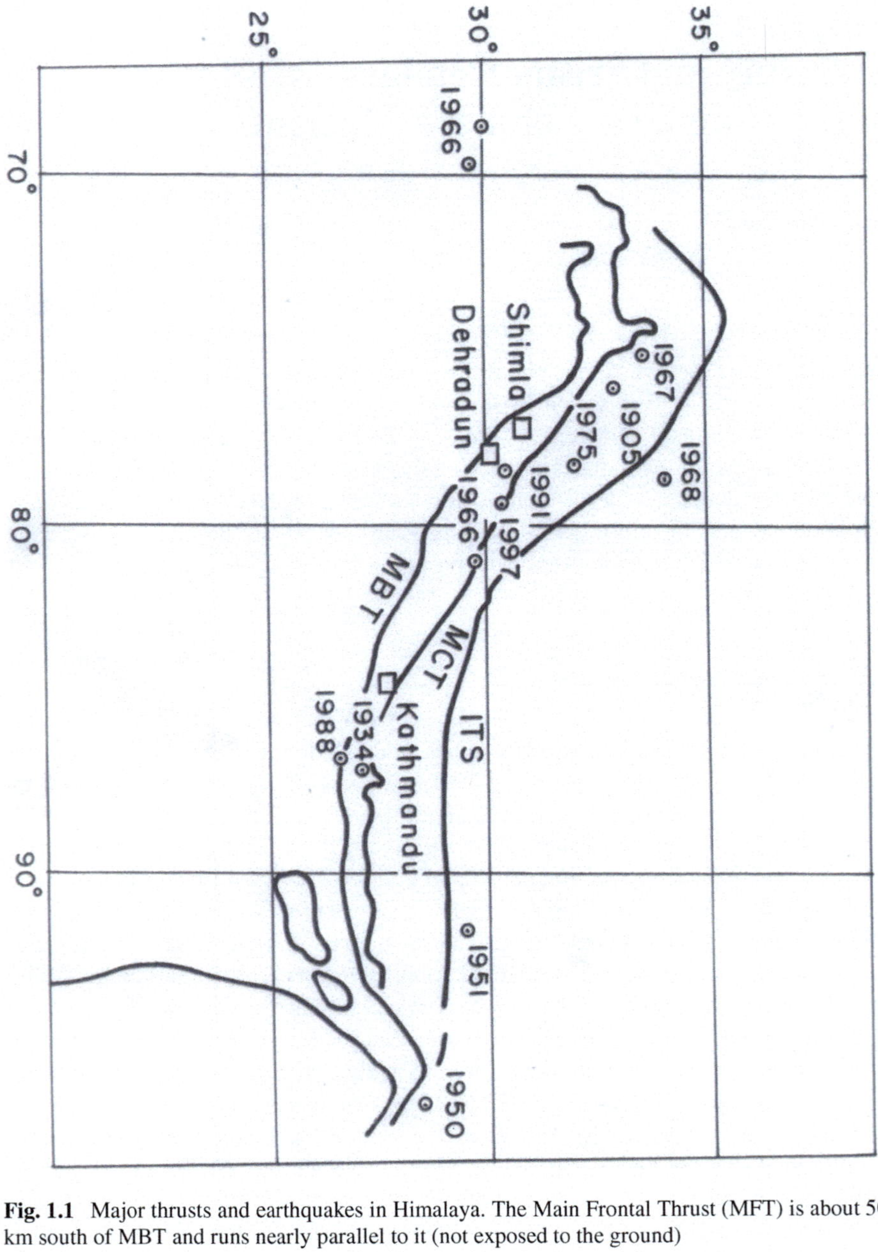

Fig. 1.1 Major thrusts and earthquakes in Himalaya. The Main Frontal Thrust (MFT) is about 50 km south of MBT and runs nearly parallel to it (not exposed to the ground)

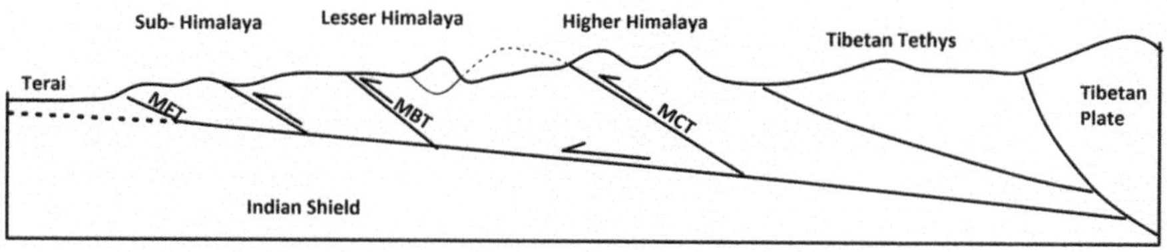

Fig. 1.2 A schematic section across Himalaya

surface or slip surface. Convergence of the Indian and Eurasian plates is accommodated through slip on the detachment surface or the Main Himalayan Thrust (MHT) causing large shallow earthquakes (Seeber and Armbruster 1981; Lave and Avouac 2001). Smaller shocks or aftershocks occur along MBT and MCT.

South of ITS and north of Higher Himalaya (HH) are Tethyan sediments of Cambrian to Eocene age which are low-grade metamorphic rocks. HH consists of high-grade metamorphic crystalline rocks (nappes and klippes of gneiss and schist, etc.) extending southward up to the Main Central Thrust (MCT). Some 80–200 km of shortening occurred between ITS and MCT. South of MCT is the Lesser Himalaya (LH) zone of less metamorphosed rocks extending south to MBT. MBT brought Lesser Himalayan rocks southwards over the Sub-Himalaya (SH) belt of Neogene to Quaternary Siwalik foreland sediments overlying the Main Frontal Thrust (MFT).

1.2 Crustal Structure in the Himalayan Region

In this section, we describe some new insights on detachment surface or Main Himalayan Thrust (MHT), thickening or doubling of the crust and Low-Velocity Zones (LVZs) or layers. Salient results of the study of Earth's structure by NGRI groups (Manglik et al. 2021) and other researchers are summarised.

1.2.1 Detachment Surface

Most of the moderate and large earthquakes like those at Uttarkashi and Chamoli are inferred to have occurred along the detachment surface MHT which coincides with the upper surface of the subducting Indian lithosphere (Caldwell et al. 2013; Mandal et al. 2000). This detachment surface is gently dipping (less than ~10–15°) under the Lesser Himalaya and south of the Vaikrita thrust (a branch of MCT). The Vaikrita thrust marks the line separating the gentle to steep-dipping MHT in the north. The steeper-dipping MHT is aseismic.

1.2.2 Thick or Doubling of the Crust

Thick crust of 65–70 km below the Himalaya and Tibet Plateau region was revealed by Gupta and Narain (1967) by surface wave dispersion studies, and now generally believed so (Mitra et al. 2015), but Lesser Himalayan zone is believed by many to form a duplex (Srivastava and Mitra 1994; Monsalve et al. 2006). Double Moho has been detected in Uttarakhand, Nepal and Sikkim and may exist for a length of about 1200 km or more.

Mandal et al. (2023) found double Moho between 35 and 60 km depths in Kumaon–Garhwal Himalaya by Seismic Imaging. Priestley et al. (2008) found Moho depth of 40–70 km in Nepal and Tibet. Monsalve et al. showed doubling of the crust for earthquakes in eastern Nepal and Tibet. The focal depths within Nepal are down to 60 km increasing to 80 km in Tibet. The structure is like two MHTs: one shallow from a depth of about 5 to 40 km and the other deeper from 45 to 75 km.

In Kashmir–Ladakh–Karakoram region, Moho depth is around 50 km at the foothills, 60–65 km beneath the Higher Himalaya and 75 km at Karakoram (from seismological methods by Kaila et al. (1984) and Rai et al. (2006) and by gravity methods by Mishra (1982) and Chamoli et al. (2011)).

In the Higher Himalaya of Sikkim area, crustal thickness is found to be 61 km forming a Moho doublet (Singh et al. 2010). In the north-eastern Himalaya region, crustal thickness is found to be 50 km or more (Priestley et al. 2008; Kumar et al. 2004).

1.2.3 Low-Velocity Zones or Layers

Mid-crustal low-velocity layers or zones (LVZ) have been detected at several sectors of Himalaya. The upper part of the subducting Indian plate is considered as the detachment surface or Main Himalayan Thrust (MHT). The LVZ lies over the MHT. Such a low-velocity layer was inferred in Kashmir–Ladakh–Karakoram region by Caldwell et al. (2009, using the data of Rai et al. (2006) from surface wave dispersion, by Rai et al. (2009) from Lg attenuation and by Gilligan et al. (2015) from joint inversion of P-RFs and surface-wave dispersion. Singh et al. (2010) using P-RF technique found a mid-crustal low-velocity layer in Sikkim area. Patro and Harinarayana (2009) using MT data delineated some high conductivity zones at the mid-crustal levels in Sikkim area and attributed these to the presence of metamorphic fluids.

Kanna and Gupta (2021) from P-RFs study detected an intra-crustal LVL in the NW Himalaya. For the Kumaun–Garhwal Himalaya, P-wave tomography indicated occurrence of saline-rich aqueous fluid/partial melt in the upper-mid crust lying over the Main Himalayan Thrust (MHT) (Sandeep Gupta et al. 2022). Mandal et al. (2023) found that the MHT in Uttarakhand is associated with a low-velocity (~10–20% low) north-dipping (~2°) zone at depths 8–20 km from seismic velocity tomography.

They found lower crust at 20–40 km having lower velocities and opine that the low-velocity pockets in the lower crust may be indicative of partial melts, while those in the upper and middle crust may be indicative of aqueous or metamorphic fluids. They also suggested that the 1803 Mw7.8 Garhwal earthquake, the 1991 Mw6.8 Uttarkashi earthquake, and the 1999 Mw6.5 Chamoli earthquake occurred in low-velocity zones on the MHT, which might be accumulating crustal stresses and the fluids or high pore-fluid pressure might have aided the nucleation of these large earthquakes.

Low-velocity layers were detected in the epicentral region of Mw 6.4 Killari earthquake at 5–6 km depth (Gupta et al. 1998) and for 2001 Mw 7.7 earthquake of Kachchh two LVL were detected one at 1–3.7 km depth and the other between 30 and 40 km depth (Mandal et al. 2004).

1.2.4 Sedimentary Basins

Sedimentary basins have high amplification of seismic wave amplitude, velocity and acceleration. The detailed studies done for Kashmir and Indo-Gangetic plains are described here. Bhukta et al. (2006) found a four-layered sedimentary basin in Jammu region having a maximum depth of 4.7 km south of the MBT and about 2.0 km thick north of the MBT. In Kashmir region, the basin has three layers. Kaila et al. (1968) found a crustal thickness of 28 ± 8.4 km, including the top sedimentary layer of 6 ± 1 km beneath the Indo-Gangetic plains and the Himalayan foothills using shallow earthquake travel time data. A 100 km long DSS profile revealed the detachment surface at 6–8 km depth in this area (Rajendra Prasad et al. 2011).

1.3 Seismotectonics of Himalaya

In this section, we describe some new insights on seismotectonics of Himalaya, role of transverse faults and influence of subducting ridges of the Indian plate as well as asperities on earthquakes.

Himalayan belt is one of the most seismically hazardous regions on Earth due to the Indian plate underthrusting northwards beneath Eurasia and producing continental collision of Indian and Eurasian plates. MHT accommodates the convergence. Beneath it lies the underthrusting Indian plate and above are the overriding mountains. The plates are converging at a relative rate of 50 mm/yr, nearly half of which (~20 mm/yr) is accommodated in Himalaya and rest north of it (Bilham 2004; Ader et al. 2012). Diffused but high seismicity occurs in a 200 km wide belt by thrusting over the Main Himalayan Thrust (MHT) in a densely populated region.

The northern deeper part of the MHT is creeping smoothly or slipping aseismically as the zone is warmer (> 350 °C) causing aseismic sliding (smooth creep). The shallower southern part of the MHT, south of MCT and lying under the Sub- and

Lesser Himalaya has a low-dip angle of about 10°, is frictionally locked as the rocks at 4–20 km depths are colder and brittle, is seismogenic and slips episodically. It accumulates strain during the inter-seismic period when it is locked and is released during the earthquakes through sudden slip (Bilham et al. 2001) when strain exceeds the rock failure stage. At this time, strain is estimated to be ~10^{-4} by Mitra et al. (2015).

From the rupture zones and amount of slip of past earthquakes, the areas of deficit convergence have been identified. It is observed that the rupture zones of large earthquakes in Himalaya had been about 150–200 km long. From a convergence rate of 2 cm/yr, a strain enough for 4 m slip good enough for a great earthquake may be accumulated every 200 years in different sectors of Himalaya. From the great earthquakes since 1500, it can be assessed that 2/3rd portion of Himalaya has potential to produce a few great earthquakes (Bilham et al. 2001). Three most potential regions are proposed for future great earthquakes, viz. Kashmir to NW Himalaya in the western sector, Uttarakhand to western Nepal in the central sector and the NE Himalaya.

Gupta et al. (1998) analysed 40 major and great earthquakes from Himalaya–Myanmar region recorded at the Gottingen observatory during 1905–1976. Several other major earthquakes have occurred afterwards in Myanmar, Kashmir, Nepal, etc. Great and major earthquakes in the Himalayan region are known to have occurred in 1803, 1833, 1897, 1905, 1934 and 1950. Little information is available on Himalayan earthquakes in the eighteenth century and before. The list compiled from limited damage description and palaeoseismology includes the earthquakes of 844, 1123, 1501, 1555, 1669, 1736, 1779, 1784, 1828, 1885 and 2005 all in Kashmir, while during 1400 in western Himalaya. In Nepal, the list includes those in 1223, 1225, 1344 and 1505 (Rastogi 2024).

The focal depths are mostly from near surface to 23 km in Uttarakhand, 5–40 km with concentration at 15–35 km close to the ramp of MHT in Nepal. LVZs lying over the MHT aid in the earthquake generation process. Earthquakes could be a little deeper in the western and eastern parts of Himalaya. Earthquakes down to depths of 60–70 km occur along transverse faults. Two such examples are 1988 Nepal earthquake and 2011 Sikkim earthquake. In Sikkim Himalaya, the depth of seismicity is more than 35 km associated with a NW–SE-oriented strike-slip motion. Kayal et al. (1993) and De and Kayal (2003) opine that these may be due to doubling of the crust or presence of transverse ridges.

Most of the fault plane solutions in Himalaya indicate thrusting along shallow plane trending along the thrust belt and slip vectors perpendicular to it (Rastogi 1974; Chandra 1977, 1978; Molnar et al. 1973, 1977; Ni and Barazangi 1984).

1.3.1 *Influence of Subducting Transverse Ridges and Transverse Faults on Seismicity*

Himalaya has some transverse strike-slip faults (up to lower crust at intermediate depth). Rastogi (1974) suggested that the transverse faults are boundaries of different underthrusting blocks. Rastogi (2018) inferred 1988 Udaipur Nepal earthquake of magnitude 6.8 to be along a transverse NE trending strike-slip Koshi River fault which might be extension of the eastern boundary fault of the Munger–Saharsa Ridge. In Sikkim Himalaya, the depth of seismicity is more than 35 km associated with a NW–SE oriented strike-slip motion. Rao et al. (2015) analysed all such transverse faults along the Himalayan arc from Kumaun to Sikkim and Bhutan and further east covering the Sagaing fault and the Burmese arc. They found a consistent right-lateral slip for these faults.

The transverse structures acting as boundaries of thrusting blocks may have segmented the lithosphere below the MHT in Himalaya, reducing rupture lengths for future great earthquakes ($M > 8$). Mandal et al. (2023) found mafic crustal thinning extending down to the lithosphere–asthenosphere boundary associated with northward extensions of the three transverse lithospheric structures of Delhi–Haridwar ridge, Moradabad fault, and Great Boundary fault. These transverse ridges/faults have possibly influenced the 1803 Mw7.8 Garhwal earthquake, the 1991 Mw6.8 Uttarkashi earthquake and the 1999 Mw6.5 Chamoli earthquake.

1.3.2 *Asperities*

Abrupt displacement of faults/thrusts or junction of faults or sudden change in structure or intrusives are the possible areas favouring accumulation of stresses and are prospective sites of large earthquakes. Low-velocity layers or zones are one major type of asperity which becomes the nucleation point for large earthquakes.

In Koyna region, junctions of faults were identified as locales of stress accumulation and Mw 5 + earthquakes (Talwani 1997). The whole Nepal Himalayan zone is protruding south and could favour accumulation of stress. Himalayan thrusts have several sharp bends. In Kangra area, the MBT is displaced by 70 km along a prominent bend. Such areas need special attention. A wide high-density mafic intrusive body in Kachchh region is attributed to be a favourable structure for stress/strain accumulation giving rise to large earthquakes (Mandal et al. 2004).

Dip of the detachment ramp has a sudden change from gentle (locked) to steep (creeping) dip at a junction near MCT. Within the gentle dip portion of the detachment also there is a junction north of MBT where dip increases. These junctions are favourable locations for higher seismicity. Aftershocks of the 2015 Nepal earthquake were more concentrated in these two zones.

1.4 Slip Modelling of Some Large Earthquakes

1.4.1 Slip Modelling of the Mw 6.8 1991 Uttarkashi Earthquake in Garhwal Himalaya, Uttarakhand

For the Mw 6.8 1991 Uttarkashi earthquake, rupture was found to have propagated towards west by forward modelling of the strong motions recorded at six stations (Cotton et al. 1996; Mandal et al. 2000). Inversion of the accelerograms gave slip distribution on an inferred fault plane of about 36 km × 48 km. The maximum slip of 1.5 m occurred 10 km west and 15 km south-west of the hypocentre. The inferred slip source–time function showed a low moment release at the beginning of the rupture and maximum 4 s after. Most of the seismic moment was released between 2 and 8 s with a total rupture duration of 9–11 s. Similar results were obtained with teleseismic records.

1.4.2 Slip Modelling of the Mw 6.5 1999 Chamoli Earthquake in Kumaon Himalaya, Uttarakhand (About 60 km East of Uttarkashi Earthquake Epicentre)

A rupture (detachment) surface dipping at 15° to NNE at depths of 10–16 km was inferred at a depth of 15 km by forward modelling of the strong motions (Rastogi, unpublished work). Rupture propagated west and southward for 35 km with maximum slip ~1 m. Rupture duration was 15 s increasing in nearly concentric rings.

1.4.3 Slip Modelling of Nepal 2015 Earthquake

Galetzka et al. (2015) modelled kinematics of source process of the 2015 earthquake with a slip pulse of width ~20 km, duration ~6 s and sliding velocity of 1.1 m/s. A large ~5 m slip-weakening distance was inferred from the smooth onset of source–time function indicating lower fault friction resulting into moderate ground shaking (~0.16 g) at high frequencies and limited damage to houses. Using the data from InSAR, strong motion accelerographs and teleseismic seismographs they suggested that the rupture initiated at the locked part of the MHT near Nepal–Tibet border at a depth ~10 km. It propagated south-eastward and up-dip along the MHT and ended at some depth north of the Main Boundary Thrust (MBT) (not breaking the surface). Slip estimated by Yagi from teleseismic broadband velocity data is 7.5 m near Kathmandu.

1.5 Geodetic Measurements

Global Positioning System (GPS) measurements show that India and Eurasia converge at 10–26 mm/year at surface in different segments of Himalaya (Bilham 2019; Jade et al. 2017). The slip rates are 13–20 mm/yr at 10–20 km depths along MHT. The slip rates are not steady as the large earthquakes are associated with sudden and large slips. About 100–150 km wide zone beneath Sub-Himalaya and Lesser Himalaya is locked and the zone further north up to the Tibet border is sliding. The locked portion slips episodically when the strain reaches elastic limit. Only half of the India–Eurasia plate convergence is accommodated in Himalaya, while the other half is accommodated in Tibet. Site velocities with reference to Indian plate in these regions vary gradually from nil in southern part of the Himalaya to about 16 mm/year in the northern part in the Higher Himalaya indicating high strain accumulation. GPS measurements from Jammu–Kashmir and Karakoram region (Mohanty and Gahalaut 2023) showed gradual increase in velocity from southwest to northeast suggesting accumulation of strain on the underlying seismically active detachment (MHT). The width of locked zone is estimated to be 175 km with an oblique convergence of 13.6 mm/yr in Kashmir Himalaya. The strike-slip Karakoram fault slips at average rate of 5 ± 1 mm/yr.

Convergence rates and strain rates are computed from GPS stations velocities in different sectors of Himalaya (Banerjee and Burgmann 2002; Bilham et al. 1997; Bettinelli et al. 2008; Gahalaut 2008; Jade 2004; Jade et al. 2017; Mukul et al. 2010; Yadav et al. 2019). The convergence rates at surface in different sectors of Himalaya are—Kashmir Himalaya: 13.6 mm/yr (Mohanty and Gahalaut 2023), NW Himalaya: 15 mm/yr (Banerjee and Burgmann 2002), Garhwal-Kumaon: 10–18 mm/yr (Jade 2004; Jade et al. 2017), Western Nepal: 20 mm/yr (Ader et al. 2012), Central & Eastern Nepal: 18 mm/yr (Ader et al. 2012), NE India: 16 mm/yr (Jade 2004), Sikkim: 10–12 mm/yr (Jade 2004) Shillong Plateau: 11 mm/yr (Banerjee et al. 2008). Slip rate along MHT is estimated from surface convergence rate through inverse modelling. Geodetic strain rate and slip rate along MHT are used to estimate the accumulated geodetic slip potential, from which the seismic slip due to past earthquakes is subtracted to get the balance slip potential (probable magnitude of a large future earthquake) for different sectors of Himalaya. Seismic potential for some sectors in near future estimated by Jade et al. (2017) are given below:

Kashmir–Ladakh–Himachal: geodetic strain rate 32.4 ns/yr and Mw 8.2 Garhwal–Kumaun: geodetic strain rate 96.2 ns/yr and Mw 8.6 Nepal: geodetic strain rate 73.29 ns/yr and Mw 7–9.

1.6 Conclusions

The new insights about structure and seismotectonics in the Himalaya–Myanmar–Andaman belt are described. The crust is found to be 60–80 km thick with indication of duplex crust in some sectors. The southern shallow and gentle dipping part of the detachment surface is locked and slips episodically in large earthquakes, while the northern deeper, steeply dipping part is creeping. Low-velocity layers or zones (LVZ) are detected over the detachment surface. LVZs detected in upper crust are attributed to saline-rich aqueous fluid resulted from metamorphism, while the LVZs detected in lower crust of a few areas may be due to partial melts. The areas of deficit convergence with possibility of future large earthquakes have been identified from the rupture zones and amount of slip of past earthquakes. Each 200 km long sector may have return period of 200–400 years if convergence rate is 2–1 cm/yr for 4 m slip. The prominent seismic gap areas are Kashmir to NW Himalaya in the western sector, Uttarakhand to western Nepal in the central sector and the NE Himalaya. Asperities like low-velocity zones, sudden structural variations and fault junctions are ascribed to be the nucleation points for large earthquakes.

References

Ader T, Avouac JP, Liu-Zeng J, Lyon-Caen H, Bollinger L, Galetzka J, Genrich J, Thomas M, Chanard K, Sapkota SM, Rajaure S, Shrestha P, Ding L, Flouzat M (2012) Convergence rate across the Nepal Himalaya and Interseismic coupling on the Main Himalayan Thrust: Implications for seismic hazard. J Geophys Res, Solid Earth, 117:B04403, https://doi.org/10.1029/2011JB009071

Banerjee P, Burgmann R (2002) Convergence across the northwest Himalaya from GPS measurements. Geophys Res Lett 29(13):1652. https://doi.org/10.1029/2002GL015184

Banerjee P, Burgmann R, Nagarajan B, Apel E (2008) Intraplate deformation of the Indian subcontinent. Geophys Res Lett 35: L18301.https://doi.org/10.1029/2008GL035468

Bettinelli P, Avouac J-P, Flouzat M, Bollinger L, Ramillien G, Rajaure S, Sapkota S (2008) Seasonal variations of seismicity and geodetic strain in the Himalaya induced by surface hydrology. Earth Planet Sci Lett 266:332–344

Bhukta SK, Sain K, Tewari HC (2006) Crustal structure along the lawrence Pur-astor profile in the Northwest Himalayas. Pure Appl Geophys 163:1257–1277

Bilham R (2004) Earthquakes in India and the Himalaya: tectonics, geodesy and history. Ann Geophys 47(2):839–858

Bilham R, Gaur VK, Molnar P (2001) Himalayan seismic hazard. Science 293:1442–1444

Bilham R, Larson K, Freymueller J (1997) GPS measurements of present-day convergence across the Nepal Himalaya. Nature, 386:61–64

Bilham R (2019) Himalayan earthquakes: a review of historical seismicity and early 21st century slip potential. In: Treloar PJ, Searle MP (eds) Himalayan tectonics: a modern synthesis, vol 483. Geological society, London, Special Publications, pp 423–482. First published online February 5, 2019. https://doi.org/10.1144/SP483.16

Caldwell WB, Klemperer SL, Rai SS, Lawrence JF (2009) Partial melt in the upper-middle crust of the northwest Himalaya revealed by Rayleigh wave dispersion. Tectonophysics 477:58–65

Caldwell WB, Klemperer SL, Lawrence JF, Rai SS, Ashish (2013) Characterizing the main Himalayan thrust in the Garhwal Himalaya, India with receiver function CCP stacking. Earth Planet Sci Lett 367:15–27

Chamoli A, Pandey AK, Dimri VP, Banerjee P (2011) Crustal configuration of the Northwest Himalaya based on modelling of gravity data. Pure Appl Geophys 168:827–844

Chandra U (1977) Earthquakes of Peninsular India- a seismotectonic study. Bull Seism Soc Am 67:1387–1413

Cotton F, Campillo M, Deschamps A, Rastogi BK (1996) Rupture history and seismotectonics of the 1991 Uttarkashi, Himalayas earthquake. Tectonophysics 258:35–51

De R, Kayal JR (2003) Seismotectonic model of the Sikkim Himalaya: Constraint from micro earthquake surveys. Bull Seism Soc Am 93(3):1395–1400

Gahalaut VK (2008) Major and great earthquakes and seismic gaps in the Himalayan arc. Mem Geol Soc India 66:373–393

Galetzka et al. (2015) Slip pulse and resonance of Kathmandu basin during the 2015 Mw7.8 Gorkha earthquake, Nepal imaged with geodesy. Science Express:9

Gupta HK, Narain H (1967) Crustal structure in the Himalayan and Tibet Plateau region from surface wave dispersion. Bull Seism Soc Am 57:235–248

Gupta HK, Rastogi BK, Rao CVRK, Mohan I, Sarma SVS, Rao RUM (1998) An investigation of Latur Earthquake of September 29, 1993 in South India. Tectonophysics 287:299–318

Gupta S, Mahesh P, Kanna N, Sivaram K, Paul A (2022) 3D seismic velocity structure of the Kumaun-Garhwal (Central) Himalaya: insight into the Main Himalayan thrust and earthquake occurrence. Geophys J Intl 229:138–149

Jade S (2004) Estimates of plate velocity and crustal deformation in the Indian subcontinent using GPS geodesy. Curr Sci 86(10):1443–1448

Jade S et al (2017) India Plate angular velocity and contemporary deformation rates from continuous GPS measurements from 1996 to 2015. Nat Sci Rep 7:11439. https://doi.org/10.1038/s41598-017-11697-w

Kaila KL, Reddy PR, Narain H (1968) Crustal structure in the Himalayan foothills area of north India from ‘P’ wave data of shallow earthquakes. Bull Seismol Soc Amer 58:597–612

Kaila KL, Tripathi KM, Dixit MM (1984) Crustal structure along Wular lake-Gulmarg-Naoshera profile across Pir Panjal range of the Himalayas from deep seismic soundings. Jour Geol Soc India 25:706–718

Kanna N, Gupta S (2021) Seismic crustal shear velocity structure across NW Himalaya and Ladakh-Karakoram using receiver function modelling: Evidence of the Main Himalayan Thrust. Phys Earth Planet Int 311:106422

Kayal JR, Reena De, and Chakraborty P (1993) Microearthquakes at the main boundary thrust in eastern Himalaya and the present-day tectonic model. Tectonophysics, 218:375–381

Kumar MR, Raju PS, Devi EU, Saul J, Ramesh DS (2004) Crustal structure variations in northeast India from converted phases. Geophys Res Lett 21:L17605. https://doi.org/10.1029/2004GL020576

Lave J, Avouac JP (2001) Fluvial incision and tectonic uplift across the Himalayas of central Nepal. J Geophys Res: Solid Earth 106(B11):26561–26591

Mahesh P, Gupta S, Rai SS, Rajagopala Sarma P (2012) Fluiddriven earthquakes in the Chamoli region, Garhwal Himalaya: evidence from local earthquake tomography. Geophysical J Intl 191:1295–1304

Mandal P, Rastogi BK, Gupta H (2000) Recent Indian earthquakes. Current Sc 79:1334–1346

Mandal P, Rastogi BK, Satyanarayana HVS, Kousalya M, Vijayraghavan R, Satyamurty C, Raju IP, Sarma ANS, Kumar N (2004) Characterization of the causative fault system for the 2001 Bhuj aftershocks. Tectonophysics 378(1–2):105–121

Mandal P, Raghavan RV, Saha S, Biswas R, Prathigadapa R (2023) Seismic imaging of the Kumaon-Garhwal Himalaya: Insights into the main Himalayan thrust and the crust-mantle boundary. Research Square

Manglik A, Gupta S, Tiwari VM (2021)Geophysical studies for the crust and upper mantle structure of the Himalaya: Contributions of CSIR-NGRI. J Geol Soc Ind 97:1190–1199
Mishra DC (1982) Crustal structure and dynamics under Himalaya and Pamir ranges. Earth Planet Sci Lett 57:415–420
Mitra SH, Paul A, Kumar SK, Singh SD, Powali D (2015) The 25 April 2015 Nepal earthquake and its aftershocks. Curr Sc 108:1938–1943
Mohanty A, Gahalaut V (2023) Seismic hazard and seismogenesis in Kashmir Himalaya and the role of Karakoram fault in the Himalayan tectonics, XXVIII General Assembly of the International Union of Geodesy and Geophysics (IUGG), GFZ OU_5011304
Molnar P, Fitch TJ, Wu FT (1973) Fault plane solutions of shallow earthquakes and contemporary tectonics in Asia. Earth Planet Sc Lett 19:101–112
Molnar P, Chen WP, Fitch TJ, Tapponnier P, Warsi WEK, Wu FT (1977) Structure and tectonics of the Himalaya: a brief summary of relevant geophysical observations. Himalaya. Science de la Terre, Paris, pp 269–294
Monsalve G, Sheehan A, Schulte-Pelkum V, Rajaure S, Pandey MR (2006) Seismicity and one-dimensional velocity structure of the Himalayan collision zone: earthquakes in the crust and upper mantle. J Geoph Res 111:B10301
Mukul M, Jade S, Bhattacharyya AK, Bhusan K (2010) Crustal shortening in convergent Orogens: insights from global positioning system (GPS) Measurements in Northeast India. J Geol Soc India 75:302–312
Ni JF, Barazangi M (1984) Seismotectonics of the Himalayan collision zone: Geometry of the underthrusting Indian Plate beneath the Himalaya. J Geophys Res 89:1147–1163
Patro PK, Harinarayana T (2009) Deep geoelectric structure of the Sikkim Himalayas (NE India) using magnetotelluric studies. Phys Earth Planet Int 173:171–176
Rai SS, Ashish, Padhi A, Sarma PR (2009) High crustal seismic attenuation in Ladakh–Karakoram. Bull Seismol Soc Am 99:407–415
Rai SS, Priestley K, Gaur VK, Mitra S, Singh MP, Searle M (2006) Configuration of the Indian Moho beneath the NW Himalaya and Ladakh. Geophys Res Lett, 33:L15308, https://doi.org/10.1029/2006GL026076
Rajendra Prasad B, Klemperer SL, Vijaya Rao V, Tewari HC, Khare P (2011) Crustal structure beneath the Sub-Himalayan fold–thrust belt, Kangra recess, northwest India, from seismic reflection profiling: Implications for Late Paleoproterozoic orogenesis and modern earthquake hazard. Earth Planet Sci Lett 308:218–228
Rao NP, Tiwari VM, Kumar MR, Hazarika P, Saikia D, Chadha RK, Bhaskar Rao YJ (2015) The Mw 6.9 Sikkim-Nepal earthquake of September 2011: a perspective for wrench faulting in the Himalayan thrust zone. Nat Hazards 77:355–366
Rastogi BK (1974) Earthquake mechanism and plate tectonics in the Himalayan region. Tectonophysics 21:47–56
Rastogi BK (2018) Intensity and source parameters study of August 20, 1988 Udaipur Nepal earthquake of magnitude 6.8, Bhoo-Kampan. Journal Ind Soc Earthq Sc 5:16–31
Rastogi BK (2020) Bihar-Nepal earthquake of 15 January 1934: isoseismals and damages, *Bhoo-Kampan.* J Ind Soc Earthquake Sc 7:1–12
Rastogi BK (2024) Nepal earthquake of 2015. Springer-Nature. ISBN 978-981-97-4683-5. https://doi.org/10.1007/978-981-97-4684-2
Seeber L, Armbruster JG (1981) Great detachment earthquakes along the Himalayan arc and long-term forecasting. Earthq Pred Int Rev 4:259–277
Singh A, Kumar MR, Raju PS (2010) Seismic structure of the underthrusting Indian crust in Sikkim Himalaya. Tectonics, TC6021. https://doi.org/10.1029/2010TC002722
Srivastava P, Mitra G (1994) Thrust geometries and deep structure of the outer and lesser Himalaya, Kumaon and Garhwal (India): implications for evolution of the Himalayan fold-and-thrust belt. Tectonics 13(1):89–109
Talwani P (1997) Seismotectonics of the Koyna-Warna area, India. Pure Appl Geophys 150:511–550

Valdiya KS (1973) Tectonic framework of India: A review and interpretation of recent structural and tectonic studies. Geophys Res Bull 11:79–114

Valdiya KS (2010) The making of India: Geodynamic evolution, Macmillon Publisher India Ltd, pp 802

Yadav RK, Gahalaut VK, Bansal AK, Sati SP, Catherine J, Gautam P, Kumar K, Rana N (2019) Strong seismic coupling underneath Garhwal-Kumaun region, NW Himalaya, India. EPSL 506:8–14

Chapter 2
Natural Hazards in India: A Review

B. K. Rastogi

Abstract The Himalayas are facing the maximum number of natural disasters like earthquakes, landslides and floods. The young age of the mountains, active continent–continent collision tectonics and geographical position are all contributory factors. The levels of different hazards in the Himalayas are identified in this chapter.

2.1 Introduction

It is important to assess hazards in Himalaya for sustainable development of the region as it faces frequent natural hazards like earthquakes, landslides, landslides accompanied with glacier slides, floods, cloud bursts, snow avalanches, etc., on account of its geological position, climate and geographical setting. According to the Institute for Environmental and Human Security, India, on an average of 5000 people are killed and about 60 million get affected annually by various types of disasters in India. More than half of this loss is accounted by the Himalayas. The Himalayan ranges are highest, youngest and still rising due to active geology caused by active continent–continent collision of Indian and Eurasian plates making it most vulnerable to erosion, landslides and earthquakes.

The country is divided into five zones on the basis of seismicity. Most of the Himalayan States, particularly north-eastern states and some of the western parts of the country lie in seismic zone 5, having the highest degree of vulnerability for earthquakes of magnitude reaching 8 or more, while zone 4 in rest of Himalaya. Four earthquakes exceeding magnitude 8 on the Richter scale occurred in this region since 1897.

Almost the entire mountain chain is vulnerable to landslides triggered by heavy rains and earthquakes in fragile and weak-unconsolidated rocks due to erosion or young age (Rastogi 2005a). Many locations experience multiple landslides year after year. Landslides and avalanches are major hydro-geological hazards that affect 15%

B. K. Rastogi (✉)
President Indian Society of Earthquake Sc. 1-2-63/3 Street No.1, Habsiguda, Hyderabad 500 007, India
e-mail: bkrastogi12@gmail.com

B. K. Rastogi et al. (eds.), *Natural Hazards and Risk Mitigation*, Springer Transactions in Civil and Environmental Engineering, https://doi.org/10.1007/978-981-97-7658-0_2

of area of India besides the Himalayas, the Northeastern Hill ranges, the Western Ghats, the Nilgiris, the Eastern Ghats and the Vindhyans are affected in that order. Average every year 75 lakh hectares of land (12% area of India) is affected, 1600 lives and 10 lakh houses are lost and economic loss of over Rs. 1800 crore is caused by floods (National Disaster Management Authority) and increasing due to encroachment of flood plains and global warming. Economic loss was about Rs. 500 cores/ annum in the beginning of this century and increasing rapidly. Perennial floods are due to the reason that 80% of precipitation is confined to four summer months of June to September, the rivers bring heavy dose of sediments from catchment areas and flood plains are encroached. Cyclones and flash floods are devastating. Steps are taken by Govt. of India for preparedness and advance warning. This chapter describes three types of hazards, viz. earthquakes, landslides and floods.

2.2 Geology and Geomorphology of the Himalayas

The Himalayas formed in the Tertiary period (starting 50 My ago) due to continent–continent collision while the Indian Plate drifted at 5 cm/y towards NE and under-thrusted Eurasia. Mountains rose at a rate of 1 cm/y. It is the world's highest and youngest mountain belt (Valdiya 1973, 2010). It is spread over 2400 km in length and 320 km in width in a crescent shape and encompasses five countries. Himalaya is a highly seismic belt experiencing periodically great and major devastating earthquakes. The Himalayas can be grouped into three roughly parallel belts of varying width, namely, Higher, Lower and Sub-Himalayas. the northern most region has Tethys Himalayas, limited (in the south) by South Tibetan Detachment System (STDS). South of it are three major groups of thrusts in its entire length from west to east, viz. the Main Central Thrust (MCT) dividing Higher and Lesser Himalayas, further south is the Main Boundary Thrust (MBT) limiting the Lesser Himalayas and the southernmost is the Main Frontal thrust (MFT) which limits Sub-Himalayas and the mountains.

The Tethyan sequence is of unmetamorphosed/slightly metamorphosed sedimentary rocks of Cambrian to Eocene age (Burchfiel et al. 1992). Higher Himalayas comprise of high-grade metamorphic rocks of the upper Proterozoic–lower Paleozoic age. Lesser Himalayas contain metasedimentary rocks of Late Precambrian and some Early Paleozoic (Proterzoic) rocks (Paudel and Arita 2000) including schist, phyllite, marble, quartzite, unmetamorphosed limestone, dolostone, shale and sandstone, etc., as well as orthogneisses, Neoproterozoic sedimentary rocks, and sedimentary rocks of Permian to Miocene. Rocks of Lesser Himalaya may be grouped into northern and southern groups divided by Munsiari Thrust/Bari Gad–Kali Gandaki fault. The southern belt contains low-grade metamorphic rocks similar to Higher Himalaya. The northern belt is characterized by sedimentary, low-grade metamorphic, mafic volcanic and crystalline rocks. MCT in Nepal had been active generating MCT I and MCT II branches, while in Kumaon–Garhwal area its branches are Munsiari and Vaikrita Thrusts. Lesser Himalaya has a series of thrusts indicating highly active tectonics, e.g. between the MBT and Munsiari thrusts are the Tons, Ramgarh and Berinag thrusts. Nepal has also similar thrusts but named differently. The presence of phyllites makes the Lesser Himalayan region highly prone to landslides (summarised from different sources by Halder et al., this volume). The Sub-Himalayas consist of Miocene and Pliocene sedimentary rocks of the Siwalik Group.

2.3 Earthquakes in India

Figure 2.1 shows the deep faults in India taken from the 1968 Tectonic Map of India prepared by the Oil and Natural Gas Commission of India and modified after Reddy et al. (1988). For most of these faults, the deep seismic sounding studies (Kaila 1982) confirm the depths of 35 km or more. Valdiya (1973, 2010) and Powar (1981) described additional tectonic features. Most earthquakes are associated with these faults and thrusts. Seismotectonic Atlas of India and its Environs is a comprehensive publication by the Geological Survey of India (2000).

Figure 2.2 gives epicentres of earthquakes in the Himalayas and the peninsular India having different magnitudes and depths and Tables 2.1 and 2.2 list them. Data is mostly from USGS modified using Gupta et al. (1995) and others for Himalaya while using Oldham (1883) who described significant Indian earthquakes from earliest time through 1869. Kelkar (1968) brought to light a number of additional events in Maharashtra for 200 years of period. Tandon and Srivastava (1974) published a catalogue of earthquakes of magnitude 5 and above up to 1970. Srivastava and Ramachandran (1985) prepared a catalogue of historical earthquakes in Peninsular India. Using all these sources, Chandra (1977) compiled quite a reliable catalogue for earthquakes of intensity ≥ VI and magnitude ≥ 5 for the Indian Peninsula. Bapat et al. (1983) prepared the catalogue of earthquakes for India and surrounding regions using the data from USGS catalogue and other sources. Most magnitudes are equivalent to moment magnitude. Data for Mw 6.2 Latur (Killari) earthquake is from Gupta et al. 1998), for Mw 5.8 Jabalpur earthquake from Acharya et al. (1997) and Gupta et al. (1997). Malik et al. (1999) listed some historical earthquakes in Kachchh. Table 2.3 gives magnitude wise number of earthquakes in Stable Continental Region (SCR) of India or Peninsular India.

2.3.1 *Seismic Hazard in the Himalaya*

Rupture zones of large earthquakes in the Himalaya are about 200 km long. From a convergence rate of 2 cm/y, a strain enough for 4 m slip and a great earthquake may be accumulated every 200 years. From the great earthquakes since 1500 it can be assessed that only 1/3rd portion of the Himalya has ruptured in known history and 2/3rd portion of the Himalaya has potential to produce a few great earthquakes (Bilham 2019, Bilham et al. 2001). Three most potential regions are proposed for

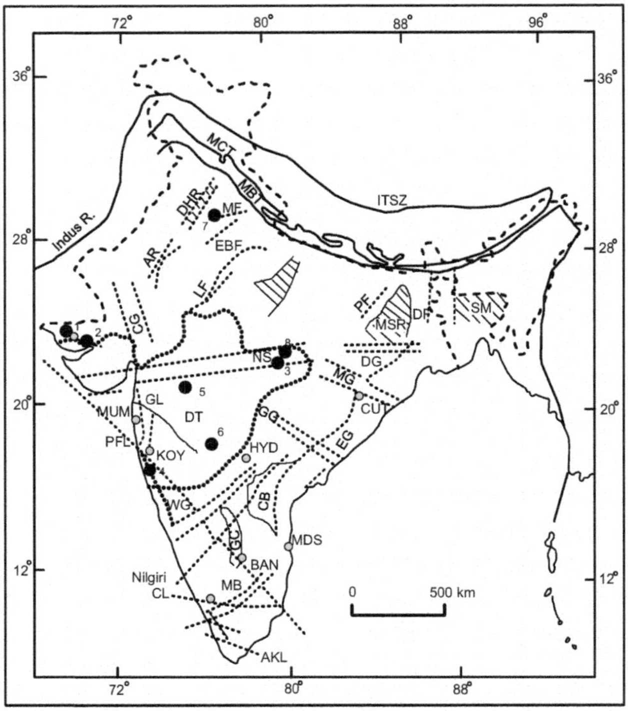

Fig. 2.1 Simplified Tectonic Map of India based on the one prepared by ONGC in 1986. Locations of M ≥ 6 earthquakes in Peninsular India are also shown, most of which are in the areas of past volcanism and rift valleys. Location no. 9 for the 2001 Bhuj earthquake is nearly same as that of no.2. Deccan Trap (DT) boundary is shown by dotted line. ISZ = Indus Suture Zone, MCT = Main Central Trust, MBT = Main Boundary Thrust (Main Frontal Thrust is nearly parallel to MBT and about 50 km south of it), CG = Cambay Graben, DHR = Delhi–Hardwar Ridge, MF = Moradabad Fault, EBF = Etah–Budaun Fault, LF = Lucknow Fault, FR = Faizabad Ridge (shown east of LF), PF = Patna Fault, MSR = Monghyr–Saharsa Ridge, DF = Dhubri Fault, SM = Shillong Massif, NS = Narmada–Son rift, MG = Mahanadi Graben, GG = Godavari Graben, EG = Eastern Ghat Belt, CB = Cuddapah Basin, CG = Closepet Granite, CL = Cauvery Lineament, AKL = Achankovil Lineament

future great earthquakes, viz. in Kashmir to NW Himalaya in the western sector, in Uttarakhand to western Nepal in the central sector and in the NE Himalaya.

In the Himalayan belt including the border regions of Nepal, Burma and China, 33 earthquakes of M ≥ 7.0 have occurred in the twentieth century including 5 of M > 8.0 (including one of M8.1 in the Chinese territory and north of the 1950 epicentral area in 1951 (Chandra 1997; Gupta et al. 1995). This means one earthquake of M > 7 every 3 years somewhere in the Himalaya and Myanmar arc. If we consider seven sectors of the Himalayan belt, roughly 350 km each, then each sector had one every 21 years. These sectors are Kashmir, Punjab, Garhwal, Nepal, Assam, Naga and Patkai. However, it has to be mentioned that the earthquakes are not uniformly distributed. There was no earthquake of M > 7 in the Kashmir sector in the twentieth

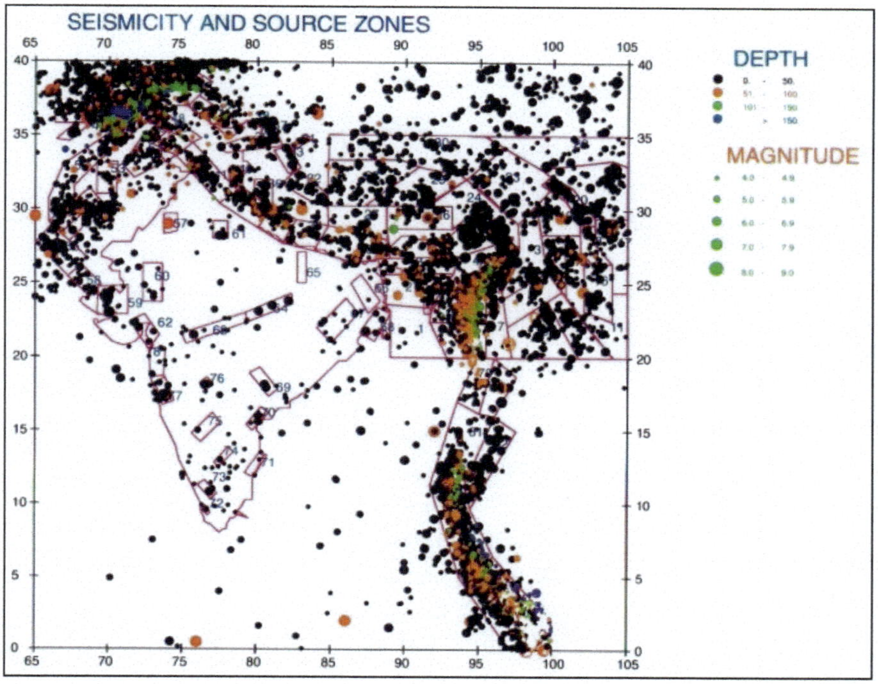

Fig. 2.2 Epicentres of earthquakes of M ≥ 4 or intensity ≥V in the Indian region from earliest time to 2015 (source USGS). As the epicentre of the 2004 Sumatra–Andaman earthquake of M9.3 is in Sumatra, it is not shown here

century or until 2005 in its western part. Also, there are 20–50 years of cycles of low and high seismicity (Rastogi and Sharma 2016).

Table 2.1 lists damaging earthquakes in the Himalayas. The four great earthquakes of M > 8 in the twentieth century in the Himalaya–Myanmar belt have been Kangra 1905, Myanmar 1912, Nepal 1934 and Assam 1950. There have been over 100 earthquakes of M 6 to 6.9, several of them, which occurred near towns, caused severe damaged. This means roughly 1 potentially damaging earthquake every year. Counting 1897 Assam earthquake there were 5 great earthquakes in 53 years period in the Himalaya from 1897 to 1950 or one every 10 years. However, no great earthquake has occurred in the last 73 years. Nevertheless, carbon dating indicates about 500 years of return period of such earthquakes in Northeast India.

Number of deaths due to earthquakes in the Himalayan region total about 1,60,000 in the last 200 yr. Number of deaths in each earthquake is increasing with time in view of increasing population and prevalent practice of constructing non-engineered dwellings. In the seismic zoning map of the Himalayas, about 40% area lies in seismic zone V where MM intensity IX or more are probable, e.g. in most of Northeast India, Sikkim, Kumaon, Himachal and Kashmir. The remaining part is placed in Zone IV where maximum MM intensity VIII is likely to occur sometime in future. However,

Table 2.1 Damaging earthquakes in Himalaya and active northern region

Date	Place	Mag	Death Toll
01.09.1803	Kumaon	8.0	300
08.06.1828	Srinagar	8.0	1000
26.08.1833	Bihar-Nepal	8.0	1000
30.05.1885	Kashmir	8.0	3000
06.12.1897	Assam	8.7	1542
04.04.1905	Kangra	8.5	20,000
23.5.1912	Myanmar	8.0	(~50), Mw7.8
02.07.1930	Dhubri	7.1	100
15.01.1934	Bihar-Nepal	8.4	11,000
31.05.1935	Quetta	7.5	25,000
30.08.1941	Andamans	8.1	100
15.08.1950	Assam	8.5	1500
28.12.1958	Kapkote	6,3	100
02.09.1963	Badgam	5.3	79
20.02.1967	Anantnag	5.6	1
19.01.1975	Kinnaur	6.8	47
29.07.1980	Dharchula	6.1	200
24.08.1980	J&K	5.2	15
30.12.1984	Cachar	5.4	11
26.04.1986	Dharmashala	5.7	10
06.02.1988	NE India	5.8	2
06.08.1988	Burma-India	7.2	5
21.08.1988	Udaipur, Nepal	6.4	382 Rastogi (2018)
20.10.1991	Uttarkashi	6.6	769
28.03.1999	Chamoli	6.4	104
08.10.2005	Kashmir	7.6	86,000
25.04.2015	Gorkha, Nepal	7.8	9,000
	Total		1,60,037

it will be more practical to consider probable intensity IX everywhere. Intensity $\geq$ X might go up to 150 km distance, intensity IX up to 200 km and intensity VIII up to 300 km distance from the rupture zone for M > 8 earthquake. Table 2.4 gives areas and radii of different intensities for different magnitude earthquakes.

Table 2.2 List of SCR India earthquakes of M ≥ 5 or strongly felt

D M Y	Lat long N E	Depth	Area	Int	Mag Mw	Deaths	Ref
04 06 1764	24.0 88.0		Banks of Ganga	VIII	6.3		OLD
16 06 1819	23.6 69.6	< 30	Kutch	XI	7.8	2000	OLD
19 04 1845	23.8 68.9		Lakhpat	VIII	6.3		OLD
26 04 1848	24.8 72.4		Mt Abu		5.7		
23 05 1866	25.0 87.0		Bengal	VIII	6.3		OLD
07 02 1900	10.8 76.8		Coimbatore		6.0		
02 06 1927	23.5 81.0		Son valley		6.4		
14 03 1938	21.5 75.7		Satpura, Khandva	VII	6.3		AJ
21 07 1956	23.0 70.0	15	Anjar		6.0	115	
10 10 1956	28.1 77.7		Khurja		6.1		AJ
13 10 1956	15.6 80.1		Ongole	VI	5		IMD
10 12 1967	17.3 73.76	4.5	Koyna	VIII	6.3	177	DDS
30 08 1970	21.7 73.0	11	Broach		5.4	26	
29 09 1993	18.09 76.5	2.8	Latur	VIII	6.2	7610	USG
22 05 1997	23.08 80.06	35	Jabalpur		5.8	39	
26 01 2001	23.44 70.31	22	Bhuj, Guj	IX	7.7	14000	ISC

Note Bellary (1843 M5.7), Coimbatore (1900 M5.8), Son Valley (1927 M6.4) and Satpura-Khandva (1938 M6.3) earthquake are likely to be of focal depth around 35 km

Table 2.3 Number of earthquakes in SCR India from historical times

Magnitude	Number of earthquakes
≥7	2
6–6.9	7
5–5.9	95
Total (M ≥ 5)	104

Table 2.4 Size of damage areas (km^2) and radius (km) for different intensities (Rastogi 2005a)

Intensity	Magnitude 6.5		7		8	
	Area	Radius	Area	Radius	Area	Radius
≥X					80,000	150
IX			700	15	120,000	200
VIII	700	15	2000	25	280,000	300
VII	2000	25	6000	45	800,00	500
VI	8000	45	30,000	100	1000,000	600

2.3.2 *Seismic Hazard in Peninsular India*

The list of earthquakes in Peninsular India includes two of Mw magnitudes 7.7 and 7.8 (which occurred in Kutch), at least 8 of Mw $\geq$ 6, 95 of Mw magnitude $\geq$ 5 and about 500 other smaller earthquakes. Twelve earthquakes of M $\geq$ 6 during the last 200 years give a return period of 14 years. The number of earthquakes causing deaths has increased after the 1950s as prior to that only the M7.8 Kutch earthquake of 1819 was known to be a killer one causing 2000 deaths. From the 1960s, a number of strong damaging earthquakes have occurred causing deaths as given in Table 2.2. These earthquakes have changed the long-held belief of the slight seismicity of Peninsular India and necessitated revision of the seismic zoning map of India.

Most focal depths are within 10 km. However, depths of the order of 35 km are in Narmada–Son and southern shear zones where the lower crust could be brittle due to under-plating, pillow lavas, mafic intrusion or dehydration. Around the hypocentre of the 2001 M7.7 Bhuj earthquake, a high velocity and rigid mafic intrusive body has been inferred in a large zone. The Bhuj earthquake nucleated from a fluid-filled zone at depths of 15–25 km. Low-velocity zones are found at shallower depths of about 5 km in Koyna and Latur. The foreshock–aftershock activity is pronounced for upper crustal earthquakes.

Peninsular India shows a preponderance of strike-slip faulting along NE–SW or NW–SE trending faults (Rastogi 1992, 2010). Reverse faulting is observed along the failed rifts of Kutch (Rastogi et al. 2014) and Narmada–Son (Rastogi 1997) indicating a change of past tensional regime to current compressional regime causing uplift of tectonic blocks. The earthquakes in the Kutch region are infrequent but of high intensity giving a high moment release. The other regions in order of decreasing moment release are Narmada–Son belt, Western Ghats and the southern mobile (shear) zone. The compression axis changes from NNW to N from west coast area to central parts of India and further changes to the NNE direction in the eastern marginal area. This indicates the effect of counterclockwise rotational tendency.

2.4 Landslides in India

India has the highest mountain chain on earth, the Himalayas, which are formed due to the collision of Indian and Eurasian plates and rising 1 cm/y. The northward movement of the Indian plate at about 5 cm/year towards China accumulates stress on the rocks causing earthquakes and rendering them friable, weak and prone to landslides in Himalaya, some of which are disastrous and pose chronic problem, causing recurring economic losses worth billions of rupees. Besides Himalaya, the hill ranges in eastern and western coast regions are also landslide-prone. The laterite cap in the Western Ghats along the steep slopes overlooking the Konkan coast poses constant problem of landslides. Most of the Indian landslides are rain-induced. Water saturation reduces the tangential stress holding the slope materials thereby resulting into reduction of frictional force and landslide. Wayanad, Kerala landslide of 30 July, 2024, death toll 400, has been caused by excessive rains. Rainfall was 400 cm during June–July 2024. Landslides are frequent in Kerala. Big and small total landslide count made by IIT, Chennai during 2018 in Kerala state was 4726, half of them in Idukki district. Western Ghats Ecology Expert Panel report (2011) identified Meppadi area where Waynad falls and nearby Churalmala area, and 16 other areas in Kerala, as ecologically sensitive localities.

Geological Survey of India has prepared landslides-hazard zonation maps on 1:50,000 scale for the country. It has been realized that a considerable improvement is required in terms of instrumentation and geotechnical and geophysical studies. So far ad hoc measures to control landslides like drainage of surface and sub-surface water, removal of slide mass and slope modification by cutting and filling, have been adopted with limited success. Use of geosynthetics and jute derived material for arresting soil erosion and soil slips has shown encouraging results. In countries like Japan and Hongkong, strict implementation of land-use regulations has considerably reduced (by a factor of 10) damage due to landslides. The same type of land-use regulations has to be reinforced in India also.

2.4.1 Factors Affecting Landslides

Some of the factors that contribute to the high frequency and intensity of landslides in Himalaya are geological factors, rainfall, seismic activity and climate change. Factors like haphazard growth, poorly planned highways and deforestation greatly influence the severity of landslides. The Himalayas have different levels of landslides: big and small or quick and creeping. Different types of landslides are debris flow, slump and debris-slides and avalanches. Certain parameters control the occurrences and aerial extent of landslides such as slope angle, nature of slope materials and the normal stress acting on the slope. When the slope angle exceeds the critical slope angle, the slope materials start to move causing landslides. Slopes > 15° are favourable for landslides.

Geological Factors for Landslides

Himalayan mountain chain is at high risk of landslides due to unstable tectonics in Lesser and Higher Himalayas in the zones of Main Central Thrust and Main Boundary Thrust while Siwaliks, south of the Main Boundary Thrust have soft rocks. Himalaya has highly complex folding and faulting pattern rendering favourable geological conditions for landslides. Severity of landslides in Himalaya can be gauged from the rate of creep movement. The creep rates on Eastern Himalayan slopes range from 5 cm in a decade on slopes less than 35°–75°. Movements of landslides and mudslides are very fast typically falling in the range 1–4 cm per day.

Landslides Due to Rainfall

Rainfall is the major factor to affect landslides (Dixit et al. 2020). Himalaya is a great climatic divide causing high rainfall and cloud bursts. The rainfall shows a progressively decreasing trend from the east to the west in Himalayas. However, the precipitation in the form of snow is more in the central and the western Himalayas, than in the eastern region. In the eastern Himalayas, the summer monsoon starts in the first week of June, i.e. almost a month earlier and leaves nearly three weeks later than it does in the western Himalayas. The eastern Himalaya gets heavy to very heavy rainfall punctuated with occasional cloudbursts acquiring rainfall intensities in the range 300 mm/day to 1100 mm/day, higher figures being for lesser duration. The high event coefficients (Ce) defined as the ratio of the rainfall of the event to the mean annual rainfall have occurred during 1902 (6%), 1915 (14%), 1950 (11%), 1968 (20% & 10%), 1969 (19% & 15%) and 1977 (10%). Their interpretation in terms of occurrence of landslides are suggested in Table 2.5 indicating fair to sure chances of triggering a landslide.

Table 2.6 lists the devastating landslides mostly caused by exceptionally heavy rainfall. For example, in Sikkim Himalayas, the annual rainfall ranges between 350 and 500 cm most of which is concentrated in the period of four monsoon months. Landslides occur within this period or immediately afterwards.

Landslides Due to Earthquakes

The second major factor is earthquake (Parkash 2013). Several Himalayan earthquakes have caused landslides. Large earthquakes have triggered severe and widespread landslides while moderate ones have caused limited number of landslides. Bihar-Nepal earthquake of January 15, 1934, triggered several landslides near Kathmandu, Garhi, Udaipur, and in eastern Nepal (Pandey and Molnar 1988; Rastogi 2020). The rocks in the region belong to the metamorphic and crystalline complex (phyllites, quartzites, granite-pegmatites) of Lesser/Higher Himalaya. Rock falls in the Siwalik Sandstones near Muksar clogged a river channel generating multiple lakes, one of which was 200 m long and about 8 m deep. Bihar-Nepal Earthquake of August, 1988 triggered rockslides at Tindharia in Darjeeling and Bansoi in North Sikkim (Parkash 2013; Rastogi 2018). Nepal earthquake of 2015 caused severe, valley-blocking and widespread landslides, but many more were of limited widths throughout the rupture area of about 150 km × 55 km (Rastogi 2016). Major

Table 2.5 Event coefficients and landslides in Eastern Himalayas (Bhandari 1988)

Event coefficient, Ce%	Probability of Landslide
>20	Will trigger landslides and other mass movements at all times
10–20	Very high probability
5–10	Fair probability
<5	Very low probability of new slides and fair probability of reactivation of old untreated slides

Table 2.6 Exceptionally heavy rainfall and devastating landslides (Media and Government sources)

Place	Date	Consequence of heavy rainfall
Darjeeling and Jalpaiguri. N. Bengal	3–5 Oct 1968	Widespread landslides and other mass movements causing death and devastation all over
Uttarakhand	July 1970	Alaknanda river caused considerable loss of life among pilgrims. Many bridges, houses and an entire village were washed away
Uttarakhand	Sep 1970	Landslides and house collapse took a toll of 223 human lives
Jammu and Kashmir	Feb 1971	Widespread landslides causing disruption of traffic and communication systems
Jammu and Kashmir	Aug 1972	Widespread landslides causing damage to life and property
Jammu and Kashmir	March 1973	Landslides cut off Kashmir valley from the rest of the country
Shimla (Himachal Pradesh)	July 1973	Landslides cut off Shimla from the rest of the country
North Bengal	July 1975	Teesta, Jaldhaka and Diana rivers were in spate. Widespread landslides and floods rendered 45,000 people homeless
Jammu and Kashmir	Sep 1975	Landslides killed 2 laborers and disrupted transportation system for 3 days
Darjeeling, North Bengal	June 1976	Teesta in floods triggering many landslides; 3 people buried alive due to caving-in of a hillock
Jammu and Kashmir	July 1977	Srinagar–Leh road blocked by landslides
Malpa (Pithoragarh, Kumaon)	Aug 11–17, 1998	Landslide killed 380 people
Ukhimath	Aug 1998	Landslide killed 68 people
Mandakini valley, Uttarakhand	16 Jul 2001	More than 200 landslides resulting in 27 fatalities Naithani et al. (2002)
Alaknanda Valley, Uttarakhand	Sep 2004	Debris flow on the left bank of Alaknanda River blocked the river for a short time, changed the river course and flooded the roads on the right bank
Mandakini valley, Uttarakhand	Sep13–14 2012	Excessive rains totaling > 20 cm, which resulted in 473 landslides and death of 51 people
Kedarnath, Uttarakhand	16 June 2013	A cloudburst in Kedarnath caused devastating floods and landslides that destroyed several villages and towns killing 6000 people. It melted the Chorabari glacier and flooded the Mandakini River Martha et al. (2015)

(continued)

Table 2.6 (continued)

Place	Date	Consequence of heavy rainfall
Rishiganga in Chamoli district, Uttarakhand	7 Feb 2021	Rishiganga flash-flood tragedy was due to landslide. A triangle-shaped landslide trigger zone of 0.34 sq km size was noticed in the Nanda Ghunti Mountain of Nanda Devi National Park near to Raini village by satellite image of Feb. 2. A huge chunk of rock mass of over 2 Mcm^3 with snow/ice that detached had a fall of 1800 m attracting additional rock mass. The impact produced dust and melted frozen glacial mass. The Rishiganga hydro project was swept away. The disaster also severely damaged another hydropower project located in the Dhauliganga River, adjacent to Tapovan village. The flood carried with it tons of mud which trapped and killed at least 204 workers in a tunnel inside the project site
Uttarakhand	29 Jul 2022	30-m section of Nainital–Bhawali road destroyed
Uttarakhand	20 Aug 2023	Roads, buildings and temples destroyed due to floods and landslides in Uttarakhand killing 46 people. Areas affected included, Uttarkashi, Joshimath, Kedarnath, Rudraprayag, Badrinath, Chamoli, Pithoragarh and Champavat
Himachal	23 Aug 2023	At least 47 deaths and 36 landslides. Flash floods triggered by heavy rainfall recked havoc in Himachal. Beas River was swollen. Debris from flash floods, about 3 m high has made half the farmland unusable
Sikkim	4 Oct 2023	Cloudburst over Lhonak glacial lake in north Sikkim triggered flash floods; 18 killed and 100 missing. Release of water from the Chungthang Dam increased water levels up to 7 m in the Teesta River basin. Dam was also completely damaged. A steel river bridge at Singtam washed away

disruptive landslides were in High Himalaya encompassing the districts of Gorkha, Dhading, Nuwakot, Rasuwa, Sindhupalchok, Dolakha, Ramechhap and Khavre. Several landslides, a km wide and landslide dams formed along two highways going northward from Kathmandu to Tibet, one being Aranico Highway through Kodari and the other west of it to Rasuwa. The highway through Dolakha also had several landslides. Huge mud-rockslide covered the town of Langtang valley on popular trekking route burying many. Hill-slope slipped in Hulakdanda, 40 km WSW of Pokhra. Some ice-rock avalanches occurred at elevations of over 3500 m. The 1991 Uttarkashi earthquake of Mw 6.6 caused some landslides. Due to it, the epicentral area is reported to have been filled with dust soon after the earthquake (Rastogi and Chadha 1995).

River Action as a Cause for Landslides

River action is a major natural destabilizing factor specially during floods. It undercuts the river banks engulfing the houses and buildings.

Developmental Activities Causing Landslides

Road-cutting, building of hydroelectric projects and quarrying–mining activity, destabilizing hill slopes, are major causative factors for landslides. Developmental activities involve deforestation, blasting and huge excavations which add to the landslide problem. Every kilometre of road construction may bring about a stress relief equivalent to about 0.2 M tons of rock mass and if the cuttings are not properly protected, landslides and rock falls become imminent adding additional of about 1,000 tons of land loss per km annually. At many major landslide locations, the debris clearance may well be of the order of 4,000–5,000 tons annually.

A number of hydro-electric power projects have been built in the Himalayas like the large Tehri Dam on Bhagirathi and many more are currently under construction. Quarrying and mining, for example, in Doon Valley, Jhiroli (Almora) and Chandhak (Pithoragarh) have inflicted heavy damages to slopes and to the environment of the area. Construction of large number of buildings due to rapid urbanisation causes instability. Some of the constructions are coming up on old landslides without adequate pretreatment and investments on hillside stability, compounding the problem. Underground water-storage tanks are being built without any precautions despite the bad experiences of slope instability from such tanks at Mussoorie and Shimla.

How Climate Affects Landslides

The Himalayan mountain belt lies in the sub-tropical zone of high pressure of the general circulation system of the globe, which, over the vast expanse of Asia, gets greatly modified by the intense heat of summer and the cold of winter. The seasonal meridional migration of the pressure and wind systems brings about considerable changes in wind and weather regimes during different months. Most part of the northern India being within 400–700 km from the Tropic of Cancer, the climate changes from the tropical of the Indian plains to the almost alpine of the higher reaches of the Himalayan region. High variations in temperature, air pressure and weather conditions influence landslides.

2.4.2 *Severity of Landslide Problems in the Himalayas*

Himalaya is the source of India's major rivers. It is also repository of treasures like forests and wild life. However, the unprecedented and unwise exploitation of Himalayas combines with its immature geology, heavy rainfall, river action and seismicity to make it World's worst landslide affected region.

Almost entire belt of Himalayas suffers from frequent landslides. NE India has perennial problem as it has high rainfall for 5 months or more. Sikkim, Mizoram,

Tripura, Meghalaya, Assam, Nagaland, Arunachal Pradesh and district of Darjeeling in West Bengal of the north-eastern region are badly affected by landslides. In the western Himalaya, J&K, Himachal and Uttarakhand have perennial landslides in rainy season (National Disaster Management Authority, Rastogi 2005b). The mean annual rate of land loss is about 120 m/km or 2,500 tons/km. At tens of places along important roads almost every year severe landslides occur. Central Building Research Institute, Roorkee, had prepared detailed maps and history of landslides at these places (Bhandari 1988, 1999).

Landslides are a serious problem in Uttarakhand, especially during the monsoon season. Some severe tragedies include Varnavat landslide in Uttarkashi District, Malpa landslide in Pithoragarh district and Ukhimath landslide in Chamoli district. In North Sikkim and Garhwal, on an average, there are two landslides every km and new locations also add every few km. Landslides have increased five times in recent years. Seventy-two per cent of Uttarakhand is landslide-prone, and had the second highest number of landslides (11,219) between 1988 and 2022 among the Himalayan sectors ('Landslide Atlas of India' by Indian Space Research Organisation). During 2023 itself, Uttarakhand had a total of about 1500 landslides killing about 100 people.

At some locations landslides are recorded to have frequently occurred since decades. And these locations have multiple hazards like earthquakes, cloud bursts, forest fires and snow avalanches. In Uttarakhand, Alaknanda Valley is one such area (Sarkar et al. 2006) which had Malpa and Ukhimath landslides in 1998. Malpa, a tiny village in the Pithoragarh district, near the Indo-China border, evidenced landslides during August 11–17, 1998 resulting in the death of 380 people (Dikshit et al. 2020). In Ukhimath area in Mandakini Valley, 466 destructive landslides killed 103 people and damaged 47 villages (Dikshit et al. 2020). Kaliasaur, 15 km east of Srinagar (Garhwal) on Haridwar–Badrinath road, has a century old record of landslides, notable in the years 1952, 1963, 1964, 1965, 1969, 1970, 1970, 1972, 1984 and frequently after that. The slide area dimension increased with every major landslide. A cloudburst in a section of the Mandakini Valley in July 2001, caused more than 200 landslides, resulting in 27 fatalities (Naithani et al. 2002). Same region during September 13–14, 2012 suffered excessive rains totalling more than 200 mm, which resulted in 473 landslides and death of 51 people (Martha et al. 2015).

Nashri, in J&K, located on the national highway 1, from Jammu to Srinagar at km 138, is a good example depicting how a stable slope develops into a recurring landslide area. Banihal cart road was made in 1914, in 1953 a forest fire robbed the stable slope of its vegetative cover which was the start of instability. In 1954, a small landslip occurred at the toe of the stable slope after undercutting. Since then, landslides have occurred almost every year with gradual increase in the slide (Bhandari 1999).

High-speed mudflows and avalanches are most hazardous which acquire speeds as high as 3–50 m/s. These are due to heavy rainfall or earthquakes or both. One such example was the devastating debris flow in the year 1880 on the slopes of Sher-Ka Danda hill in Nainital, which travelled 1 km in 30 s, killed 150 people and swept away 'Victoria Hotel', 'Naina Temple' and other buildings and filled a part of the lake. Similarly, the flow slide of Kuekhola trestle point, Sikkim (Bhandari 1999) generated large movements in a few minutes destroying the huge retaining wall, the road and the ropeway trestle platform.

A multi-institutional program on landslides in fault zones is currently underway with support from MoES. This is an all-inclusive program incorporating experts in structural geology, geomorphology, geophysics, remote sensing, computational sciences, sensor networking and design engineers for optimization efforts and will be first such effort. It is being coordinated by Prof. Malay Mukul of IITb (Mukul et al. 2018).

2.5 Floods in India

Floods are the most common natural disaster in India. The heaviest south-west summer monsoon from June to September causes flooding of most of the rivers resulting in flooding of surrounding areas. Many a time, the change in river course floods large habited and agriculture land, for example, the Kosi River in Bihar which has changed the course several times causing devastation. NE Winter monsoon causes flooding of some rivers in southern India. Every year on average 1,600 lives is lost due to floods. Table 2.7 lists notable recorded floods that have occurred in India.

Table 2.7 Notable recorded floods in India (Media and government sources)

Lucknow	1923	Heavy rainfall in the monsoon season made the Gomti River overflow its banks and inundate many parts of the city. It damaged property, crops and infrastructure
Chennai	Oct 1943	Flood occurred due to excessive rains that lasted for 6 days and overflowed Coovum and the Adyar rivers. Damage caused to life and property was immense. However, estimate figure is unknown. The flood left thousands of people homeless
Lucknow	13 Oct 1960	Gomti River overflowed due to heavy rainfall in the Himalayan region, submerged half of the city, killing 16 people and making thousands homeless
Lucknow	8 Sep 1971	Gomti River overflowed due to heavy rainfall by 2 m above the danger mark. The flood submerged many parts of the city, affecting the main shopping area, some fashionable residential areas, and the zoological gardens. The flood also disrupted the electricity supply, the communication and transportation systems, and the public health and safety. At least 13 deaths due to drowning were reported from Uttar Pradesh
Morbi, Rajkot dt., Guj	11 Aug 1979	Machchu Dam burst due to overflowing of Machchu River causing flooding of Morbi town that killed about 2500 people
Bihar State	1987	One of the worst flood occurred due to overflow of Kosi River, which claimed lives of 1,399 humans and loss of Rs 68 billion
Punjab	1988	All the rivers in Punjab overflowed

(continued)

Table 2.7 (continued)

Eight states of India	July 1993	Flash floods killed 530 people
Mumbai and Maharashtra	26 July 2005	Heavy rains in Maharashtra including 1500 cm on a day in Mumbai killed at least 1,094 people. Airport and highways closed for a day. Public property loss was estimated at Rs. 550 crore (US $69 million)
Lucknow	Aug 2008	Heavy rainfall caused overflow of Gomti River (8.5 m above the danger mark) causing flooding of 11 districts and several parts of Lucknow affecting more than 100,000 people and killing at least 15
Uttarakhand	June 2013	A cloud burst caused severe floods mainly in Uttarakhand but at lesser level in nearby states. More than 5,700 people were presumed dead
Gujarat	June 2015	Heavy rain resulted into flooding of Saurashtra region of Guj. Causing more than 70 deaths. The wildlife of Gir forest National Park and nearby areas was also affected
Gujarat	July 2015	Heavy rain caused flooding in north Gujarat and 70 deaths
Tamilnadu	Nov–Dec 2015	Heavy rain resulted in flooding of Adyar and Cooum rivers in Chennai causing financial loss and lives
Assam	Jul–Aug 2016	Heavy rains caused floods affecting 1.8 million people and flooding the Kaziranga National Park killing about 200 wild animals
Gujarat	Jul 2017	Heavy rains caused severe floods killing 200 people
Kerala	Aug 2018	Floods due to heavy rains caused floods resulting into 445 deaths
Nine states of India	Jul–Aug 2019	Following high rain in late July and early August 2019, series of floods affected over nine states in India. The states of Kerala, Madhya Pradesh, Karnataka, Maharashtra and Gujarat were the most severely affected
Assam	2020	Floods
Hyderabad	2020	Flash flood in Oct due to a north Indian cyclone causing 98 fatalities
Maharashtra	22 Jul 2021	Widespread flooding in Mahad and Chiplun caused by exceptionally heavy rainfall
Assam	May 2022	Heavy flooding
Balrampur	2022	Over 1300 villages affected by floods
Himalaya	2023	Heavy rains caused flooding and landslide killing at least 50 people and washed away numerous buildings in HP & UTK
Bihar	Aug 2023	Floods in Bihar destroyed 7.54 lakh Hectares of agricultural land in 2023. Almost every year Kosi, Gandak, Boorhi Gandak, Bagmati, Lakhandei and other rivers are flooded. Moreover, Kosi sometimes changes its course causing havoc. Since independence millions of hectares of agricultural land has been affected precipitating huge losses

2.6 Conclusions

As prediction of natural hazards is not possible, preparedness measures need to be undertaken. Risk due to hazards is multiplied because of anthropological reasons; hence, there has to be a well-planned developmental activity. Mitigation and management of landslides need to be tackled by mapping hazard zones and specific slides are to be stabilized and managed in addition to monitoring and early warning systems to be placed at selected sites.

References

Acharya SK, Kayal JR, Roy A, Chaturvedi RK (1997) Jabalpur earthquake of May 22, 1997: Constraints from aftershock study. J Geol Soc Ind 50:1–12
Bapat A, Kulkarni RC, Guha SK (1983) Catalogue of earthquakes in India and neighbourhood from historical period up to 1979. Ind Soc Earthq Tech, Roorkee:211
Bhandari RK (1988) Landslides in the Himalayas and protection facilities, Landslides & Mudflows, Chapter 12, vol 2. A Joint Publication of UNESCO and UNEP, Moscow, p 95
Bhandari RK (1999) Natural disasters in Asia and the IDNDR Stimulus. In: XI Asian regional conference on IDNDR (1999–2000). Seol, Korea, p 24
Bilham R, Gaur VK, Molnar P (2001) Himalayan seismic hazard. Science 293:1442–1444
Bilham R (2019) Himalayan earthquakes: a review of historical seismicity and early 21st century slip potential, In: Treloar PJ, Searle MP (eds) 2019 Himalayan tectonics: A modern synthesis. Geological Society, London, Special Publications, 483, pp 423–482. First published online February 5, 2019, https://doi.org/10.1144/SP483.16
Burchfiel BC et al (1992) The South Tibetan detachment system, Himalayan orogen: Extension contemporaneous with and parallel to shortening in a collisional mountain belt. GSA Special Papers, USA
Chandra U (1997) Earthquakes of Peninsular India—A seismotectonic study. Bull Seism Soc Am 67(5):1387–1413
Dikshit A, Sarkar R, Pradhan B, Segoni S, Alamri AM (2020) Rainfall induced landslide studies in Indian Himalayan region: A critical review. Appl Sci 10:2466
Geological Survey of India (2000) Seismotectonic Atlas of India and its environs
Gupta HK, Rajendran K, Singh HN (1986) Seismicity of the northeast India region: Part I: The data base. J Geol Soc India 28:345–365
Gupta HK, Khanal KN, Upadhyay SK, Sarkar D, Rastogi BK, Duda SJ (1995) Verification of magnitudes of Himalayan region earthquakes of 1903–1985 from Gottingen observatory. Tectonophys 244:267–284
Gupta HK, Chadha RK, Rao MN, Narayana BL, Mandal P, Ravikumar M, Kumar N (1997) Jabalpur earthquake of May 22, 1997. J Geol Soc Ind 50:85–91
Gupta HK, Rastogi BK, Rao CVRK, Mohan I, Sarma SVS, Rao RUM (1998) An investigation of Latur earthquake of September 29, 1993 in South India. Tectonophys 287:299–318
Kaila KL (1982) Deep seismic sounding studies in India. Geophys Res Bull (India) 20:309–328
Kelker YN (1968) Earthquakes in Maharashtra in last 300 years. Kesari Daily (a Marathi language newspaper), Poona
Malik JN, Sohoni PS, Karanth RV, Merh SS (1999) Modern and historic seismicity of the Kachchh Peninsula, western India. J Geolog Soc in 54:545–550
Martha TR et al (2015) Landslides triggered by the June 2013 extreme rainfall event in parts of Uttarakhand state, India. Landslides 12:135–146

Mukul M, Jade S, Ansari K, Matin A, Joshi V (2018) Structural insights from geodetic global positioning system measurements in the Darjiling-Sikkim Himalaya. J Struct Geol 114:346–356. https://doi.org/10.1016/j.jsg.2018.03.007

Naithani AK, Kumar D, Prasad C (2002) The catastrophic landslide of 16 July 2001 in Phata Byung area, Rudraprayag District, Garhwal Himalaya, India. Curr Sci 82:921–923

Oldham TA (1883) Catalogue of Indian earthquakes. Mem Geol Survey India 19:163–215

Pandey MR, Molnar P (1988) The distribution of intensity of the Bihar-Nepal earthquake of 15 January 1934 and bounds on the extent of the rupture zone. J Nepal Geol Soc 5:22–44

Parkash S (2013) Earthquake related landslides in the Indian Himalaya: Experiences from the past and implications for the future. Landslide Sci Pract: Complex Environ 5:327–334

Paudel LP, Arita K (2000) Tectonic and polymetamorphic history of the Lesser Himalaya in central Nepal. J Asian Earth Sci 18:561–584

Powar KB (1981) Lineament fabric and dyke pattern in the western part of the Deccan Volcanism and related basalts provinces in other parts of the world. Mem GSI, Ind 3:46–57

Rastogi BK (1992) Seismotectonics inferred from earthquakes and earthquake sequences in India during the 1980s. Curr Sci, Spec Issue, Seism India—Overv, 62 (1&2):101–108

Rastogi BK (1997) Seismotectonics along Narmada-Son rift zone, GSI. Misc Publ 63:37–49

Rastogi BK (2005a) Earthquake hazard, risk and disaster management in India. In: Rastogi BK, Rai SN (eds) Proc intl seminar on natural hazards disaster mitigation. NGRI, Hyderabad, India, pp 1–19

Rastogi BK (2005b) Landslide hazard in India. In: Rastogi BK, Rai SN (eds) Proc intl seminar on natural hazards disaster mitigation. NGRI, Hyderabad, India, pp 61–68

Rastogi BK (2010) Causative mechanisms of the intraplate earthquakes occurring in India. Adv Geosci 27:97–114

Rastogi BK (2016) Salient aspects of the 2015 Mw 7.8 Gorkha Nepal Earthquake in Central Himalaya. J Earthquake Sc and Engineering (JoESE) (renamed Bhoo-Kampan: J Ind Soc Earth Sc, 3:18–28

Rastogi BK (2018) Intensity and source parameters study of August 20, 1988 Udaipur Nepal earthquake of magnitude 6.8, Bhoo-Kampan. J Ind Soc Earthquake SC 5:16–31

Rastogi BK (2020) Bihar-Nepal earthquake of 15 January 1934: Isoseismals and damages, bhoo-Kampan. J Ind Soc Earthquake Sc 7:1–12

Rastogi BK, Chadha RK (1995) Intensity and isoseismals of Uttarkashi earthquake of October 20, 1991. Mem Geol Soc India (30):19–24

Rastogi BK, Sharma J (2016) Global seismic temporal pattern and enhanced seismicity since 2000. J Ind Geophys Union 20(3):316–324

Rastogi BK, Mandal P, Biswas SK (2014) Seismogenesis of earthquakes occurring in the ancient rift basin of Kachchh, Western India. In: Earthquakes I (ed) Pradeep Talwani). Cambridge Univ, Press, pp 126–161

Reddy AGB, Mathew MP, Singh B, Naidu PS (1988) Aeromagnetic evidence of crustal structure in the Granulite terrain of Tamil Nadu-Kerala. J Geol Soc Ind 32:368–381

Sarkar S, Kanungo D, Patra A (2006) Landslides in the Alaknanda valley of Garhwal Himalaya, India. Q J Eng Geol Hydrogeol 39:79–82

Srivastava HN, Ramachandran K (1985) New catalogue of earthquakes for Peninsular India during 1839–1900. Mausam 36:351–358

Tandon AN, Srivastava HN (1974) Earthquake occurrence in India, In: Earthquake Engineering (Jai Krishna vol.), Ind Soc Earthq Tech, Roorkee, pp 1–48

Valdiya KS (1973) Tectonic framework of India: A review and interpretation of recent structural and tectonic studies. Geophys Res Bull 11:79–114

Valdiya KS (2010) The making of India: Geodynamic evolution. Macmillon Publisher India Ltd, pp 802

Chapter 3
GPS Deformation and Earthquakes in Himalaya

Sridevi Jade and T. S. Shrungeshwara

Abstract The 2500 km Himalayan arc spanning Kashmir, Ladakh, in the west to Eastern syntaxis in the east is a tectonically complex and seismically active northern subduction boundary of the Indian plate. Three decades of GPS data give well-constrained surface convergence rates ranging from 10 to 26 mm/yr in the various segments of the Himalayas (Kashmir, Ladakh, Himachal, Garhwal, Kumaun, Nepal, Sikkim, Bhutan, Arunachal and Eastern syntaxis). Arc parallel rates of 3–10 mm/yr is the manifestation of locked curvature of the central Himalayan arc and the E–W extension rate of Tibet. Inverse modelling of surface convergence rates is used to estimate oblique slip rate of 13–20 mm/yr along Main Himalayan Thrust (MHT) at a depth of 15–20 km and locking width of 100–150 km from the frontal Himalayas suggesting that each segment of the Himalaya is unique in nature. Geodetic strain rates derived from the GPS-derived surface convergence rates suggest that the Himalayan region is predominantly under compression with high strain rate coinciding with the northern boundary of sub surface basal decollement (MHT) along which Indian plate subducts below Tibet. Seismic strain rates for each segment of the Himalaya are computed using the instrumental and historical earthquake catalogue. Seismic potential of each segment of the Himalaya is estimated by a combined analysis of geodetic and seismic strain rates and the corresponding moment rates. Further, strain budget and accumulated slip since the last devastating earthquake was used to estimate the recurrence interval and probable magnitude of impending earthquake in the Himalayan segments.

Keywords Himalayan arc · Slip models · Geodetic strain · Seismic strain · Seismic potential

S. Jade (✉) · T. S. Shrungeshwara
Bangalore CSIR-4PI, India
e-mail: head@csir4pi.in

T. S. Shrungeshwara
e-mail: shrungeshsringeri@gmail.com

B. K. Rastogi et al. (eds.), *Natural Hazards and Risk Mitigation*, Springer Transactions in Civil and Environmental Engineering, https://doi.org/10.1007/978-981-97-7658-0_3

3.1 Introduction

Seismically active northern convergent boundary of the Indian tectonic plate is the 2500 km Himalayan arc extending from Kashmir Himalaya in the west to Eastern syntax is in the east (Fig. 3.1). Two-dimensional boundary of the Himalayan arc can be represented by a circle of radius 1696 ± 15 km with centre located at 42.4 ± 2.1 °N and 91.6 ± 1.6 °E (Bendick and Bilham 2001). This definition of curved tectonic boundary is consistent with micro-seismicity, surface deformation, strain rate and is a prerequisite to model the convergent arcs. Tectonics, seismicity and GPS surface deformation are not uniform throughout the arc (Jade 2004; Jade et al. 2014, 2017) from west to east and hence each segment of the arc behaves uniquely warranting an in-depth study. Seismic hazard of the Himalaya was broadly quantified (Bilham et al. 2001) by dividing the Himalayan arc into 10 segments of 220 km length equivalent to the rupture length of great Himalayan earthquakes.

Indian tectonic plate subducts below the Eurasian plate along the Main Himalayan thrust with a slip rate of 13–20 mm/yr at a depth of 12–20 km in various segments of the Himalaya. This subduction rate is not steady as during the major earthquakes the slip along MHT is sudden and large (Bilham et al. 2001). MHT marks the upper surface boundary of the subducting Indian plate over which Tibet together with the Himalayan wedge moves south. Surface expression of MHT coincides with Main Frontal Thrust (MFT) which is the southern boundary of the Himalaya and marks the beginning of the Himalayan foreland. Major faults with surface expressions along the Himalayan fold and thrust belt from south to north are Main Frontal Thrust (MFT), Main Boundary Thrust (MBT), Main Central Thrust (MCT), South Tibet Detachment System (STDS) and Indus Suture Zone (ISZ) (Figs. 3.1 and 3.2) which demarcate the Himalayas into frontal, lesser, higher, Tethyan and Trans Himalayan litho-tectonic units.

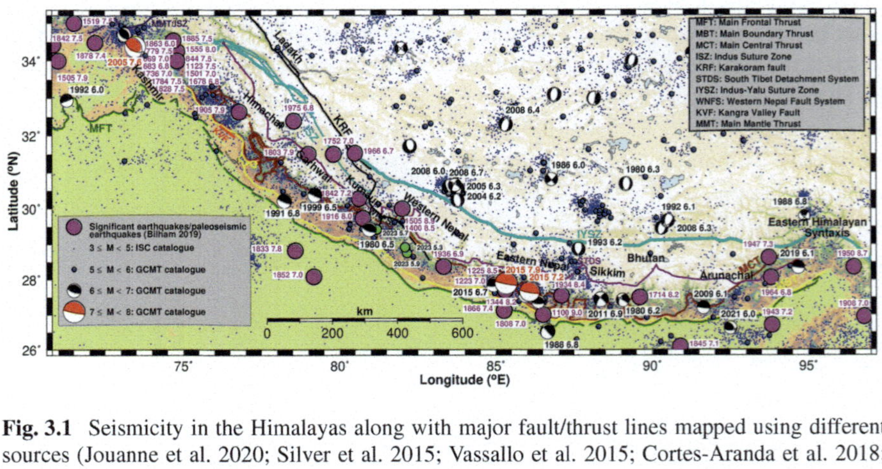

Fig. 3.1 Seismicity in the Himalayas along with major fault/thrust lines mapped using different sources (Jouanne et al. 2020; Silver et al. 2015; Vassallo et al. 2015; Cortes-Aranda et al. 2018; Mohadjer et al. 2016; Dasgupta et al. 2000). The figure was created using GMT (Generic Mapping Tool) software version 6.0.0 (Wessel et al. 2019)

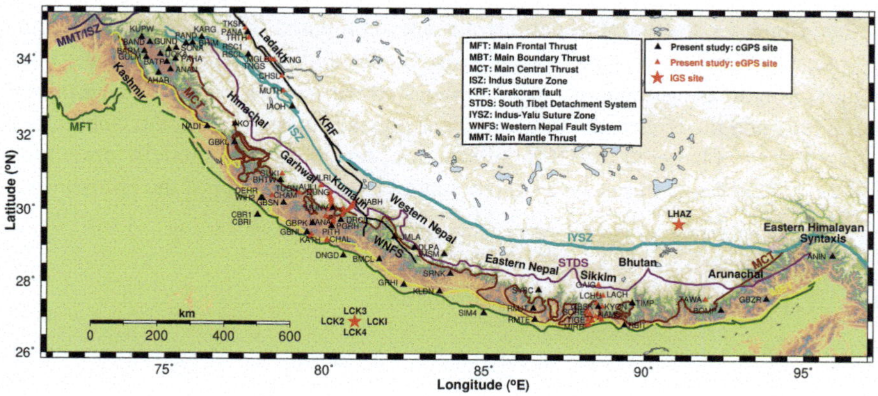

Fig. 3.2 Location map of GPS stations analysed in the present study with new extended span of GPS data. The figure was created using GMT (Generic Mapping Tool) software version 6.0.0 (Wessel et al. 2019)

Himalayan seismicity can be demarcated into the Historical era since 800 AD–1964 followed by the Instrumental period with well-documented earthquakes starting from 1964 (Fig. 3.1). Major historical earthquakes in the Himalayas since 800 AD were comprehensively listed in Biham 2019 from uneven, sparse and fragmented data availability as shown in Fig. 3.1. It was noted that the earthquake data before 1800 is remarkably sparse and is from felt records and before 1500 there is no or insufficient data of many major earthquakes. However, reasonable data of earthquakes since 1800 through historical records and paleo-seismic evidences is available in many parts of the Himalaya. Devastating Himalayan earthquakes in the instrumental period from west to east are M_w 7.6 2005 Kashmir earthquake, M_w 6.8 1991 Uttarkashi earthquake, M_w 6.5 1999 Chamoli earthquake, M_w 6.6 1980 Nepal earthquake, M_w 7.9 & 7.2 2015 Nepal earthquake, M_w 6.8 1988 Nepal Bihar earthquake and M_w 6.9 2011 Sikkim earthquake.

GPS measurements in the Himalayas were first initiated by us in 1994 (Fig. 3.2) when GPS became fully operational and was made available for civilian use. Subsequently, continuous and episodic GPS measurements were carried out by several researchers in the Himalayan arc giving GPS-derived surface deformation rates (Freymueller et al. 1996; Paul et al. 2001; Bilham et al. 2001; Banerjee and Burgmann 2002; Jouanne et al. 2004, 2011, 2020; Bettinelli et al. 2006; Mukul et al. 2018; Vernant et al. 2014; Kundu et al. 2014; Schiffman et al. 2013; Jade 2004; Jade et al. 2011, 2014, 2017, 2020). GPS rates indicate primarily convergence in various segments of the Himalaya from west to east which is being accommodated differentially in the litho-tectonic units of each segment from south to north. Inverse modelling of the GPS surface convergence rates estimates the locking depth, width and slip along MHT in each Himalayan segment thus providing the subduction rate along the northern convergent boundary of the Indian tectonic plate. Slip accumulation estimated from these slip rates gives the seismic potential of each segment of the Himalaya. Further, Geodetic strain rates computed from GPS velocities and seismic strain rates computed using instrumental and historical earthquake catalogues provide an estimate of strain budget, which in turn provides the probable magnitude and the recurrence interval of earthquakes in the Himalayas. Subsequent sections discuss in detail the GPS deformation, seismicity, slip models, strain rates and associated moment rates, seismic potential in each segment of the Himalayas.

3.2 GPS Displacements

Since 1994, we made continuous (cGPS) and episodic (eGPS) GPS measurements in the Himalayas and Table 3.1 gives the site locations, brief site description, data span of all our GPS sites along with Nepal cGPS and IGS (International GNSS Service) sites. At continuous GPS sites, antenna is mounted on a concrete pillar grouted to bedrock exposure with clear sky visibility and, at episodic sites, antenna is mounted on bipods placed over 2–3 mm diameter holes drilled into bedrock exposures marked prominently for subsequent measurements. GPS data at our sites collected with new GPS measurements till 2021 is analysed using GAMIT/GLOBK (Herring et al. 2018) along with the IGS sites listed in Table 3.1 to obtain loosely constrained daily solutions. ITRF 2014 (Altamimi et al. 2016) velocities and uncertainties of our GPS sites (Table 3.1) are estimated by combining these daily solutions and stabilizing the IGS sites to their predetermined ITRF precise locations and velocities with uncertainties. India fixed velocities (Table 3.1) of all the GPS sites are estimated using pole of rotation of the Indian tectonic plate (DeMets et al. 2020) located at $51.51° \pm 0.003$ N, $-4.00° \pm 0.013$ E with an angular velocity of $0.509°\ \text{Myr}^{-1}$. In addition to site velocities listed in Table 3.1, published velocities (Jouanne et al. 2020; Schiffman et al. 2013; Kundu et al. 2014; Banerjee et al. 2008; Vernant et al. 2014; Devachandra et al. 2014; Kumar et al. 2023) of all the Himalayan sites are plotted in Fig. 3.3 to give a complete picture of GPS deformation in the Himalayas. Figure 3.3 is subdivided into Fig. 3.3a and b for better visualization. Published velocities are first converted to ITRF2014 and then to India fixed velocities using the angular velocity of Indian tectonic plate (DeMets et al. 2020). Velocity vector at a given location from one reference frame A is converted to another reference frame B using three translational rates of reference frame B relative to frame A, corresponding velocities in reference frame B is calculated using the equations given by Snay (1999). Horizontal Time-Dependent Positioning (HDTP) utility (geodesy.noaa.gov/TOOLS/Htdp/Htdp.shtml) is used to convert published velocities in different reference frame to ITRF 2014 reference frame. Arc-normal and arc-parallel components of India fixed velocities of a total of 255 Himalayan GPS sites are estimated using the two-dimensional boundary of the Himalayan Arc defined by Bendick and Bilham (2001) and plotted in Fig. 3.5a and b. The arc of a circle with centre at latitude $42.4° \pm 2.1°$ N, longitude $91.6° \pm 1.6°$ E (Bendick and Bilham 2001) is used as reference point to calculate azimuth of each GPS site. North and East velocities of each GPS site are rotated into arc-parallel and arc-normal velocities using GAMIT/GLOBK. We use latitude, longitude, azimuth, 1/2 width of profile swath (km) and velocities with its associated uncertainty of a GPS site as input to generate arc-normal and arc-parallel velocity components.

Southward velocities of all the Himalayan sites (Figs. 3.3a and b) indicate predominantly convergence of 13–28 mm/yr along the northern boundary of Indian tectonic plate. This convergence is being accommodated differentially in the litho-tectonic units of each segment. GPS surface velocities of 255 GPS sites give well constrained arc normal convergence rate of 14 mm/yr in Kashmir, 16 mm/yr in Himachal, 16 mm/yr in Ladakh, 13 mm/yr in Garhwal, 18 mm/yr in Kumaun, 14 mm/yr in western Nepal, 13 mm/yr in central Nepal, 17 mm/yr in eastern Nepal, 10 mm/yr in Sikkim, 12 mm/yr in Bhutan and 18 mm/yr in Arunachal Himalaya (Fig. 3.4). Arc-normal convergence rates of GBZR cGPS site in Arunachal Himalaya is 18 mm/yr and ANIN cGPS site in Eastern Himalayan syntaxis is 25 mm/yr (Fig. 3.5a). These convergence rates are inverted for estimating the locking depth, width and slip along the MHT in each Himalayan segment which is explained in detail in the subsequent section.

Segments of the Himalaya from Himachal in north-west Himalaya to Eastern Himalayan syntaxis record arc parallel rate (Fig. 3.5b) of about 3 mm/yr which reflects elastic deformation from the locked curvature of Main Himalayan Thrust fault. Ladakh Himalayan sites to the west and Lhasa IGS site located in south Tibet indicate arc parallel extension rate of 6–9 mm/yr which is a component of east–west GPS extension rates recorded in Tibet (Mukul et al. 2010; Jade et al. 2014, 2017). The reported arc parallel extension rates are consistent with the 3D block models of the Tibet (20–45 °N; 76–108 °E) proposed considering inter-seismic GPS deformation, mapped fault geometry, earthquake cycle and moment release (Chen et al. 2004; Meade 2007).

Kashmir Himalaya and adjoining regions record arc normal convergence rate of 5–14 mm/yr and arc parallel extension rate of 7 mm/yr suggesting that this region is currently experiencing oblique deformation pattern. Our broadband seismic data indicates moho depth variations in this region suggesting the presence of active unmapped deep-rooted faults which may be contributing to the oblique deformation in this region. Post seismic displacement of about 10–60 mm/yr was observed within the 100 km radius of the 2005 earthquake (Fig. 3.1) soon after the event, reducing through time (4–5 years after the event) in the hanging wall of BBT (Jouanne et al. 2011; Bendick et al. 2015). High post-seismic displacement values are observed at the stations located within the 50 km radius of the epicentre. Our GPS measurements in Kashmir Himalaya were made almost 10–16 years post the 2005 event and only four Kashmir valley GPS sites (KUPW, BAND, BARM and GULM) are located beyond 50 km and within 100 km radius. (Table 3.1 and Figs. 3.1 and 3.2). However, only Kupwara site located at a radial distance of ~75 km to the extreme north-west of the valley recorded high westward velocity 14 mm/yr, i.e. ~9 mm/yr relative to the rest of the valley sites suggesting that this region may be recording a component of the post-seismic deformation even 16 years after the 2005 event. However, we cannot rule out the possibility that Kupwara velocity may have a component contributed by active deep-rooted fault in Kanzalwan and Nangaparbhat to further north (Shrungeshwara et al. 2023; Jade et al. 2020; Jouanne et al. 2020).

Table 3.1 ITRF 14 and India Fixed rates of cGPS & eGPS sites with location description and the data span

Site code	Lat (°N)	Lon (°E)	Epoch	ITRF14 velocities (mm/yr)				India fixed velocities (mm/yr)				Description
				N	σN	E	σE	N	σN	E	σE	
Kashmir–Himachal/Ladakh sites												
Kupwara KUPW	34.6	74.3	Permanent (2015–2019)	32.26	1.07	18.21	1.09	–2.23	1.09	–14.29	1.35	North–western part of Kashmir Valley
Baramulla BARM	34.2	74.3	Permanent (2015–2019)	28.33	1.05	29.12	1.07	–6.17	1.07	–3.60	1.35	Located on the bank of Jhelum River, also near NW edge of the valley
Gulmarg GULM	34.1	74.4	Permanent (2015–2019)	32.25	1.12	29.52	1.14	–2.26	1.13	–3.31	1.39	Located on foothills of Pir Panjal Range, SE of Baramulla
Bandipora BAND	34.4	74.5	Permanent (2015–2019)	26.34	1.12	28.20	1.13	–8.18	1.13	–4.46	1.38	Located on the NW of Wular Lake
Aharbal AHAR	33.6	74.8	Permanent (2015–2019)	30.94	1.06	29.79	1.07	–3.61	1.08	–3.38	1.33	South–western part of Kashmir valley, on foothills of Pir Panjal
Srinagar UOKA	34.1	74.8	Permanent (2015–2019)	29.15	1.15	29.29	1.16	–5.41	1.16	–3.64	1.41	Located in centre of Kashmir Valley, on western bank of Dal Lake
Batpal BATP	33.9	75.0	Permanent (2015–2017)	28.99	1.41	27.23	1.42	–5.59	1.42	–5.88	1.63	NE edge of the valley, NNW of ANAN
Gund GUND	34.3	75.1	Permanent (2015–2019)	26.65	1.09	26.86	1.10	–7.94	1.10	–6.09	1.36	Located on bank of Sind river
Anantnag ANAN	33.7	75.2	Permanent (2015–2019)	28.98	1.06	28.44	1.07	–5.62	1.07	–4.81	1.33	South–eastern edge of the basin
Pahalgam PAHA	34.0	75.3	Permanent (2015–2019)	27.38	1.06	28.10	1.08	–7.23	1.07	–5.05	1.34	Located near Lidder river, Tethys Himalaya
Sonamarg SONA	34.3	75.3	Permanent (2015–2019)	25.92	1.06	27.73	1.08	–8.70	1.07	–5.28	1.34	Located on the bank of Sind river, NE of the valley

(continued)

Table 3.1 (continued)

Site code	Lat (°N)	Lon (°E)	Epoch	ITRF14 velocities (mm/yr)				India fixed velocities (mm/yr)				Description
				N	σN	E	σE	N	σN	E	σE	
Pandrass PAND	34.4	75.6	Permanent (2015–2019)	25.35	1.11	26.70	1.12	–9.30	1.12	–6.35	1.38	Indus suture zone, Zanskar
Bhimbat BHIM	34.4	75.8	Permanent (2017–2020)	23.77	1.39	24.88	1.40	–9.79	1.40	–9.34	1.62	Indus suture zone, Zanskar
Kargil KARG	34.6	76.1	Permanent (2017–2019)	23.97	1.52	26.79	1.53	–10.73	1.53	–6.34	1.73	Indus suture zone, Zanskar
Hanle IAOH	32.8	79.0	Permanent (2008–2019)	18.78	0.63	28.18	0.64	–16.18	0.65	–6.81	1.00	Mt Saraswathi, Ladakh Himalaya
Leh RSCL	34.1	77.6	Permanent (2008–2012)	20.36	1.07	26.84	1.07	–14.49	1.08	–7.03	1.34	Ladakh Himalaya
Leh RSC1	34.1	77.6	Permanent (2016–2019)	21.61	1.27	26.55	1.28	–13.24	1.28	–7.32	1.51	Ladakh Himalaya
MUTH	33.2	78.7	Episodic (1998–2008)	18.39	0.76	26.75	0.77	–16.55	0.76	–7.95	0.77	North of Hanle in Ladakh granites
CHSL	33.6	78.7	Episodic (1998–2008)	18.22	0.77	27.42	0.81	–16.72	0.77	–7.08	0.81	Pangong range of the reunited strands of Karakoram fault
LKNG	34.0	78.4	Episodic (1998–2008)	17.78	0.76	26.88	0.78	–17.14	0.76	–7.33	0.78	NW of Pangong lake
MGLB	34.0	78.3	Episodic (1998–2008)	18.16	0.78	26.31	0.80	–16.75	0.78	–7.85	0.80	Pangong strand of Karakoram fault
TNGS	34.0	78.2	Episodic (1998–2008)	18.95	0.77	25.53	0.79	–15.95	0.77	–8.58	0.79	Tangste strand of Karakoram fault

(continued)

Table 3.1 (continued)

Site code	Lat (°N)	Lon (°E)	Epoch	ITRF14 velocities (mm/yr)				India fixed velocities (mm/yr)				Description
				N	σN	E	σE	N	σN	E	σE	
Panamik PANA	34.7	77.6	Permanent (2004–2006)	20.67	0.67	26.07	0.68	–14.17	0.67	–7.49	0.68	Nubra valley, Karakoram terrane
Panamik PNMK	34.7	77.6	Episodic (1997–2008)	20.35	0.67	25.61	0.68	–14.49	0.67	–7.94	0.68	Nubra valley, Karakoram terrane
Tirth TRTH	34.6	77.6	Episodic (1997–2008)	19.96	0.74	26.27	0.75	–14.89	0.74	–7.38	0.75	Nubra valley, Karakoram terrane
Staksha TKSH	34.8	77.2	Episodic (2004–2006)	19.79	0.78	26.63	0.78	–15.05	0.78	–6.86	0.78	Nubra valley, Karakoram terrane
Naddi NADI	32.3	76.3	Permanent (2004–2005)	30.47	1.52	24.41	1.52	–4.26	1.52	–9.98	1.52	MCT zone south of Leh
Kothi KOT1	32.3	77.2	Permanent (2004–2005)	24.53	1.61	25.45	1.61	–10.28	1.61	–9.19	1.61	MBT zone south of Leh
Kullu GBKL	31.8	77.2	Permanent (2008–2014)	29.30	0.81	28.88	0.82	–5.51	0.83	–5.99	1.12	MCT Zone, on the banks of Beas River
Garhwal Himalaya												
Roorkee CBRI	29.9	77.9	Permanent (2014–2015)	34.68	1.74	33.68	1.75	–0.20	1.75	–2.34	1.90	Frontal Himalaya
Roorkee CBR1	29.9	77.9	Permanent (2015–2019)	35.02	1.07	34.56	1.08	0.14	1.08	–1.46	1.31	Frontal Himalaya
Dehradun DEHR	30.3	78.1	Permanent (2009–2011)	34.44	1.24	34.17	1.24	–0.45	1.25	–1.69	1.44	South of MBT in Garhwal Himalaya

(continued)

Table 3.1 (continued)

Site code	Lat (°N)	Lon (°E)	Epoch	ITRF14 velocities (mm/yr)				India fixed velocities (mm/yr)				Description
				N	σN	E	σE	N	σN	E	σE	
Dehradun WIH2	30.3	78.0	Permanent (2002–2005)	34.90	1.04	34.14	1.04	0.01	1.04	−1.70	1.04	South of MBT in Garhwal Himalaya
Nagoli GBSN	30.2	78.7	Permanent (2008–2014)	32.90	0.82	34.82	0.83	−2.05	0.84	−1.30	1.11	Garhwal Himalaya
Chamba CHAM	30.4	78.4	Episodic (1996–2016)	33.31	0.49	33.90	0.49	−1.61	0.49	−2.02	0.49	Jhandi Daar north of MCT
Bhatwari BHTW	30.8	78.6	Permanent (2004–2005)	30.79	2.08	47.93	2.09	−1.74	2.08	−4.15	2.09	In Higher Himalayas
Sukhi SUKI	31.0	78.7	Episodic (1996–2016)	24.38	0.49	29.70	0.49	−10.56	0.49	−6.04	0.49	Sukhi top mountain in the Higher Himalayas
Lansdowne LANS	29.9	78.7	Episodic (1996–2006)	34.75	0.73	34.25	0.75	−0.20	0.73	−2.02	0.75	North of MBT in Lesser Himalaya
Tungnath TUGN	30.5	79.2	Episodic (1996–2016)	27.30	0.49	29.45	0.49	−7.69	0.49	−6.69	0.49	Close to the 1999 Chamoli earthquake in MCT zon
Auli AULI	30.5	79.6	Episodic (1996–2016)	32.84	0.49	33.78	0.49	−2.17	0.49	−2.45	0.49	North of MCT zone in Higher Himalaya
Malari MLRI	30.7	79.9	Episodic (1996–2016)	22.19	0.52	29.62	0.52	−12.84	0.52	−6.64	0.52	In Higher Himalaya
Almora GBPK	29.6	79.6	Permanent (2008–2014)	34.03	0.80	35.23	0.81	−0.99	0.82	−1.41	1.09	Lesser Himalaya of Garhwal
Kumaun Himalaya												

(continued)

Table 3.1 (continued)

Site code	Lat (°N)	Lon (°E)	Epoch	ITRF14 velocities (mm/yr)				India fixed velocities (mm/yr)				Description
				N	σN	E	σE	N	σN	E	σE	
Nainital GBNL	29.4	79.4	Permanent (2008–2014)	34.16	0.80	35.76	0.81	–0.84	0.82	–0.94	1.09	Kumaun Himalaya
Almora KTML	29.6	79.6	Permanent (1997–1999)	29.89	1.53	38.88	1.53	–5.13	1.53	2.24	1.53	In Lesser Himalaya
Chaukadi CHAU	29.8	80.0	Episodic (1996–2015)	30.72	0.50	34.11	0.50	–4.32	0.50	–2.57	0.50	Hard rock exposure NE of Almora in Lesser Himalaya
Chalti CHAL	29.2	80.1	Episodic (2004–2007)	21.35	2.29	30.46	2.54	–13.70	2.29	–6.51	2.54	In Lesser Himalaya
Kathgodam KATH	29.3	79.5	Episodic (2004–2007)	27.57	1.76	24.73	1.87	–7.44	1.76	–12.05	1.87	In Lesser Himalaya
Sukhi Dhang SUKH	29.2	80.1	Episodic (2004–2015)	35.56	0.68	36.98	0.70	0.51	0.68	–0.01	0.70	Sukhi Dhang Mountain in Lesser Himalaya
BHIM	29.4	79.6	Episodic (2004–2015)	34.33	0.67	40.68	0.67	–0.68	0.67	3.91	0.67	In Lesser Himalaya
Pithoragarh PITH	29.6	80.2	Episodic (1999–2007)	32.65	0.81	34.82	0.82	–2.40	0.81	–2.01	0.82	Hard rock exposure E of Almora in Lesser Himalaya
Pithoragarh PIHT	29.6	80.2	Episodic (1999–2007)	33.12	0.71	33.66	0.75	–1.93	0.71	–3.17	0.75	Hard rock exposure E of Almora in Lesser Himalaya
Pithoragarh PGRH	29.6	80.2	Permanent (2005–2006)	35.22	0.71	32.70	0.75	0.17	0.71	–4.15	0.75	Hard rock exposure E of Almora in Lesser Himalaya
Bala BALA	30.0	80.2	Episodic (2002–2014)	25.97	0.63	32.41	0.63	–9.08	0.63	–4.22	0.63	Southernmost point of the Gori valley transect

(continued)

Table 3.1 (continued)

Site code	Lat (°N)	Lon (°E)	Epoch	ITRF14 velocities (mm/yr)				India fixed velocities (mm/yr)				Description
				N	σN	E	σE	N	σN	E	σE	
Khaliatop KHAL	30.1	80.2	Episodic (1999–2006)	25.74	0.75	31.04	0.75	–9.31	0.75	–5.59	0.75	In MCT Zone N of Bala in Higher Himalayas
Munsyari MUNS	30.1	80.2	Episodic (1996–2002)	26.39	0.49	32.85	0.49	–8.66	0.49	–3.78	0.49	In MCT zone N of Bala in Higher Himalayas
Lilam LILM	30.2	80.2	Episodic (2002–2012)	24.21	0.70	32.56	0.72	–10.85	0.70	–4.05	0.72	In higher Himalayas
Munsyari MUNT	30.1	80.2	Episodic (1999–2014)	25.21	0.49	31.67	0.49	–9.85	0.49	–4.98	0.49	In MCT zone N of Bala in Higher Himalayas
Munsyari MUNV	30.1	80.2	Permanent (2005–2007)	23.35	0.78	32.36	0.78	–11.71	0.78	–4.29	0.78	In MCT zone N of Bala in Higher Himalayas
Railgadi RALG	30.2	80.2	Episodic (2002–2014)	25.75	0.63	31.68	0.63	–9.30	0.63	–4.91	0.63	In Higher Himalayas
Bogdiar BOGD	30.2	80.2	Episodic (2002–2012)	24.41	0.75	30.44	0.76	–10.64	0.75	–6.13	0.76	In Martoli fault region of Higher Himalaya
Martoli MART	30.3	80.2	Episodic (2003–2014)	22.30	0.65	30.55	0.65	–12.75	0.65	–5.97	–12.75	In Martoli fault region of Higher Himalaya
Laspa LASP	30.3	80.2	Episodic (2000–2014)	24.41	0.59	30.08	0.59	–10.64	0.59	–6.45	0.59	Hard rock exposure in Martoli fault region of Tethyan Himalaya
Burfu BURF	30.4	80.2	Episodic (1998–2014)	20.04	0.54	24.74	0.55	–15.01	0.54	–11.76	0.55	Tethyan Himalaya
Belju BEL1	30.4	80.2	Episodic (2003–2007)	25.30	1.31	29.13	1.37	–9.75	1.31	–7.34	1.37	Tethyan rocks of Tethyan Himalaya

(continued)

Table 3.1 (continued)

Site code	Lat (°N)	Lon (°E)	Epoch	ITRF14 velocities (mm/yr)				India fixed velocities (mm/yr)				Description
				N	σN	E	σE	N	σN	E	σE	
Milam MILA	30.4	80.2	Episodic (2000–2014)	21.15	0.59	29.54	0.60	−13.90	0.59	−6.91	0.60	Tethyan rocks of Tethyan Himalaya
Dung DUNG	30.5	80.2	Episodic (2002–2007)	23.29	1.13	29.69	1.25	−11.76	1.13	−6.75	1.25	Northernmost point of the transect in Tethyan rocks of Tethyan Himalaya
Navighang NABH	30.2	81.0	Episodic (2004–2015)	22.38	0.65	30.78	0.66	−12.72	0.65	−6.01	0.66	Tethyan rocks of Tethyan Himalaya
Kalapani KALA	30.2	80.9	Episodic (2003–2008)	25.17	1.27	31.80	1.28	−9.92	1.27	−4.98	1.28	Tethyan rocks of Tethyan Himalaya
Gunji GUNJ	30.2	80.9	Episodic (2004–2015)	23.00	0.69	31.58	0.71	−12.09	0.69	−5.21	0.71	Martoli Schist of Tethyan Himalya
Garbyang GARB	30.1	80.9	Episodic (2004–2015)	22.04	0.67	31.49	0.67	−13.05	0.67	−5.32	0.67	In Martoli fault region of Tethyan Himalaya
Chialekh CHIA	30.1	80.8	Episodic (2004–2015)	21.65	0.65	32.37	0.66	−13.44	0.65	−4.43	0.66	In Martoli fault region of Tethyan Himalaya
Lamari LAMB	30.1	80.8	Episodic (2003–2008)	19.67	1.43	31.08	1.51	−15.42	1.43	−5.73	1.51	In Martoli fault region of Tethyan Himalaya
Jipti JIPT	30.0	80.7	Episodic (2003–2008)	22.05	1.35	29.53	1.37	−13.04	1.35	−7.29	1.37	MCT zone in Higher Himalayas
Pangu PANG	30.0	80.6	Episodic (2004–2006)	22.99	1.71	28.45	1.90	−12.09	1.71	−8.35	1.90	MCT zone in Higher Himalayas
Sobla SOBL	30.1	80.6	Episodic (2004–2007)	23.00	1.06	32.84	1.17	−12.08	1.06	−3.91	1.17	MCT zone in Higher Himalayas

(continued)

Table 3.1 (continued)

Site code	Lat (°N)	Lon (°E)	Epoch	ITRF14 velocities (mm/yr)				India fixed velocities (mm/yr)				Description
				N	σN	E	σE	N	σN	E	σE	
Tapovan TAPO	29.9	80.6	Episodic (2003–2008)	32.50	1.71	26.84	1.79	–2.57	1.71	–9.98	1.79	MCT zone in Higher Himalayas
Askot ASKO	29.7	80.3	Episodic (2004–2015)	32.48	0.69	33.37	0.69	–2.58	0.69	–3.43	0.69	In Lesser Himalayas NE of Pithoragarh
Kanalichhina KANA	29.7	80.3	Episodic (2003–2008)	34.22	1.65	28.22	1.67	–0.84	1.65	–8.60	1.67	Southern most point of the transect in Lesser Himalaya NE of Pithoragarh
Nepal Himalaya												
Syangboche SYBC	27.8	86.7	Permanent (2009–2010)	24.57	1.46	36.50	1.47	–10.66	1.46	–2.95	1.47	North–eastern Nepal
Ramite RMTE	27.0	86,6	Permanent (2009–2010)	35.89	1.58	36.31	1.59	0.65	1.58	–3.39	1.59	South–eastern Nepal
Rumjartar RMJT	27.3	86,6	Permanent (2009–2010)	30.55	1.95	36.34	1.96	–4.69	1.95	–0.3.24	1.96	Mid–eastern Nepal
Simara SIM4	27,2	84.9	Permanent (2004–2005)	37.77	2.04	35.53	2.07	2.54	2.04	–3.66	2.07	South–eastern Nepal
Sarangkot SRNK	28.3	83.9	Permanent (2009–2010)	29.16	1.55	34.20	1.56	–6.05	1.55	–4.29	1.56	Sarangkot Mountain
Jomsom JMSM	28.8	83.7	Permanent (2004–2010)	25.82	0.81	35.11	0.81	–9.39	0.81	–3.11	0.81	On the banks of Kali–Gadanki river
Koldana KLDN	27.8	83.6	Permanent (2009–2010)	33.70	1.46	35.49	1.46	–1.51	1.46	–3.09	1.46	Central Nepal

(continued)

Table 3.1 (continued)

Site code	Lat (°N)	Lon (°E)	Epoch	ITRF14 velocities (mm/yr)				India fixed velocities (mm/yr)				Description
				N	σN	E	σE	N	σN	E	σE	
Dolpa DLPA	29.0	82.8	Permanent (2009–2010)	23.70	1.46	32.02	1.47	–11.48	1.46	–5.85	1.47	Karnali zone of North–western Nepal
Ghorahi GRHI	28.0	82.5	Permanent (2009–2010)	33.47	1.46	35.07	1.47	–1.70	1.46	–3.12	1.47	South–west Nepal
Jumla JMLA	29.3	82.2	Permanent (2009–2010)	25.50	1.46	31.63	1.46	-9.66	1.46	–5.94	1.46	Western Nepal
Bhimchula BMCL	28.7	81.7	Permanent (2009–2010)	32.31	1.48	32.95	1.49	-2.83	1.48	-4.73	1.49	Western Nepal
DNGD	28.8	80.6	Permanent (2009–2010)	32.82	1.46	34.92	1.47	-2.26	1.46	–2.39	1.47	Western Nepal
Darchula DRCL	29.7	80.5	Permanent (2009–2010)	27.12	1.48	30.00	1.48	-7.95	1.48	–6.87	1.48	Western Nepal
Other Himalayan sites												
Panthang GBSK	27.4	88.6	Permanent (2008–2014)	28.01	0.80	39.19	0.81	-7.19	0.85	–0.94	1.07	Sikkim Himalaya
Mirrik MIRR	26.8	88.2	Episodic (2006–2012)	30.20	1.15	39.13	1.22	-5.01	1.15	–1.07	1.22	Foreland of Darjiling–Sikkim–Eastern Nepal Himalaya
Giddpahar GIDD	26.9	88.3	Episodic (2006–2012)	34.57	0.92	36.57	0.94	–0.64	0.92	–3.64	0.94	Foreland of Darjiling–Sikkim–Eastern Nepal Himalaya
TIGE	27.0	88.3	Episodic (2005–2012)	36.39	0.88	37.68	0.90	1.18	0.88	–2.49	0.90	Sikkim Himalaya

(continued)

Table 3.1 (continued)

Site code	Lat (°N)	Lon (°E)	Epoch	ITRF14 velocities (mm/yr)				India fixed velocities (mm/yr)				Description
				N	σN	E	σE	N	σN	E	σE	
Mungpu MUNG	27.0	88.4	Episodic (2000–2012)	35.12	0.66	38.04	0.67	–0.09	0.66	–2.17	0.67	Foreland of Darjiling–Sikkim–Eastern Nepal Himalaya
Nim NIMC	27.0	88.7	Episodic (2002–2012)	34.01	0.74	37.94	0.76	–1.19	0.74	–2.34	0.76	Foreland of Darjiling–Sikkim–Eastern Nepal Himalaya
Labha LAVA	27.1	88.7	Episodic (2000–2012)	33.21	0.65	37.64	0.66	–1.99	0.65	–2.61	0.66	Lesser Himalaya–Darjiling–Sikkim–Tibet Wedge
GAIG	28.0	88.6	Episodic (2005–2009)	24.13	1.11	38.45	1.12	–11.07	1.11	–1.48	1.12	Sikkim Himalaya
Delo DELO	27.1	88.5	Episodic (1997–2012)	33.41	0.61	37.20	0.63	–1.79	0.61	–3.00	0.63	Lesser Himalaya–Darjiling–Sikkim–Tibet Wedge
Namchi NAMC	27.2	88.3	Episodic (2000–2012)	33.16	0.65	36.59	0.74	–2.05	0.65	–3.54	0.67	Lesser Himalaya–Darjiling–Sikkim–Tibet Wedge
Soreng SORE	27.2	88.2	Episodic (2005–2012)	31.10	0.87	36.74	0.89	–4.11	0.87	–3.34	0.89	Lesser Himalaya–Darjiling–Sikkim–Tibet Wedge
Peling PELI	27.3	88.2	Episodic (2005–2012)	29.06	1.55	38.75	1.59	–6.15	1.55	–1.30	1.59	Lesser Himalaya–Darjiling–Sikkim–Tibet Wedge

(continued)

Table 3.1 (continued)

Site code	Lat (°N)	Lon (°E)	Epoch	ITRF14 velocities (mm/yr)				India fixed velocities (mm/yr)				Description
				N	σN	E	σE	N	σN	E	σE	
Tsangu KYON	27.4	88.7	Episodic (2000–2009)	28.83	0.74	38.55	0.77	–6.37	0.74	–1.62	0.77	Lesser Himalaya–Darjiling–Sikkim–Tibet Wedge
Lachung LCHU	27.7	88.7	Episodic (2005–2009)	26.10	1.10	38.72	1.13	–9.10	1.10	–1.35	1.13	Higher Himalaya–Darjiling–Sikkim–Tibet Wedge
Lachen LACH	27.7	88.6	Episodic (2005–2012)	26.47	0.89	36.31	0.91	–8.73	0.89	–3.70	0.91	Higher Himalaya–Darjiling–Sikkim–Tibet Wedge
RBIT	26.9	89.4	Permanent (2003–2005)	35.15	1.42	42.84	1.43	–0.03	1.42	2.32	1.43	Bhutan
TIMP	27.5	89.6	Permanent (2003–2005)	31.74	1.41	41.51	1.42	–3.43	1.41	1.12	1.42	Bhutan
Lhasa LHAZ	29.7	91.1	IGS (2001–2020)	15.54	0.60	46.34	0.61	–19.55	0.69	6.23	0.95	IGS station located in south–eastern Tiber, China
Bomdilla BOMP	27.3	92.4	Permanent (2008–2016)	19.67	0.73	43.31	0.73	–15.35	0.82	2.08	1.01	Arunachal Himalaya
GBZR	27.6	93.8	Permanent (2010–2014)	16.92	0.89	43.78	0.89	–17.99	0.89	2.25	0.89	Arunachal Himalaya
TAWA	27.6	91.9	Episodic (2002–2011)	15.26	1.42	48.23	1.98	–19.79	1.42	7.22	1.98	Arunachal Himalaya
Anini ANIN	28.80	95.90	Permanent (2007–2008)	9.84	0.64	45.58	0.66	–24.87	0.64	3.79	0.66	Arunachal Himalaya

(continued)

Table 3.1 (continued)

Site code	Lat (°N)	Lon (°E)	Epoch	ITRF14 velocities (mm/yr)				India fixed velocities (mm/yr)				Description
				N	σN	E	σE	N	σN	E	σE	
Stable Indian plate sites												
Lucknow LCKI	26.9	81.0	IGS (2012–2014)	34.59	1.38	37.74	1.39	–0.51	1.39	–0.43	1.55	IGS station, Uttar Pradesh
Lucknow LCK2	26.9	81.0	IGS (2012–2014)	34.75	1.40	37.79	1.41	–0.35	1.41	–0.38	1.57	IGS station, Uttar Pradesh
Lucknow LCK3	26.9	81.0	IGS (2015–2020)	35.32	0.94	36.01	0.94	0.22	0.96	–2.16	1.17	IGS station, Uttar Pradesh
Lucknow LCK4	26.9	81.0	IGS (2015–2020)	35.32	0.94	36.01	0.94	0.22	0.96	–2.16	1.17	IGS station, Uttar Pradesh
Hyderabad HYDE	17.4	78.6	IGS (2002–2021)	34.96	0.55	40.54	0.55	0.01	0.57	–0.42	0.78	IGS station in Central India located on deccan Plateau
Bangalore IISC	13.0	77.6	IGS (1996–2021)	34.89	0.51	42.18	0.52	0.01	0.53	0.13	0.73	IGS station in South India located on Bedrock exposure
Bangalore BAN2	13.0	77.5	IGS (2004–2012)	34.52	0.75	43.38	0.76	–0.35	0.77	1.34	0.92	IGS station
Colombo SGOC	6.9	79.9	IGS (2013–2019)	34.71	0.91	43.87	0.91	–0.35	0.91	0.30	1.02	IGS station, Narahenpita, Colombo, Srilanka
IGS sites												
Maldives MALD	4.2	73.5	IGS (2000–204)	34.26	0.97	43.87	1.01	–0.17	0.99	0.20	1.11	IGS station, Southernmost of the Indian Plate
Portblair PBR2	11.6	92.7	IGS (2012–2016)	18.85	1.05	8.00	1.06	–16.17	1.12	–36.26	1.17	Andaman and Nicobar Islands

(continued)

Table 3.1 (continued)

Site code	Lat (°N)	Lon (°E)	Epoch	ITRF14 velocities (mm/yr)				India fixed velocities (mm/yr)				Description
				N	σN	E	σE	N	σN	E	σE	
Karratha KARR	-21.0	117.1	IGS (1996–2020)	59.33	0.61	39.51	0.63	29.15	1.06	4.57	0.85	Australian tectonic plate
Wuhan WUHN	30.5	114.4	IGS (1996–2016)	–11.04	0.48	33.28	0.48	–42.04	0.95	–13.40	0.83	China in Eurasian tectonic plate
Bakosurtanal BAKO	–6.5	106.8	IGS (2002–2021)	–6.17	0.54	24.80	0.55	–39.12	0.87	–17.85	0.73	Located in Indonesia
Nanyang NTUS	1.3	103.7	IGS (1997–2020)	–6.83	1.17	30.30	1.18	–40.43	1.31	–14.28	1.27	Located in Singapore
Kunming KUNM	25.0	102.8	IGS (1998–2013)	–16.40	1.51	33.75	1.50	–50.14	1.62	–10.73	1.63	China in Eurasian tectonic plate
Cocos COCO	–12.2	96.8	IGS (1996–2020)	50.88	0.59	44.28	0.60	16.25	0.75	2.32	0.81	Coco island, western Australian tectonic plate
Selezaschita SELE	43.2	77.0	IGS (1999–2013)	4.20	0.54	28.29	0.54	–30.58	0.56	–0.35	1.08	Kazakhstan in Tean Shan tectonic plate
Poligan POL2	42.7	74.7	IGS (1996–2021)	4.80	0.63	27.82	0.66	–29.73	0.66	–0.16	1.14	Kyrghyzstan in Eurasian tectonic plate
Diego Garcia DGAR	–7.3	72.4	IGS (1996–2021)	32.80	0.71	46.70	0.77	–1.47	0.74	1.67	0.92	Diego Garcial Island
Kitab KIT3	39.1	66.9	IGS (1996–2021)	4.99	0.42	27.84	0.42	–28.28	0.52	0.65	0.96	Uzbekistan in Eurasian tectonic plate
Seychelles SEY1	–4.7	55.5	IGS (2003–2016)	12.03	0.53	24.22	0.54	–18.35	0.75	–21.42	0.73	Mahe Island, East African tectonic plate

(continued)

Table 3.1 (continued)

Site code	Lat (°N)	Lon (°E)	Epoch	ITRF14 velocities (mm/yr)				India fixed velocities (mm/yr)				Description
				N	σN	E	σE	N	σN	E	σE	
Bahrain BHR1	26.2	50.6	IGS (2000–2009)	30.35	0.62	30.72	0.62	1.62	0.88	–0.12	0.86	Located in Arabian tectonic plate
Bahrain BHR2	26.2	50.6	IGS (2005–2009)	31.02	0.99	29.76	0.98	2.29	1.17	–1.08	1.15	Located in Arabian tectonic plate

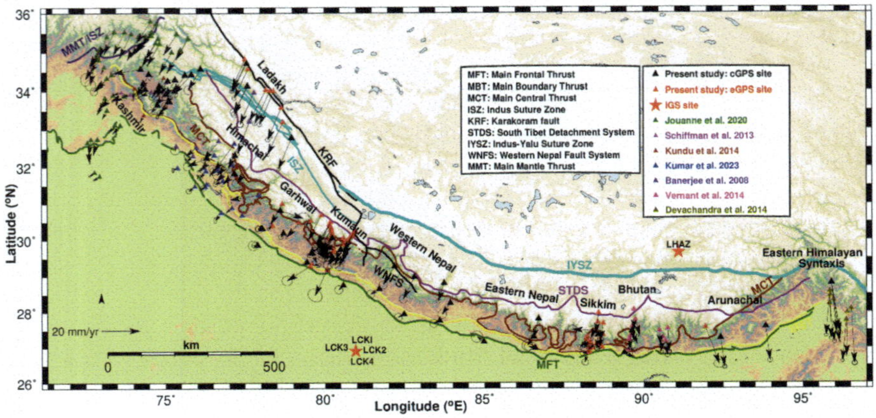

Fig. 3.3 India fixed velocities estimated using the Euler pole of Indian tectonic plate given by DeMets et al. (2020) of all GPS sites (Table 3.1) and published (Banerjee et al. 2008; Schiffman et al. 2013; Kundu et al. 2014; Vernant et al. 2014; Jouanne et al. 2020 and Devachandra et al. 2014; Kumar et al. 2023). The figure was created using GMT (Generic Mapping Tool) software version 6.0.0 (Wessel et al. 2019). **a** India fixed velocities estimated using Euler pole of Indian tectonic plate given by DeMets et al. (2019) of Kashmir, Ladakh, Himachal, Garhwal and Kumaun Himalaya. The figure was created using GMT (Generic Mapping Tool) software version 6.0.0 (Wessel et al. 2019). **b** India fixed velocities estimated using Euler pole of Indian tectonic plate given by DeMets et al. (2019) of Nepal, Sikkim, Bhutan, Arunachal and Eastern Himalayan syntaxis. The figure was created using GMT (Generic Mapping Tool) software version 6.0.0 (Wessel et al. 2019)

3.3 Decollement Slip Models

Okada dislocation theory and weighted least square was used for inverse modelling of India fixed velocities to determine the subsurface decollement parameters and the associated slip. We assume that the observed GPS-derived surface deformation is primarily caused due to subsurface slip along a buried dislocation and for the Himalayas the decollement is MHT along which Indian tectonic plate subducts below the Eurasian plate. We consider the decollement (MHT) in each segment of the Himalaya as a finite rectangular fault of Length L and width W in an elastic half-space and the slip m is estimated using weighted least square inversion of GPS velocities (Okada 1985) as $m = (G^T \cdot W_e \cdot G)^{-1}\ G^T \cdot W_e \cdot d$, where W_e is the weight matrix (inverse of covariance matrix of GPS velocities), d is GPS velocities and G is the function of fault parameters (length, width, locking depth, dip and strike). Total slip m has two components dip slip and strike slip in two directions perpendicular to each other and the equations are written accordingly. We use a range of apriori fault parameters derived from fault mapping and earthquake focal mechanism as input to the inversion program along with GPS velocities. Output is the slip *m* and fault parameters that best fit the observed surface deformation recorded at the GPS sites by minimizing the residual velocities.

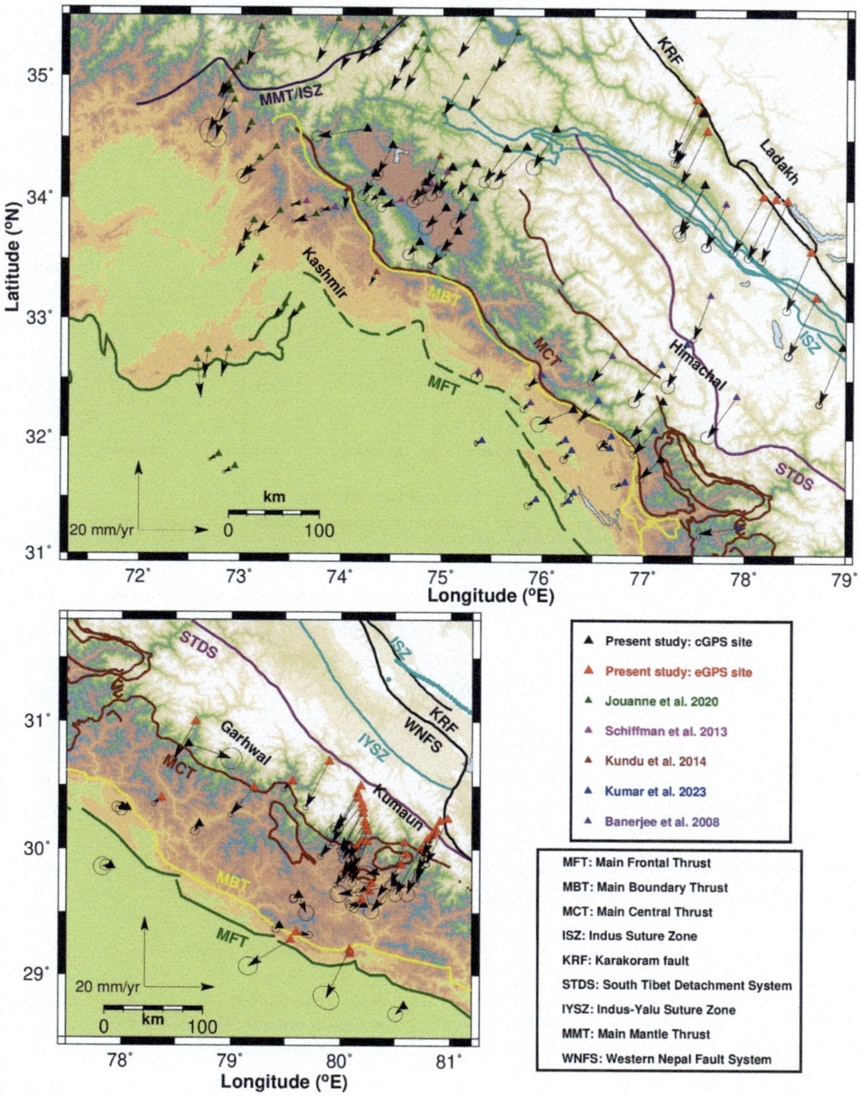

(a)

Fig. 3.3 (continued)

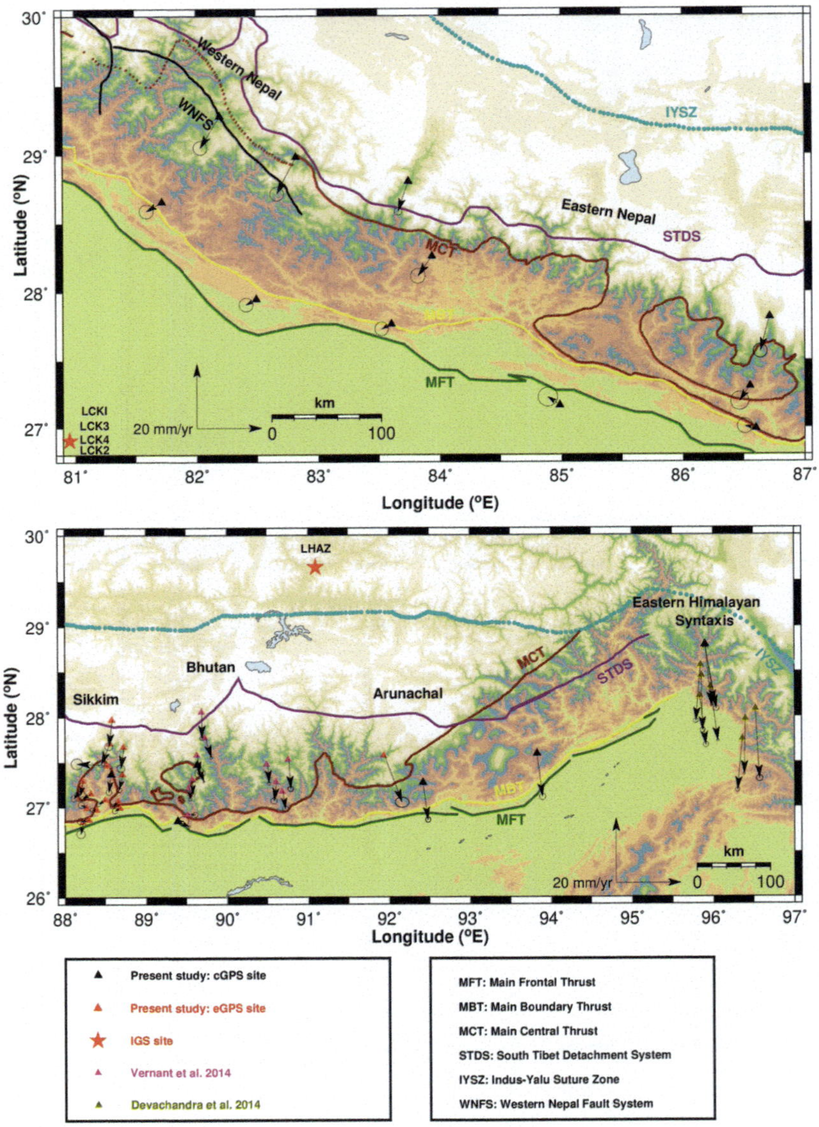

(b)

Fig. 3.3 (continued)

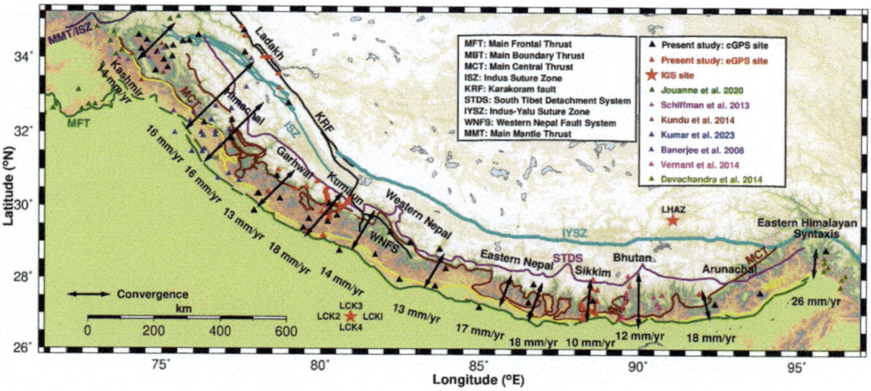

Fig. 3.4 GPS-derived arc-noral surface convergence rates across the Himalayan arc from west to east. The figure was created using GMT (Generic Mapping Tool) software version 6.0.0 (Wessel et al. 2019)

GPS velocities (Table 3.1) along with the published velocities of sites located (Figs. 3.3a and b) in each segment of Himalaya are inverted to estimate the dip slip, strike slip, length, locking depth and width of the decollement and given in Table 3.2 and Fig. 3.6. Estimated geometry of buried dislocation is consistent with micro-seismicity recorded in the Himalayas. Total slip along MHT in different segments of Himalaya varies from 13 to 20 mm/yr locked from surface to a depth of 15–20 km over a width of 100–150 km indicating high strain accumulation in the 2500 km Himalayan arc. This stored strain energy increases the inter-seismic stress which would be released by earthquakes at regular intervals whenever the strain crosses the threshold value. Well-constrained variable deformation and slip rates along the long Himalayan arc are used to derive the geodetic strain rates in each segment of Himalaya. These rates and the associated scalar geodetic moment rates precisely quantify the stored strain energy in each segment as well as slip accumulation. Taking into account the seismicity of the region which quantifies released strain energy in terms of seismic strain and associated seismic moment, seismic potential of each segment of the Himalayas is estimated as detailed in the subsequent sections.

3.4 Geodetic Strain Rates

Two-dimensional XY grid is used to estimate crustal strain rates from India fixed velocities and positions of GPS sites using Teza and Galgaro (2008) formulation. Strain tensor I is expressed as U = AI + e, in matrix form for a two-dimensional grid where matrix A contains the position of GPS points, U are the GPS velocities and e is the residual vector. Modified least square inversion is used to minimize the residual vector. In this approach, uncertainty in the GPS data points are weighed by a distance-dependent scale factor to ensure major contribution from the nearest GPS

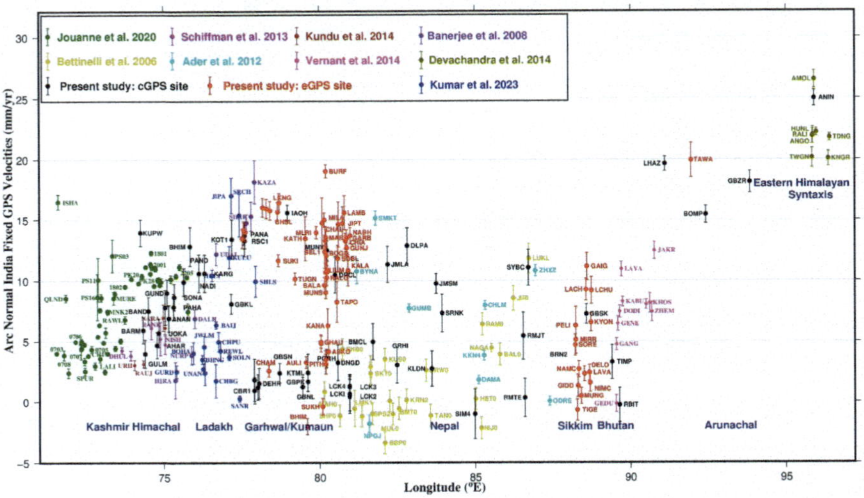

(a)

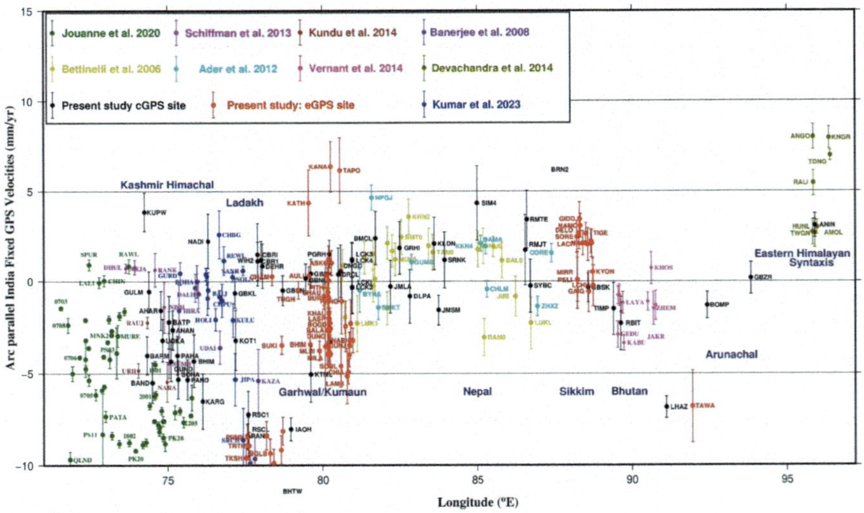

(b)

Fig. 3.5 **a** Arc-normal rates of GPS sites in the Himalayan arc with error bars determined by rotating the site velocities on to the direction locally orthogonal to the arc using arc geometry defined by Bendick and Bilham (2001). The figure was created using GMT (Generic Mapping Tool) software version 6.0.0 (Wessel et al. 2019). **b** Arc-parallel rates of GPS sites in the Himalayan arc with error bars determined by rotating the site velocities on to the direction locally orthogonal to the arc using arc geometry defined by Bendick and Bilham (2001). The figure was created using GMT (Generic Mapping Tool) software version 6.0.0 (Wessel et al. 2019)

Table 3.2 Surface convergence rates, dislocation model parameters and associated slip rates

Location	Dislocation	Source	Convergence rate (mm/yr) present study	Length (km)	Width (km)	Depth (km)	Dip (°)	Dip slip (mm/year)	Strike slip (mm/year)
Kashmir	MHT	Present study	14	145	145	15	7	16	4
Ladakh		Present study	16	–	–	–	–	–	–
Himachal	MHT	Present study	16	180	100	20	6	14	3.4
Garhwal	MHT	Present study	13	160	109	16	5	16 ± 1	1 ± 1
Kumaun	MHT	Present study	18	200	110	20	8	18 ± 1.5	1 ± 1.5
Western Nepal	MHT	Bettinelli et al. (2006)	14	400	150	12	4.5	13.4 ± 5	–
		Jouanne et al. (2004)	–	–	–	20–21	9–10	19	0–1
Central Nepal	MHT	Jouanne et al. (2004)	–	–	–	17.21	9–10	19–20	0–2
Central + Eastern Nepal	MHT	Bettinelli et al. (2006)	13	500	115	20	10	19 ± 2.5	–
Sikkim	MHT	Present study	10	200	105	20	9	18 ± 1.5	–
Bhutan	MHT	Present study	12	–	–	–	–	–	–
Arunachal	MHT	Present study	18	250	130	17	10	16 ± 2	–
Eastern Syntaxis		Devachandra et al. (2014)	26	–	–	–	–	17 ± 2	10 ± 3

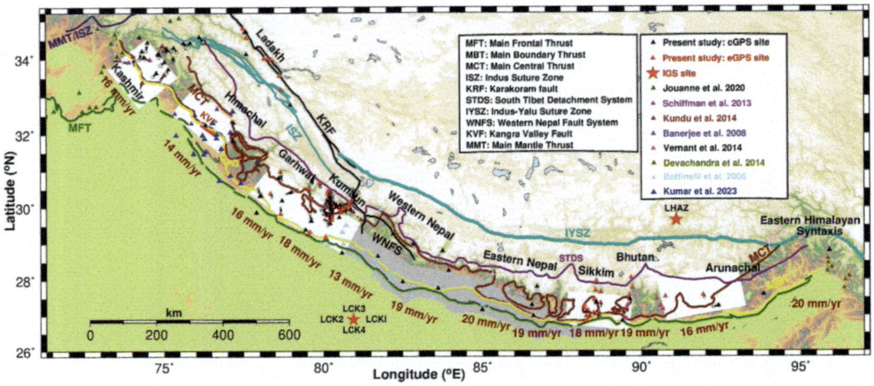

Fig. 3.6 Rectangular dislocations (white solid rectangles) estimated by inverting the GPS velocities from the present study are plotted along with dislocation model (solid grey patch) of Bettinelli et al. (2006) for Nepal Himalaya along with associated slip rates. GPS sites are denoted by triangles. The figure was created using GMT (Generic Mapping Tool) software version 6.0.0 (Wessel et al. 2019)

point for strain computations at each grid point. This enables to evaluate the scale-dependent behaviour of the grid system using a range of scale factors and grid sizes to arrive at optimal value that best fits the system. It also allows to estimate the effect of specific GPS point and gives a choice to include or exclude the point for computation of strain which is beneficial for unevenly distributed GPS points for a specific two-dimensional XY grid system. Grid strain program (Teza and Galgaro 2008) with optimized grid size and scale factor is used for geodetic strain computations with GPS velocities and positions as input. Programme performs computation at each node XY of the two-dimensional grid and the output is given as eigenvectors, eigenvalues and normalized shear of the strain field tensor. Principal strain rates at a XY grid point are considered to be of high significance, if in a grid plane, each of the three equal sectors with a 120 aperture centred at the reference grid point contains at least one GPS point at a distance less than the scale factor. If only two out of three sectors contain the GPS point at a distance less than the scale factor, the strain rate at the grid point is considered to be of mean significance.

Himachal, Kashmir and Ladakh Himalaya: Geodetic strain rates are estimated from GPS velocities of 58 GPS sites located in the region (32°–35.5° N; 72.5°–77° E) covering the 2005 Muzaffarabad earthquake and 1905 Kangra earthquake (Shrungeshwara et al. 2023). GPS velocities with uncertainty of less than 3 mm/yr were used for strain analysis with a grid size of 25 km × 25 km grid size and optimal scale factor of 80 km (Fig. 3.7). Principal strain rates of high and medium significance (Fig. 3.7) and the dilatation rates suggest high compression in the region with the mean value of 32.4 ± 7 ns/yr, with a mean orientation angle of ~ 44.7 °N. In the south-west, the presence of massive salt layer with southward horizontal flow (Jouanne et al. 2020) acts as a decollement between the Potwar Plateau and Salt ranges resulting in extension strain rates. Southward GPS velocities of the sites in Salt Range result in the compression strain rates oriented in the direction perpendicular to

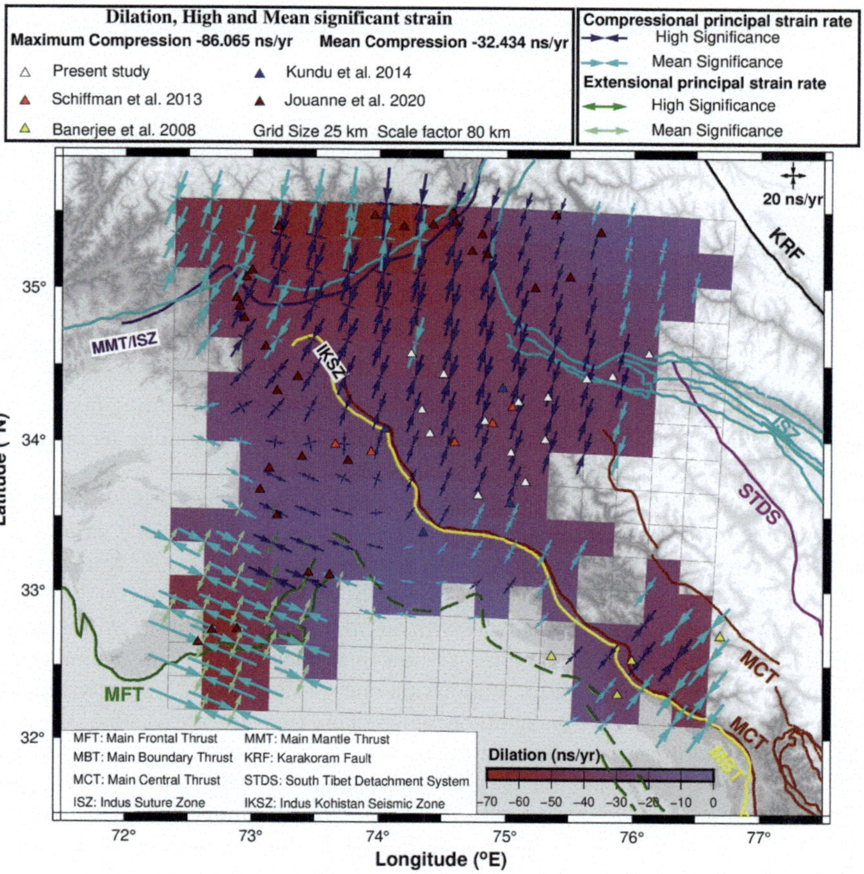

Fig. 3.7 Geodetic strain rates in the Himachal, Kashmir and Ladakh Himalaya at high and mean significant grid points. cGPS site velocities used for determining the strain rates are denoted by solid triangles. The figure was created using GMT (Generic Mapping Tool) software version 6.0.0 (Wessel et al. 2019)

the direction of the thrusting of the Mirpur event to the east. Long span of GPS data covering large spatial scale is required to minimize the non-linear strain accumulation during the earthquake cycle on individual faults so that the GPS-derived geodetic strain rates can be used to reliably compute the earthquake recurrence intervals as the accumulated strain includes both seismic and aseismic deformation. Further, we computed geodetic strain rates for the Kashmir Seismic gap and Himachal separately in the subsequent sections to estimate the seismic potential in Kashmir Seismic gap and 1905 Kangra earthquake region, respectively.

Kashmir seismic gap: About 25 cGPS site velocities with less than 3 mm/yr uncertainty and positions are used for computing geodetic strain rates in the Kashmir valley and the adjoining regions, i.e. between the 2005 Muzaffarabad earthquake and

1905 Kangra earthquake (Jade et al. 2020). Optimal two-dimensional grid size of 10 × 10 km and a scale factor of 50 km is used for strain computation.

Principal strain rates at high and mean significant grid points are plotted in Fig. 3.8 along with GPS points and the image map of dilatation. Maximum and Minimum compression rate of high and mean significance is about –75.92 and –30.12 ns/yr (nanostrain/year) indicating high compression with mean compression rate of –47.79 ns/yr. Orientation of principal strain axis ranges from 13 °N to 55 °N with mean orientation angle of 35°N. Maximum extension rate of 98.91 ns/yr with a mean extension rate of 17.15 ns/yr is due to oblique GPS convergence rate recorded in this region as detailed in the earlier section. Compression rates are higher than the extension rates and the large negative value of dilatation rate of –64.15 ns/yr also points towards high compression in this region. High compression rates in this region coincide with the northern edge of locked decollement indicating high strain accumulation along the MHT which is consistent with the seismicity recorded by our broadband observation network (Jade et al. 2020).

Himachal Himalaya: GPS measurements in Himachal (30.5–33 °N; 75–78.5 °E) covering the 1905 Kangra earthquake were used to compute geodetic strain rates in this region. GPS velocities of 24 GPS sites (Banerjee et al. 2008; Kumar et al. 2023) with a data span of about 16 years with velocity uncertainty of less than 3 mm/yr were used for strain analysis with an optimal scale factor of 95 km and a grid size of 30 km. High- and medium-significance principal strain rates along with dilatation are plotted in Fig. 3.9. Compression rates (–12.28 to –133.26 ns/yr) in this region are almost 10 orders of magnitude higher than extension rate (0.30–6.01 ns/yr) suggesting predominantly high compression in this region with a mean compression rate of –77.15 ns/yr. Orientation angle of compression strain is 26.2–61.6 °N with a mean orientation angle of 42 °N. Dilatation rates in this region range from –10.2 to –147.2 suggesting that this region is currently experiencing active high compression.

Garhwal-Kumaun Segment: cGPS and eGPS sites from Table 3.1 and the published velocities (Ghavri and Jade 2021; Yadav et al. 2019; Dumka et al. 2018; Bettinelli et al. 2006) with velocity uncertainty of less than 3 mm/yr were used for geodetic strain computation in this region (28.7–31.3 °N; 77.8–81.2 °E) (Fig. 3.10). Optimal scale factor of 45 km with a grid size of 15 × 15 km was used for the strain analysis of this region with good spatial spread of GPS sites. Geodetic strain rates at high and significant grid nodes are plotted in Fig. 3.10 along with the image map of dilation and GPS points. Principal strain rates of high- and mean-significance indicate compression of –142.77 to –31.60 ns/yr with an average rate of –96.21 ns/yr with an orientation angle of 24°N. Extension strain rate is ten orders of magnitude lower than the compression strain rate and the mean dilatation is –92 ns/yr pointing to active compression regime. Compression strain rate increases northward with a maximum value at the northern edge of basal decollement. Geodetic strain rates in this segment of Himalaya indicate high strain accumulation along the MHT which is consistent with micro-seismicity observed (Mahesh et al. 2013) along this belt. If we consider, accumulated strain as the average geodetic strain rate in Higher Himalaya, north of MCT, the geodetic strain rate in the MHT region is ~113 ns/year towards 25 °N.

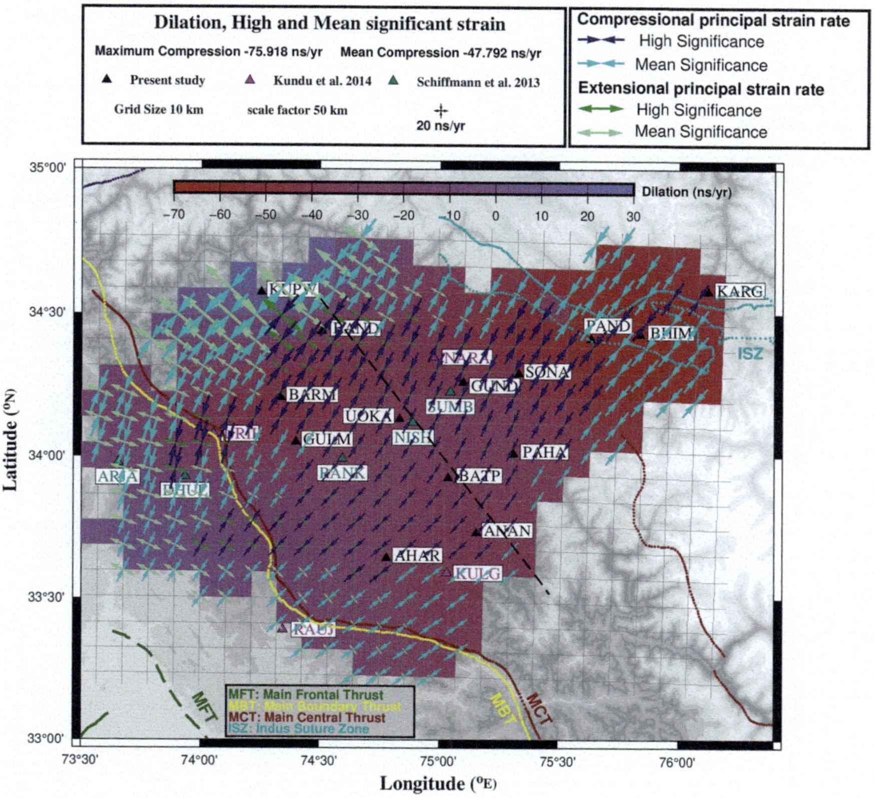

Fig. 3.8 Geodetic strain rates in the Kashmir Himalaya at high and mean significant grid points. cGPS site velocities used for determining the strain rates are denoted by solid triangles. Black dashed line is the northern edge of locked decollement estimated from dislocation model. The figure was created using GMT (Generic Mapping Tool) software version 6.0.0 (Wessel et al. 2019)

Nepal Himalaya: GPS velocities of 14 cGPS sites located in Nepal Himalaya (Table 3.1) and published velocities of 52 GPS sites (Bettellini et al. 2006; Ader et al. 2012; Zheng et al. 2017) and their positions were used for geodetic strain computation of western, central and eastern Nepal Himalayas from 26.5–30 °N; 75–88 °E (Fig. 3.11). We took GPS sites with velocity uncertainty of less than 3 mm/yr for the strain analysis and chose an optimal grid size 25 km and a scale factor of 90 km. Maximum and minimum compression strain rates are –100.58 ns/yr and –39.03 ns/yr with mean compression rate of –73.29 ns/yr and orientation angle of 14 °N. It can be seen from Fig. 3.11 that currently Nepal Himalaya is in active compressive regime indicating high strain accumulation.

Sikkim, Bhutan, Arunachal and Eastern Himalayan syntaxis: Geodetic strain rates of the rest of the Himalayan segments (Sikkim, Bhutan, Arunachal and Eastern Himalayan syntaxis) were not computed due to insufficient GPS data. GPS-derived surface convergence rates and the slip rates along MHT in these Himalayan segments

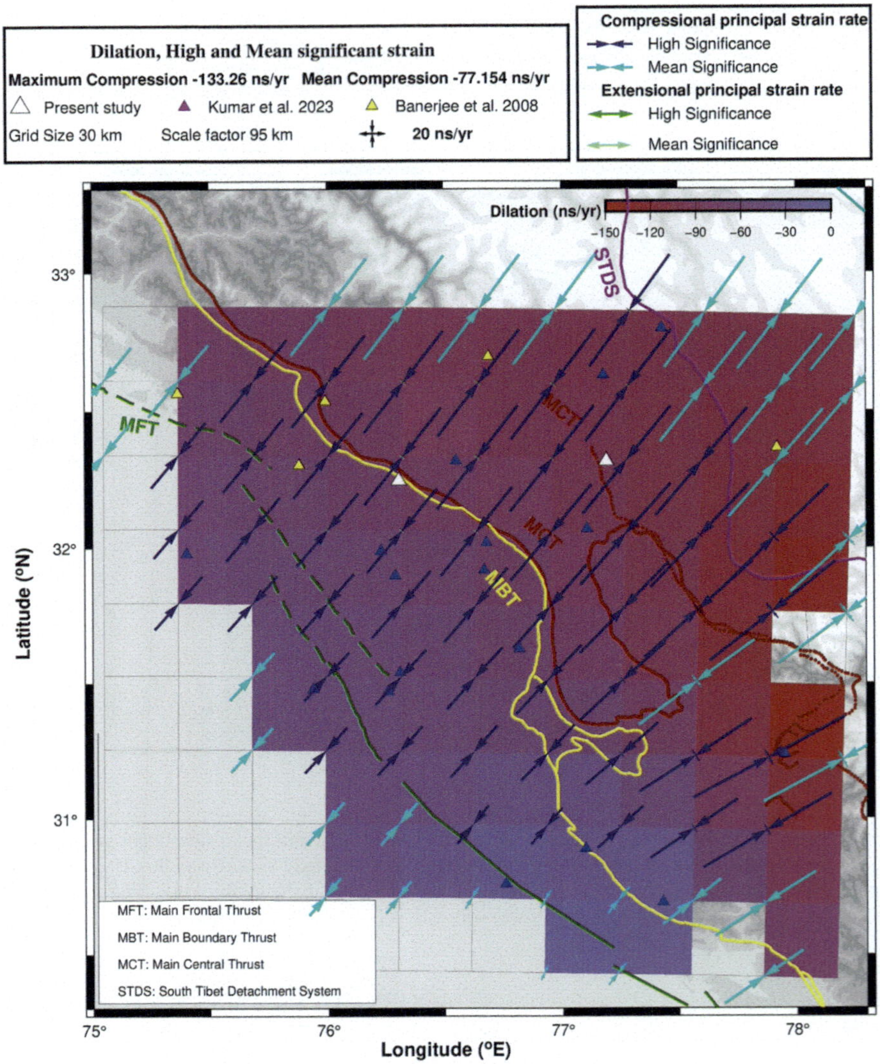

Fig. 3.9 Geodetic strain rates in the Himachal Himalaya region. GPS sites used for strain computation are denoted by solid triangles. The figure was created using GMT (Generic Mapping Tool) software version 6.0.0 (Wessel et al. 2019)

(Figs. 3.4 and 3.6) indicate active compressive regime with high strain accumulation along the decollement (MHT). This is intermittently being released by seismic events of magnitude 5 and above (Fig. 3.1). Micro-seismicity also suggests high strain accumulation in the entire Himalayan arc. Considering that the geodetic strain is the stored strain energy and seismic events release this stored energy intermittently,

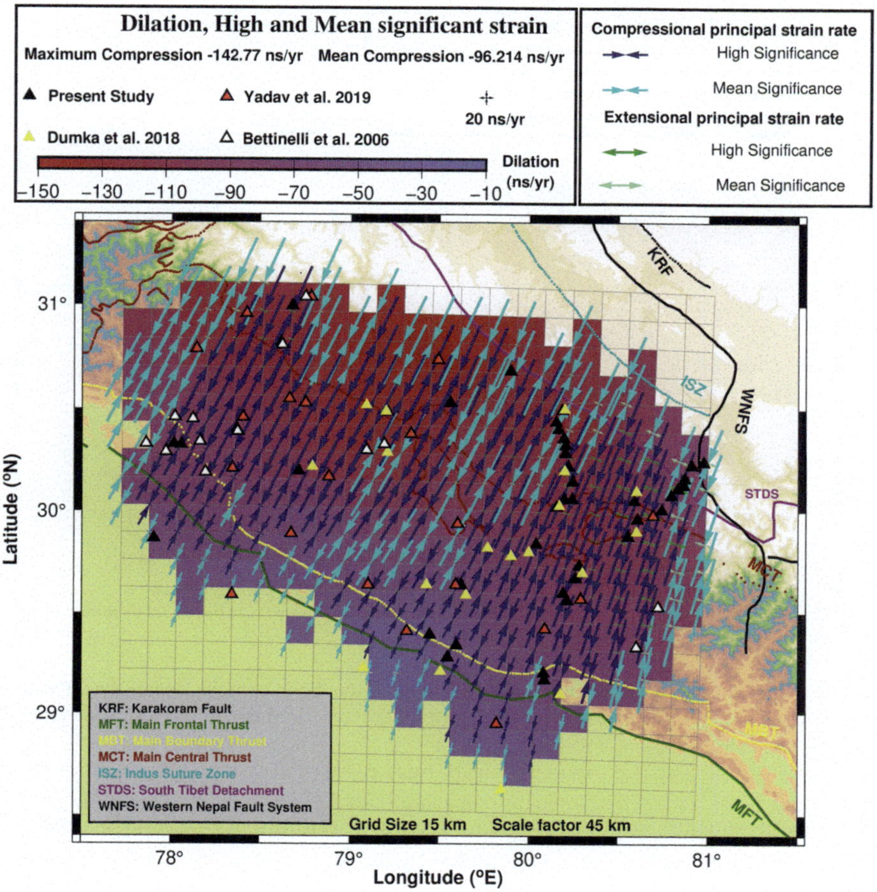

Fig. 3.10 Geodetic strain rates in the Garhwal–Kumaun region. GPS sites used for strain computation are denoted by solid triangles. The figure was created using GMT (Generic Mapping Tool) software version 6.0.0 (Wessel et al. 2019)

strain budget and slip accumulation is estimated for each segment of Himalaya to determine the seismic potential in the entire Himalayan arc from west to east.

3.5 Seismic Potential

Seismic strain rates are computed from the earthquake catalogue using (Kostrov 1974) formulations. Magnitude completeness and seismogenic thickness required for seismic strain computation are derived using ZMAP (Wiemer 2001) for the seismicity catalogue of the instrumentational period from 1964 till date. Geodetic strain rates contain both elastic and anelastic (seismic and aseismic) strain of the region

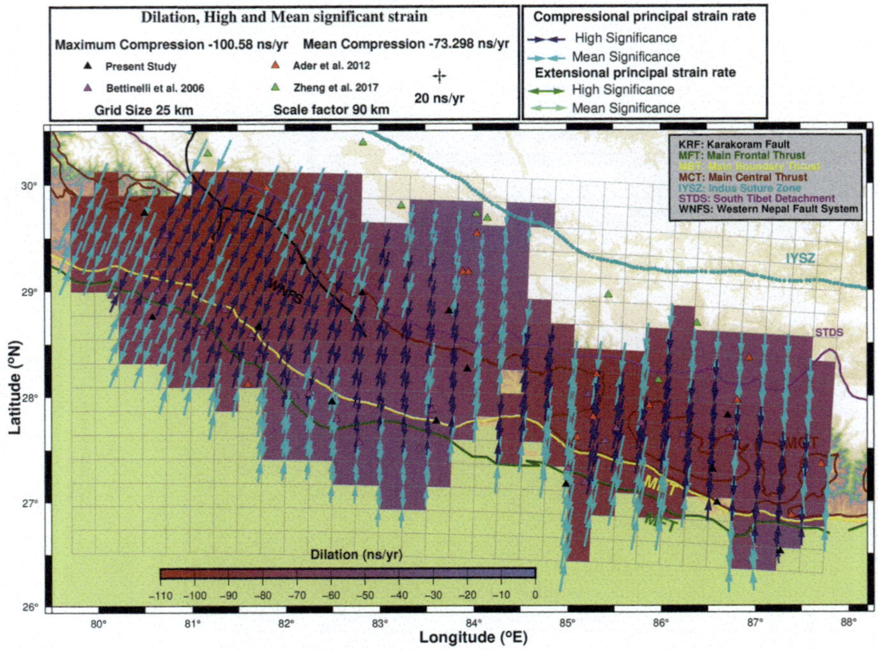

Fig. 3.11 Geodetic strain rates in the Nepal Himalaya region. Solid triangles denote the GPS sites used for strain analysis. The figure was created using GMT (Generic Mapping Tool) software version 6.0.0 (Wessel et al. 2019)

where as the seismic strain rates give a record of brittle deformation. Incomplete historical earthquake records compounded by the large uncertainty in the epicentre and magnitude of the available historical seismic events contribute to overestimation or underestimation of seismic strain rates depending on the earthquake occurrence during the catalogue interval relative to the average recurrence interval of large earthquakes in the region. To estimate reliable strain rates, the average recurrence interval must be shorter than the historical record. Further, while computing seismic moment rates using (Kanamori 1977), uncertainties associated in estimating the long-term seismic moment rate from historical record must be accounted for. This includes catalogue length, seismogenic layer thickness and the uncertainty associated with the equations used to calculate the seismic moment. Combined analysis of seismic and geodetic strain rates provide the current strain budget which is used to estimate the seismic potential of the region.

Himachal, Kashmir and Ladakh Himalaya: For Himachal, Kashmir and Ladakh Himalaya covering the rupture zones of the 2005 Muzaffarabad earthquake and the 1905 Kangra earthquake, seismic strain rates (Fig. 3.12) are calculated for instrumental (1964–2021) and historical period since 1500 AD, respectively (Shrungeshwara et al. 2023). Seismogenic thickness of 20 km and magnitude completeness of $M_w > 3.9$ are computed from instrumentation period earthquakes using ZMAP.

Seismic strain rate computed from 1964 to 2021 is –28.1 ns/yr with an orientation of N 42.5° (Fig. 3.12) which is lower than the geodetic strain rate of 32.4 ns/yr with an orientation of 45 °N. Orientation of seismic strain is same as geodetic strain indicating uniform deformation in this region. Seismic strain rate is the released energy and geodetic strain rate is the stored energy, hence higher geodetic strain rate suggests that stored strain is partially released or alternatively catalogue length is insufficient compared to recurrence interval of the earthquakes. Notable earthquake during the instrumentation period is the 2005 Muzzafarabad event which released the strain accumulated for the past 200–300 years partially and the residual strain may be released by future earthquakes. Historical records (Bilham 2019) suggest about 15 damaging earthquakes in this region since 844 AD, however reliable magnitude and epicentre location were not available. Previous studies indicate recurrence interval of ~500 years in north-west Himalaya and hence seismic strain rate since 1500 AD computed from the historical catalogue is –21.70 ns/yr which is lower than the 32.4 ns/yr geodetic strain (Fig. 3.12) suggesting aseismic deformation. Two major earthquakes with M_w ~8 during the historical period are 1905 Kangra event and 1555 Kashmir valley event. Reasonable documentation is available on the 1905 Kangra event where as the 1555 event has a large uncertainty in the magnitude (Bilham 2019; Jouanne et al., 2020). Hence, if we consider M_w of 8.2 for the 1555 event, seismic strain rate is 32.04 ns/yr which is comparable to the geodetic strain rate indicating that this region is currently poised to generate M_w 8.2 earthquake in immediate future. Following the expression given by Savage and Simpson (1997), geodetic strain rate gives the geodetic scalar moment rate of 7.5 E + 18N m/yr and the seismic moment rate calculated following (Kanamori, 1977) for the past 520 years is 5.2 E + 18 Nm/yr which gives the current accumulated seismic moment of 1E + 21Nm for the past 520 years indicating that M_w~ 8 earthquake is impending in the near future.

Kashmir seismic gap: However, if we consider Kashmir seismic gap i.e. region between the 2005 Muzaffarabad earthquake and 1905 Kangra earthquake, historical record suggests about 14 damaging earthquakes in this region from 844 to 1885 though there is a large uncertainty in the magnitude and epicentre of these events (Bilham 2019). Reasonably researched Kashmir earthquake of September 1555, 7.6 $< M_w <$ 8.0 is considered to be the largest earthquake in this region. Considering that 1555 event released all the accumulated strain energy and fresh strain accumulation started since then, geodetic strain and slip rates give the estimate of strain budget and accumulated slip in this region. Horizontal principal strain rates of the region give a scalar geodetic moment accumulation rate of ~8.91e24 dyne-cm/year (~ 4.14e27 dyne-cm for 465 years) using the analytical expression of Savage and Simpson (1997), which is capable of generating M_w 7.7 earthquake. Slip deficit since 1555 earthquake is about 7.4 m (16 mm/yr for 465 yrs) and assuming that the probable future earthquake has similar rupture parameters of 2005 Muzaffarabad earthquake, seismic moment calculated using the expressions given by Wells et al., 1994 is ~ 5.3e27 dyne-cm suggesting M_w 7.8 earthquake is long due in this region (Jade et al. 2020).

Himachal Himalaya: In Himachal segment of Himalaya, 1905 Kangra earthquake of M_w 7.9 (Fig. 3.1) is most devastating earthquake with reasonably well documented

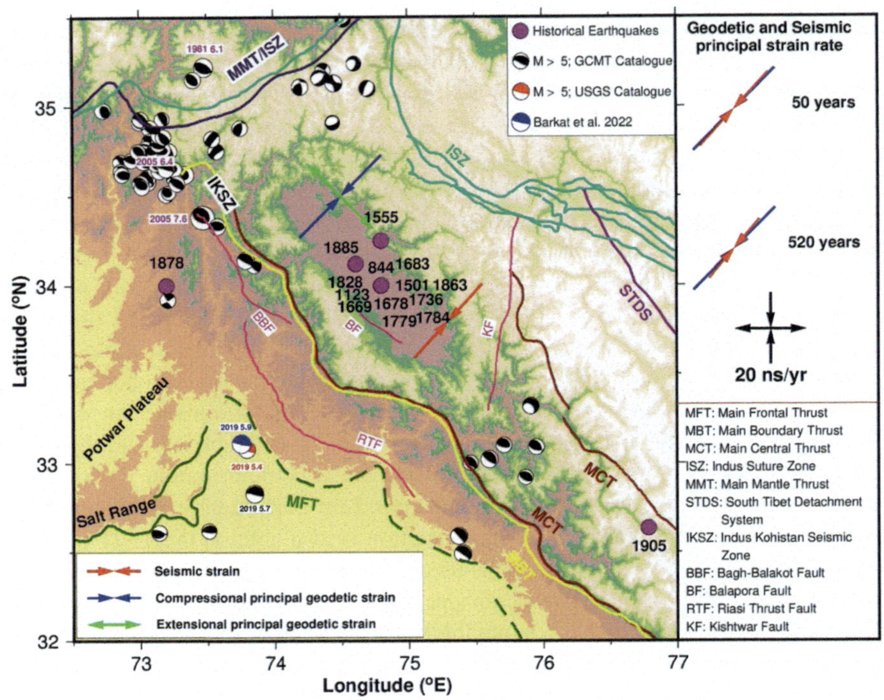

Fig. 3.12 Combined analysis of the geodetic and seismic strain rates of Himachal, Kashmir and Ladakh Himalaya. The mean geodetic and seismic strain rates are represented by blue and red colours, respectively. The figure was created using GMT (Generic Mapping Tool) software version 6.0.0 (Wessel et al. 2019)

slip, rupture, magnitude and epicentre (Bilham 2019). Geodetic data of Kangra valley fault (Figs. 3.1 and 3.6) estimated the probable mean thrust slip of up to 5.1 m due to this earthquake. It was also observed that Kangra earthquake was not followed by subsequent up dip earthquakes. Further, available seismic record suggests that no major earthquake of M_w 5 and above was recorded in this region either during instrumental or historic period. Seismogenic thickness of 15 km and Magnitude completeness of $M_w > 2.7$ is computed using ZMAP from the earthquake catalogue of instrumentation period. Since 1905 Kangra event in the only historical event, seismic strain is computed for past 120 years as –74.84 ns/yr with orientation angle of 41.8 °N (Fig. 3.13) which is comparable to the geodetic strain rate of –77.15 ns/yr with orientation angle of 42 °N suggesting recurrence interval of ~120 years for earthquake with similar magnitude of 1905 Kangra earthquake. Similar orientation angle for geodetic and seismic strain indicates uniform deformation in this region. Geodetic moment rate of 7.6E + 18Nm/yr is comparable to seismic moment rate of 7.4E + 18Nm/yr for the past 120 years indicating that this region currently has the potential to generate M_w 7.9 earthquake.

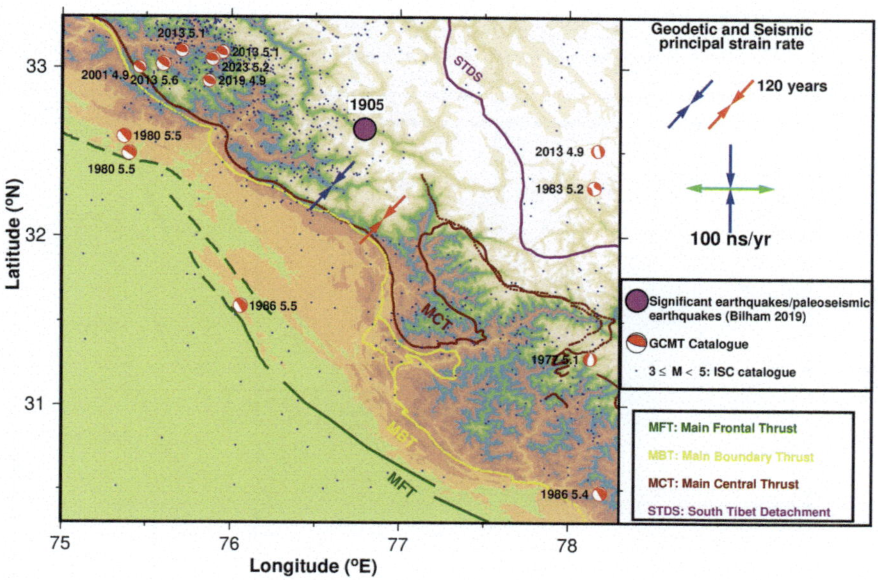

Fig. 3.13 Combined analysis of the geodetic and seismic strain rates of Himachal Himalaya. The mean geodetic and seismic strain rates are represented by blue and red colours, respectively. The figure was created using GMT (Generic Mapping Tool) software version 6.0.0 (Wessel et al. 2019)

Garhwal–Kumaun Segment: Seismic strain rates for Garhwal–Kumaun are calculated using the available earthquake catalogue of 50, 220 and 700 years using the (Kostrov 1974) expressions (Ghavri & Jade, 2021). Magnitude completeness of $M_w \geq 3.6$ and seismogenic thickness of 20 km derived using ZMAP (Wiemer 2001) for the seismicity catalogue of the instrumentational period from 1964 for seismic strain computation. All the seismic events with M_w equal and greater than 3.6 in 50 years catalogue give a strain rate of –10 ns/yr towards 26 °N suggesting the region in under compression in the NNE direction and is consistent with the geodetic strain rates. However, it is very low compared to the mean geodetic strain rate of 96.2 ns/yr suggesting that the length of the catalogue is too short for comparison. Seismic strain rates computed from 220 and 700 years of catalogue are about –38 ns/yr and –115 ns/yr which are comparable to the geodetic strain rates with consistent orientation angles suggesting uniform deformation pattern in the region justifying combined analysis of geodetic and seismic strain rates (Fig. 3.14).

These rates indicate that the stored strain energy is not completely released during the past 220 years whereas for the past 700 years seismic strain is comparable to geodetic strain rate suggesting recurrence interval of about 700 years for earthquakes with magnitude similar to the 1344 (1222–1422) events. Seismic moment rate for the past 700 years is 8.3E + 18 Nm/yr which is consistent with the scalar geodetic moment rate of 8.13E + 18 Nm/yr suggesting that this region is ready for a $M_w > 8.0$ earthquake in the near future. Magnitude of 1344 (1222–1422) events range from M_w 8.4 to 8.6 with a rupture of 8–16 m (Kumar et al., 2006; Kumahara and

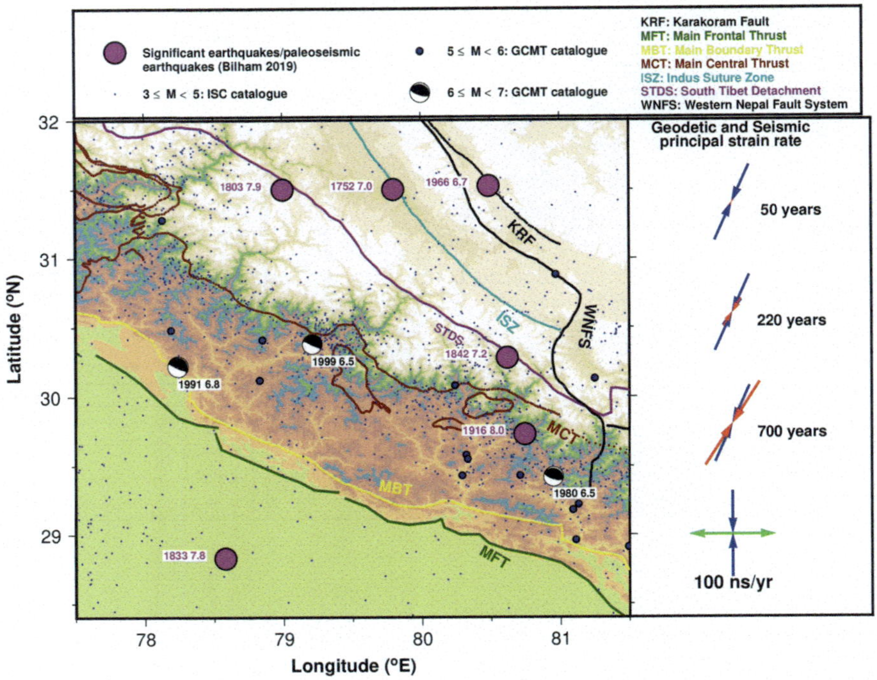

Fig. 3.14 Combined analysis of the geodetic and seismic strain rates of Garhwal–Kumaun Himalaya. The mean geodetic and seismic strain rates are represented by blue and red colours, respectively. The figure was created using GMT (Generic Mapping Tool) software version 6.0.0 (Wessel et al. 2019)

Jayangondaperumal 2013) and considering that the stored strain energy is completely released during this event, slip rate of 16–18 mm/yr in the Garhwal–Kumaun gives a slip accumulation of 11–12 m/yr for the past 700 years which is capable of generating M_w 8.6 earthquake in the near future considering fresh strain accumulation started after the 1344(1222–1422) events.

Nepal Himalaya: Several historical earthquakes at regular intervals (100–120 years) with magnitude M_w ~7–9 in Nepal Himalaya (Fig. 3.1) from 1100 to 1934 with large uncertainty in location and magnitude indicate that this region is intermittently releasing the stored energy. 1934 Bihar–Nepal earthquake is the largest and most devastating earthquake with reasonably documented magnitude (M_w of 8.4), location and rupture. During instrumental era several major earthquakes of magnitude 5 and above occurred from 1980 till date in this region which are well documented. In eastern Nepal Himalaya, 2015 Gorkha earthquake with M_w 7.9 was the most devastating earthquake which ruptured the MHT partially which progressed eastward causing the largest aftershock (M_w 7.2) about 150 km east of the main shock (Sreejith et al. 2018). Recently, in November 2023 significant earthquake of M_w 5.9 occurred in western Nepal Himalaya followed by two aftershocks of M_w 5.7 and 5.3 within 2 days of the main event. Seismogenic thickness of 15 km and Magnitude completeness of M_w 3.6 is computed using the earthquake data of the instrumentation period. Seismic strain for instrumentation period is computed as 43.76 ns/yr which is less than the geodetic strain rate of 73.29 ns/yr suggesting the insufficient length of seismic catalogue (Fig. 3.15). Seismic strain of 71.26 ns/yr is calculated for historical period from 1100 AD by taking the average magnitude of 11 historical events due to large uncertainty in the magnitudes of historical events. Considering that this region is experiencing high magnitude M_w 7–9 every ~100 years, comparable geodetic and seismic strain rates suggest that the stored strain energy is being released partially at regular intervals by intermittent seismic events suggesting that Nepal Himalaya continues to be in the phase of active seismicity and hence highly vulnerable to earthquakes. Seismic moment rate is 3.4E + 26Nm/yr which is comparable to the geodetic scalar moment rate of 3.4E + 26Nm/yr indicating the high vulnerability of this region to M_w 7–9 earthquakes currently.

Sikkim Himalaya: Recent well instrumented major earthquake in Sikkim Himalaya is the 2011 M_w 6.9 Sikkim earthquake of 18 September 2011. Further, 1980 event of M_w 6.2 also occurred during the instrumental period. Historical record suggests an earthquake of magnitude M_w 8.2 occurred in 1714 in Sikkim–Bhutan Himalaya. Slip rate of 18–19 mm/yr in this region suggest accumulated slip of about ~5.7 m since the 1714 event in this region suggesting that the recent 2011 and 1980 events partially released the stored strain energy. Current GPS convergence rates and the observed micro-seismicity indicate high rate of strain accumulation. However,

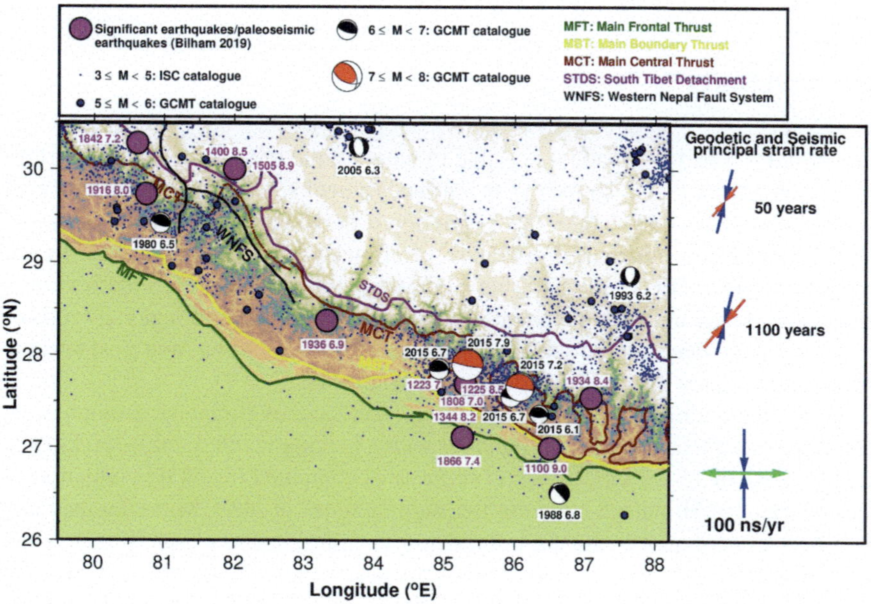

Fig. 3.15 Combined analysis of the geodetic and seismic strain rates of Nepal Himalaya. The mean geodetic and seismic strain rates are represented by blue and red colours, respectively. The figure was created using GMT (Generic Mapping Tool) software version 6.0.0 (Wessel et al. 2019)

we cannot estimate the magnitude and recurrence interval of a probable earthquake in this region due to insufficient geodetic and seismic data.

Arunachal Himalaya: Arunachal Himalaya recorded three events of magnitude ~6.0 during the instrumental era and the historical record suggests three earthquakes of magnitude 6.8–7.2 from 1943 to 1964. Since there is no earthquake data available priori to 1943 in this region, we assume that there were no earthquakes in the last 200 years which gives accumulated slip of about 4 m in this region. This region is currently under high compression similar to the rest of Himalayan segments storing strain energy at an alarming pace pointing to high seismic potential in this region which is hard to quantify due to insufficient geodetic and seismic data.

Eastern Himalayan Syntaxis: The 1950 Assam earthquake of M_w 8.6 is the largest and the latest earthquake that occurred in Eastern Himalayan Syntaxis (EHS) though it is not clear whether it occurred in adjoining Himalayan region or both. Further, the epicentre and the rupture cannot be pinpointed with the available focal mechanisms and aftershock location. However, we can safely assume that some part of the rupture occurred in EHS based on focal mechanism solution of (Devachandra et al. 2014). Slip rate in EHS is about 20 mm/yr (Fig. 3.6) gives a slip deficit of 1.3 m since the 1950 event. However, based on available historical record there was no earthquake in the past 200 years in this region other than the 1950 event, hence the slip deficit is about 4 m. Due to lack of seismicity, the seismic potential of this region cannot be estimated. However, geodetic rates and slip deficit suggest that the stored strain energy is currently high in this region.

3.6 Summary

Advent of GPS in 1994 made it possible to measure the precise arc normal surface convergence rates of 10–26 mm/yr in various segments of Himalaya. Arc parallel convergence and extension rates of ~3–9 mm/yr observed in the 2500 km Himalayan arc are consistent with locked curvature of Himalayan arc and east–west extension rate of Tibet. However, in Kashmir Himalaya the oblique surface deformation rate recorded may have a component of active deformation along regional faults of Kashmir valley, deep rooted faults in Kanzalwan, Nangaparbhat and post seismic deformation related to 2005 Muzaffarabad earthquake.

Inverse modelling of GPS velocities in each segment of Himalaya estimate variable slip along MHT from 13 to 20 mm/yr with a locking width of 100–150 km and locking depth of 15–20 km. Geodetic strain rates computed from GPS velocities in each segment of Himalaya covering the rupture zones of 2005 Muzzafarabad and 1905 Kangra earthquakes, Kashmir Seismic gap, Himachal, Garhwal–Kumaun and Nepal Himalaya suggest active compression regime with NNE orientation and strain rates increasing from frontal Himalaya to northern edge of MHT which is consistent with the micro-seismicity observed in Himalayas.

Consistent NNE strain orientation of geodetic and seismic strain suggests uniform stress regime in Himalayas justifying combined analysis. Seismic and geodetic strain rates in the segment of Himalaya covering the rupture zones of 2005 Muzaffarabad and 1905 Kangra event indicate that M_w~8 earthquake may occur in the near future. Strain budget and accumulated slip is the Kashmir seismic gap currently has the potential to generate magnitude M_w~7.7–7.8 earthquake. For Himachal segment of Himalaya, geodetic and seismic strain rates estimate the recurrence interval of 120 years for earthquake with magnitude same as the 1905 Kangra earthquake. Combined analysis of seismic and geodetic strain in Garhwal–Kumaun Himalaya suggest that magnitude M_w~8.2 earthquake is probable in near future. Nepal Himalaya is seismically active since 1100 AD with occurrence of M_w 7–9 earthquakes at regular interval of 100–120 years. Geodetic and seismic strain rates in Nepal Himalaya suggest that this region continues to be seismically active and is highly vulnerable to M_w~8 earthquakes currently. Further, significant insights in to the seismic potential of Sikkim–Bhutan, Arunachal and Eastern syntaxis segment of Himalaya were given based on the seismicity, accumulated slip and strain budget with the available GPS and seismic data.

Each segment of Himalaya from west to east behaves differently with variable GPS surface convergence rates, variable locking width, depth and slip along the MHT. GPS convergence rates in each segment of Himalaya is accommodated differently in the litho-tectonic units from south to north. Seismic potential of each segment of Himalaya varies considerably based on the geodetic strain rates and associated moments, accumulated slip and seismicity and seismic moment of the region.

Acknowledgements First and foremost, we thank Dr. B.K. Rastogi, President: Indian Society of Earthquake Science and Ex-Director General: Institute of Seismological Research, Gandhinagar, for his time and effort in making this book chapter a reality. We sincerely thank reviewers for their time and valuable comments which helped to improve the quality of the book chapter. The authors thank all the former Heads and the current Head of CSIR-4PI for supporting the GNSS and broadband network projects for past three decades. We thank Mr. Chiranjeevi G Vivek for his contribution to cGNSS network and maintenance of GNSS servers. We thank Mrs. Bhavani Narukula for technical inputs related to seismic data analysis. We acknowledge all the concerned scientific and technical personnel involved in GPS program from 1994 till date.

References

Ader T, Avouac J-P, Liu-Zeng J, Lyon-Caen H, Bollinger L, Galetzka J, Genrich J, Thomas M, Chanard K, Sapkota SN, Rajaure S, Shrestha P, Ding L, Flouzat M (2012) Convergence rate across the Nepal Himalaya and inter-seismic coupling on the Main Himalayan Thrust: implications for seismic hazard. J Geophys Res 117:B04403. https://doi.org/10.1029/2011JB009071

Altamimi Z, Rebischung P, Métivier L, Collilieux X (2016) ITRF2014: A new release of the International Terrestrial Reference Frame modelling nonlinear station motions. J Geophys Res Solid Earth 121:6109–6131. https://doi.org/10.1002/2016JB013098

Banerjee P, Bürgmann R (2002) Convergence across the northwest Himalaya from GPS measurements. Geophys Res Lett 29(13). https://doi.org/10.1029/2002GL015184

Banerjee P, Bürgmann R, Nagarajan B, Apel E (2008) Intraplate deformation of the Indian subcontinent. Geophys Res Lett 35(18):1–5. https://doi.org/10.1029/2008GL035468

Bendick R, Bilham R (2001) How perfect is the Himalayan arc? Geology 29:791–794
Bendick R, Khan SF, Bürgmann R, Jouanne F, Banerjee P, Khan MA, Bilham R (2015) Postseismic relaxation in Kashmir and lateral variations in crustal architecture and materials. Geophys Res Lett 42:4375–4383. https://doi.org/10.1002/2015GL064670
Bettinelli P, Avouac JP, Flouzat M et al (2006) Plate motion of India and Interseismic strain in the Nepal Himalaya from GPS and DORIS measurements. J Geodesy 80:567–589. https://doi.org/10.1007/s00190-006-0030-3
Bilham R et al (2001) Himalayan seismic hazard. Science 293(1442–1444):2001. https://doi.org/10.1126/science.1062584
Bilham R (2019) Himalayan earthquakes: A review of historical seismicity and early 21st century slip potential. Geol Soc Spec Publ 483:423–482. https://doi.org/10.1144/SP483.16
Chen Q, Freymueller JT, Yang Z, Xu C, Jiang W, Wang Q, Liu J (2004) Spatially variable extension in southern Tibet based on GPS measurements. J Geophys Res 109:B0940. https://doi.org/10.1029/2002JB002350
Cortés-Aranda J et al (2018) Late quaternary out-of-sequence deformation in the innermost Kangra Re-entrant, NW Himalaya of India: Seismic potential appraisal from 10Be dated fluvial terraces. J Asian Earth Sci 158:140–152. https://doi.org/10.1016/j.jseaes.2018.01.027
Dasgupta S et al (2000) Seismotectonic atlas of India and Its environs. Special Publication Geological Survey of India, Calcutta
DeMets C, Merkouriev S, Jade S (2020) High-resolution reconstructions and GPS estimates of India-Eurasia and India-Somalia plate motions: 20 Ma to the present. Geophys J Int 220(2):1149–1171. https://doi.org/10.1093/gji/ggz508
Devachandra M, Kundu B, Catherine J, Kumar A, Gahalaut VK (2014) Global Positioning System (GPS) measurements of crustal deformation across the frontal eastern Himalayan syntaxis and seismic-hazard assesment. Bull Seism Soc Am 104(3):1518–1524. https://doi.org/10.1785/0120130290
Dumka RK, Kotila BS, Kothyari GC et al (2018) Detection of high and moderate crustal strain zones in Uttarakhand Himalaya, India. Acta Geod Geophys 53:503–521
Freymueller J et al (1996) Global positioning system measurements of Indian plate motion and convergence across the lesser Himalaya. Geophys Res Lett 23:3107–3110
Ghavri S, Jade S (2021) Seismic potential of megathrust in the Kumaun-Garhwal region of NW Himalaya: implications from geodetic and seismic strain rates. Int J Earth Sci (Geol Rundsch) 110:1439–1452. https://doi.org/10.1007/s00531-021-02023-x
Herring TA, King RW, Floyd MA, McClusky SC (2018) Introduction to GAMIT/GLOBK Release 10.7. Department of Earth, Atmospheric, and Planetary Sciences, Massachusetts Institute of Technology, Cambridge. http://geoweb.mit.edu/gg/Intro_GG.pdf
Jade S (2004) Estimates of plate velocity and crustal deformation in the Indian subcontinent using GPS geodesy. Curr Sci 10:1443–1448
Jade S et al (2011) GPS-derived deformation rates in northwestern Himalaya and Ladakh. Int J Earth Sci 100(6):1293–1301. https://doi.org/10.1007/s00531-010-0532-3
Jade S et al (2014) Contemporary deformation in the Kashmir-Himachal, Garhwal and Kumaon Himalaya: significant insights from 1995–2008 GPS time series. J Geod 88:539–557. https://doi.org/10.1007/s00190-014-0702-3
Jade S et al (2017) India Plate angular velocity and contemporary deformation rates from continuous GPS measurements from 1996 to 2015. Nat Sci Rep 7:11439. https://doi.org/10.1038/s41598-017-11697-w
Jade S et al (2020) Crustal deformations rates in Kashmir valley and adjoining regions from continuous GPS measurements from 2008 to 2019. Sci Rep 10:17927. https://doi.org/10.1038/s41598-020-74776-5
Jouanne F, Mugnier JL, Gamond JF, Le Fort P, Pandey MR, Bollinger L (2004) Current shortening across the Himalayas of Nepal. Gephys J Int 157:1–14. https://doi.org/10.1111/j.1365-246X.2004.02180.x

Jouanne F et al (2011) Postseismic deformation in Pakistan after the 8 October 2005 earthquake: Evidence of after slip along a flat north of the Balakot-Bagh thrust. J Geophys Res Solid Earth 116:B07401. https://doi.org/10.1029/2010JB007903

Jouanne F et al (2020) Seismic coupling quantified on inferred decollements beneath the western syntaxis of the Himalaya. Tectonics 39:1–20. https://doi.org/10.1029/2020TC006122

Kanamori H (1977) The energy release in great earthquakes. J Geophys Res 82(20):2981–2987. https://doi.org/10.1029/JB082i020p02981

Kostrov B (1974) Seismic moment and energy of earthquake and seismic flow of rock. Izv Acad Sci USSR, Phys Solid Earth 1:23–40

Kumahara Y, Jayangondaperumal R (2013) Paleoseismic evidence of a surface rupture along the Northwestern Himalayan Frontal Thrust (HFT). Geomorpholgy 180:47–56

Kumar S, Wesnousky SG, Rockwell TK, Briggs RW, Thakur VC, Jayagonda Perumal R (2006) Paleoseismic evidence of great surface rupture earthquakes along the Indian Himalaya. J Geophys Res 111:B03304. https://doi.org/10.1029/2004JB003309

Kumar P, Malik JN, Gahalaut VK, Yadav RK, Singh G (2023) Evidence of strain accumulation and coupling variation in the Himachal region of NW Himalaya from short term geodetic measurements. Tectonics, 42:e2022TC007690. https://doi.org/10.1029/2022TC007690

Kundu B, Yadav RK, Bali BS, Chowdhury S, Gahalaut VK (2014) Oblique convergence and slip partitioning in the NW Himalaya: implications from GPS measurements. Tectonics 33(10):2013–2023. https://doi.org/10.1002/2014TC003633

Mahesh P, Rai SS, Sivaram K, Paul A, Gupta S, Sarma R, Gaur VK (2013) One dimensional reference velocity model and precise locations of earthquake hypocenters in the Kumaon-Garhwal Himalaya. Bull Seismol Soc Am 103(1):328–339. https://doi.org/10.1785/0120110328

Meade BJ (2007) Present day kinematics at the India-Asia collision zone. Geology 35(1):81–84. https://doi.org/10.1130/G22924A.1

Mohadjer S, Alan Ehlers T, Bendick R, Stübner K, Strube TA (2016) Quaternary fault database for central Asia. Nat Hazards Earth Syst Sci 16:529–542. https://doi.org/10.5194/nhess-16-529-2016

Mukul M, Jade S, Bhattacharyya AK et al (2010) Crustal shortening in convergent orogens: insights from global positioning system (GPS) measurements in northeast India. J Geol Soc India 75:302–312. https://doi.org/10.1007/s12594-010-0017-9

Mukul M, Jade S, Ansari K, Matin A, Joshi V (2018) Structural insights from geodetic global positioning system measurements in the Darjiling-Sikkim Himalaya. J Struct Geol 114:346–356. https://doi.org/10.1016/j.jsg.2018.03.007

Okada Y (1985) Surface deformation due to shear and tensile faults in a half-space. Bull Seismol Soc Am 75(4):1135–1154

Paul J, Bürgmann R, Gaur VK, Bilham R, Larson KM, Ananda MB, Jade S, Mukul M, Anupama TS, Satyal G, Kumar D (2001) Themotion and active deformation of India. Geophys Res Lett 28(4):647–650

Savage JC, Simpson RW (1997) Surface strain accumulation and the seismic moment tensor. Bull Seismol Soc Am 87:1345–1353

Schiffman C, Bali BS, Szeliga W, Bilham R (2013) Seismic slip deficit in the Kashmir Himalaya from GPS observations. Geophy Res Lett 40:5642–5645

Shrungeshwara TS, Narukula B, Jade S et al (2023) Implications of seismic and GNSS strain rates in Himachal, Kashmir and Ladakh. Sci Rep 13:21652. https://doi.org/10.1038/s41598-023-48997-3

Silver CRP, Murphy MA, Taylor MH, Gosse J, Baltz T (2015) Neotectonics of the Western Nepal fault system: Implications for Himalayan strain partitioning. Tectonics 34:2494–2513. https://doi.org/10.1002/2014TC003730

Snay R (1999) Using the HTDP software to transform spatial coordinates across time and between reference frames. Surv Land Inf Sys 59(1):15–25

Sreejith KM, Sunil PS, Agarwal R et al (2018) Audit of stored strain energy and extent of future earthquake rupture in central Himalaya. Sci Rep 8:16697. https://doi.org/10.1038/s41598-018-35025-y

Teza G, Galgaro A (2008) Grid_strain and grid_strain3: software packages for strain field computation in 2D and 3D environment. Comput Geosci 34(9):1142–1153

Vassallo R et al (2015) Distribution of the Late-Quaternary deformation in Northwestern Himalaya. Earth Planet Sci Lett 411:241–252. https://doi.org/10.1016/j.epsl.2014.11.030

Vernant P, Bilham R, Szeliga W, Drupka D, Kalita S, Bhattacharyya AK, Gaur VK, Pelgay P, Cattin R, Berthet T (2014) Clockwise rotation of the Brahmaputra Valley relative to India: Tectonic convergence in the eastern Himalaya, Naga Hills, and Shillong Plateau. J Geophys Res Solid Earth 119:6558–6571. https://doi.org/10.1002/2014JB011196

Wells DL, Coppersmith KJ (1994) New empirical relationships among magnitude, rupture length, rupture width, rupture area, and surface displacement. Bull Seismol Soc Am 84(4):974–1002. https://doi.org/10.1785/BSSA0840040974

Wessel P, Luis JF, Uieda L, Scharroo R, Wobbe F, Smith WHF, Tian D (2019) The generic mapping tools version 6. Geochem Geophys Geosyst 20:5556–5564. https://doi.org/10.1029/2019GC008515

Wiemer S (2001) A software package to analyze seismicity: ZMAP. Seismol Res Lett 72(3):373–382. https://doi.org/10.1785/gssrl.72.3.373

Yadav RK, Gahalaut VK, Bansal AK, Sati SP, Catherine J, Gautam P, Kumar K, Rana N (2019) Strong seismic coupling underneath Garhwal-Kumaun region, NW Himalaya, India. Earth Planet Sci Lett 506:8014

Zheng G, Wang H, Wright TJ, Lou Y, Zhang R, Zhang W, Wel N (2017) Crustal deformation in the India-Eurasia collision zone from 25 years of GPS measurements. J Geophys Res: Solid Earth 122:9290–9312. https://doi.org/10.1002/2017JB014465

Chapter 4
Assessing the Influence of Clay Minerals on Landslides in the Lesser Himalayas

Piyal Halder, Matsyendra Kumar Shukla, Kamlesh Kumar, and Anupam Sharma

Abstract Landslide is one of the deadliest hazards in the tectonically active Himalayan Region, particularly in the Lesser Himalayas. Rainfall-induced weathering and the hydrothermal alteration during the formation of the MCT have resulted in the production of hydrophilic clay minerals, which reduces the frictional strength of the slide planes by absorbing water in their interlayer spaces and thus triggers landslides. However, smectite, illite and vermiculite-like clay minerals with higher interlayer spacing enhance the shear strength with an increase in sliding velocity which gives rise to short-term but large-scale landslides, whereas chlorite crystals retain the strain energy within its lattice due to the presence of hydrogen bonds, which joins the talc-like T-O-T layer and brucite like-octahedral sheet and facilitates viscoelastic behaviour even at high-stress regime inhibiting the occurrences of large-scale landslides. Hence, in the NaMu highway section in Nepal, more active landslides are evidenced in the areas where the soil dominantly contains illite rather than those containing chlorite and kaolinite with little or no illite. Moreover, unlike smectite or other clay minerals, the shear strength of chlorite diminishes with increasing slide velocity and results in slow-motion landslides for longer duration or creep, as also observed frequently in parts of the Lesser Himalayas.

Keywords Hydrothermal alteration · Chlorite · Ripplocations

P. Halder · K. Kumar · A. Sharma (✉)
Birbal Sahni Institute of Palaeosciences (DST, Govt. of India), Lucknow, Uttar Pradesh 226007, India
e-mail: anupam110367@gmail.com

Academy of Scientific and Innovative Research (AcSIR), Ghaziabad 201002, India

M. K. Shukla
Goverment of India, Ministry of Earth Sciences (MoES), Borehole Geophysics Research Laboratory (BGRL), Karad 415105, India

B. K. Rastogi et al. (eds.), *Natural Hazards and Risk Mitigation*, Springer Transactions in Civil and Environmental Engineering, https://doi.org/10.1007/978-981-97-7658-0_4

4.1 Introduction

Landslides are one of those complicated detrimental events occurrences of which are completely beyond the control of humans but can be worse due to human interventions. It encompasses the downslope movement of the rocks and/or soil in a coherent or semi-coherent state with a little degree of internal deformation along the curved or planar surface. The rapid onset of landslides brings adverse impacts on the natural environment along with severe socio-economic losses. Although the socio-economic losses of the landslides are partially recoverable, the natural losses can't be recovered or restored, including morphological changes, changes in land cover (forests and grasslands), etc. Hence, considering the long-term as well as short-term calamitous impact of the landslides, they are considered the seventh largest natural disaster accounting for 17% of fatalities (Alimohammadlou et al. 2013).

A landslide takes place when the frictional force, holding the rocks and/or soil in place, is dominated by the gravitational force acting on the hill slopes. A landslide initiates with the cracking at its top part, i.e. crown, due to the shearing stress, and forms a semi-circular surface with a steep slope along the surface of failure perpendicular to the steepest section of the slide. This is called the main scarp, which often transforms into the flanks. The displaced materials get crowded at the end of the sliding plane beyond the accumulation zone which is called its foot. Overall, the concave-shaped depletion zone controls the movement and nature of the landslides. The surface roughness is generally found higher on the erosional surfaces of a landslide, i.e. on the depletion as well as on the accumulation zone rather than in nearby stable terrains. A landslide advances through the formation of several transverse cracks along the slope, whereas radial cracks can be formed at its toe, a curved part created by the displaced materials. Figure 4.1 shows the different morphological features formed during a landslide.

Certain parameters control the occurrences and aerial extent of landslides, such as slope angle, nature of slope materials and the normal stress acting on the slope. When the slope angle exceeds the critical slope angle, the slope materials start to move causing landslides. The critical slope angle has been found 15° for the debris flow while it is 9°–27° for the slumps, 10°–35° for the slow earth flows and >25° for the debris slides and avalanches (Mann et al. 2012). This means that slopes <9° are not favourable for any kind of landslide. On the other hand, the materials with less frictional resistance are more prone to induce landslides. It also has been found that incoherent soil favours shallow but elongated landslides with no size restriction, while cohesive soil favours deep-seated landslides with relatively smaller sizes.

The 2400 km long Himalayan Mountain chain passing through the five nations is one of the high-risk areas for landslides due to its existence in the unstable tectonic region. This youngest and sixth longest mountain in the world is believed to have been formed as a result of the collision of the Indian plate with the Eurasian Plate. However, a recent isotopic study by Ibarra et al. (2023) in the Palaeocene Gangdese Arc in southern Lhasa has challenged this well-accepted theory of the Himalaya formation. They have compared the triple oxygen isotopes of the modern meteoric waters with

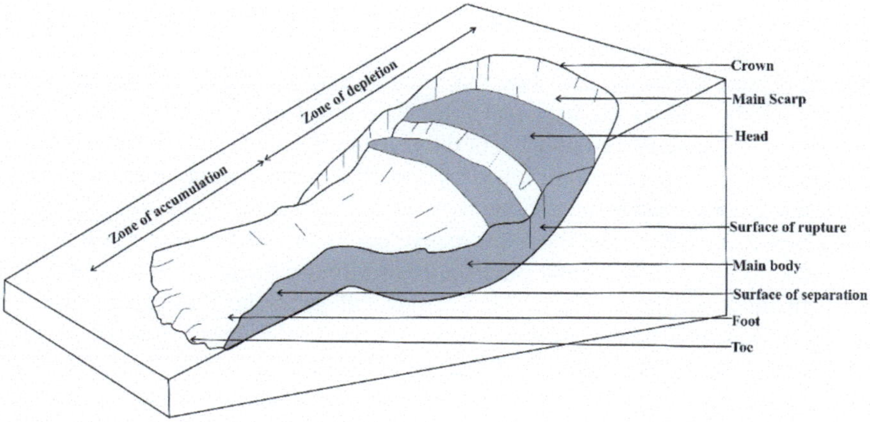

Fig. 4.1 Conceptual model of landslide morphology showing how a landslide initiates with the cracking at its crown and creates the main scarp along with several transverse cracks and lastly forms the foot with the displaced materials along with the formation of radial cracks at the toe. The zone of depletion and the zone of accumulation have been marked separately

epithermal Ag–Pb–Zn deposit quartz veins from this volcanic arc, and found that the substantial uplift took place around 63–61 million years ago, before the collision of the Indian plate with the Eurasian, probably due to the crustal shortening as a consequence of low-angle subduction of Neo-Tethyan oceanic lithosphere (Ibarra et al. 2023). However, the continuous growth of this mountain chain at a rate of about 1 cm/year does not allow us to rule out the influence of the compressional stress caused by the drifting of the Indian Plate at 5 cm/yr towards the north-east and that of the Eurasian Plate towards the north at 2 cm/yr (Bilham et al. 1998).

The mighty Himalayas are bounded by three major thrusts—Main Frontal Thrust (MFT), Main Boundary Thrust (MBT) and the Main Central Thrust (MCT) and they consist of several litho-tectonic units like Sub-Himalayas, Lesser Himalayas, Higher Himalayas and Tethys Himalayas. The Sub-Himalayas consist of Miocene and Pliocene sedimentary rocks of the Siwalik Group analogous to the modern Indo-Gangetic foreland basin and are surrounded by the Main Boundary Thrust (MBT) in the north and the Main Frontal Thrust (MFT) in the South; the Lesser Himalayas bordered by the Main Central Thrust (MCT) in the north and Main Boundary Thrust (MBT) in the south contain metasedimentary rocks of Late Precambrian ± Early Paleozoic; and the Higher Himalayas which have been thrust along the MCT zone over internally imbricated rocks of the Lesser Himalayas is comprised of high-grade metamorphic rocks (up to upper amphibolite facies) of upper Proterozoic–lower Paleozoic (Paudel and Arita 2000). The Greater Himalayas also include the Tethyan Himalayan Sequence (THS), a series of unmetamorphosed and very slightly metamorphosed sedimentary rocks of Cambrian to Eocene age, which is differentiated from the Higher Himalayas by the South Tibetan Detachment system (STDS) (Burchfiel et al. 1992). So, the occurrence of two thrusts, MBT and MCT and the presence of landslide-prone phyllites have made the Lesser Himalayas extremely susceptible to landslide-like vulnerabilities.

Most of the Himalayan landslides are rainfall-induced. Water saturation of the slope materials reduces the frictional force as a result, it comes down under the influence of the tangential component of the gravitational force and landslide occurs. In 2009, around 50 people were killed by landslides during monsoonal rainfall in Nepal. Landslides at Darbang, 200 km west of Kathmandu, took almost 200 lives in 1988 (Yagi et al. 1988). The most destructive one in the history of the Himalayan landslides occurred in the same place in 1926 killing around 500 people (Dahal 2012). In India, Malpa in Uttarakhand, a tiny village in the Pithoragarh district, near the Indo-China border, evidenced likewise landslides from 11 to 17 August in 1998 resulting in the death of 380 people (Dikshit et al. 2020). A similar sort of 466 destructive landslides in the Okhimath area in Mandakini Valley, killed 103 people and damaged 47 villages (Dikshit et al. 2020). In July 2001, a cloudburst in a section of the Mandakini Valley caused more than 200 landslides, resulting in 27 fatalities and affecting about 4000 people (Naithani et al. 2002). During the 13–14th of September 2012, the same region suffered excessive rains totalling more than 200 mm, which resulted in 473 landslides and the deaths of 51 people (Martha et al. 2015).

However, instances of earthquake-induced landslides are not also little in number in the Himalayan terrains. Bihar–Nepal earthquake of 15 January 1934, triggered such landslides in the metamorphic and crystalline complex (phyllites, quartzites, granite-pegmatites) near Kathmandu, Garhi, Udaipur, and in eastern Nepal. The falls were identified in the Siwalik Sandstones near Muksar which clogged the local channel and generated multiple lakes, one of which was 600 ft long with an approximate depth of about 25 ft. The rockslides at Tindharia in Darjeeling and Bansoi in North Sikkim were also triggered by the Bihar–Nepal Earthquake of August 1988 (Parkash 2013).

The tendency for landslides in the Himalayan region is also increased by anthropogenic activities such as excavation for buildings, roads, or other features which make the slope steeper than they were before construction. Thus, the slope angle becomes higher than the angle of repose as a result of which the slope materials get destabilised inducing landslides. Besides, the construction of high-rise buildings on the hilly slopes increases the normal stress and triggers landslides.

Apart from these classical rationales, clay minerals formed due to rainfall-induced weathering and the hydrothermal alteration during the formation of the Main Central Thrust (MCT) may have a role in landslides in the Lesser Himalayas. The clay minerals being hydrophilic in nature significantly contribute to the evolution of fluid pressure and help to reduce the frictional resistance acting on the slope surface. Thus, apart from the tectonic elements present in the Lesser Himalayas, the role of clay minerals on the different types of landslides in this highly vulnerable region can't be ruled out. Hence, the present study envisages this interfingering relationship between the clay minerals and landslides in the Lesser Himalayan perspective. The current chapter has also given an overview of how the lesser Himalayas have been surrounded by different thrusts which makes it tectonically unstable and susceptible to landslides.

Further, it discusses the different types of landslides evidenced in different parts of the Lesser Himalayas section and varied mechanisms of clay mineralisation. Finally, light has been thrown on the physics behind the landslides and the role of chemistry on landslides i.e., how the internal atomic structure of these clay minerals controls the motion of slope materials and results in different types of landslides along the stretch of the Lesser Himalayas.

4.2 Geology and Tectonics of the Lesser Himalayas

The Lesser Himalayan sequence, bounded by the Main Central thrust (MCT) and the Main Boundary thrust (MBT) in the North and South, respectively, is observed significantly in three countries—India, Nepal and Bhutan (Fig. 4.2). Its lithology is dominated by the rocks of Paleo- and Mesoproterozoic age, metamorphosed up to greenschist facies such as schist, phyllite, marble, quartzite, and unmetamorphosed limestone, dolostone, shale and sandstone, etc., as well as orthogneisses, Neoproterozoic sedimentary rocks, and sedimentary rocks of Permian to Miocene (Kohn et al. 2010; Martin et al. 2011; DeCelles et al. 2002).

A fault, recognised as the Bari Gad–Kali Gandaki Fault (BKF), has divided the Lesser Himalayas into two belts (Arita et al. 1982; Sakai 1985). The northern part is the inner belt and the southern part, which is a parautochthonous unit, is the outer belt. In India, the outer belt is distinguished from the inner belt based on ε_{Nd} values. The outer belt comprising low-grade metamorphic rocks is characterised by ε_{Nd} values similar to the Higher Himalayas (–13 to –18), whereas the inner belt has more negative ε_{Nd} values (–22 to –25) (Najman 2009). The inner belt contains sedimentary, low-grade metamorphic and mafic volcanic rocks along with a group of medium- to high-grade metamorphic rocks which is also designated as the Lesser Himalayan Crystalline Series (LHCS). LHCS is separated from the unmetamorphosed Inner Himalaya by the Munsiari Thrust (Najman 2009). Thus, each of the belts in the Lesser Himalaya contains a series of thrusts revealing its tectonically active nature. Between the Main Boundary Thrust (MBT) and Munsiari Thrust in India, there are Tons Thrust, Ramgarh Thrust, and Berinag Thrust (Célérier et al. 2009). Similarly in Nepal, the inner belt also consists of thrust sheets like TS I and TS II, divided by the Phalebas Thrust (PT) (Paudel and Arita 2000). Many of these thrusts were folded later and the Higher Himalayan rocks were thrust along the UMCT near the MBT (Paudel and Arita 2000). Thus, the Lesser Himalaya section has formed a foreland propagating duplex structure in many places where the UMCT acts as the roof thrust and the MBT as the floor thrust. The presence of several klippen are also observed overlying these thrust sheets such as the Almora–Dadeldhura klippe in Kumaun and Garwhal Himalaya, Palpa Klippe in Nepal, etc., which reveals evidence of active faulting (Célérier et al. 2009). The north-bounding Main Central Thrusts have also been found as the zone of maximum shear strain (Martin 2017; Searle et al. 2008). This MCT is bordered in the north by Upper MCT (UMCT) and in the south by Lower MCT (LMCT), which are known as MCT I & MCT II, respectively, in Nepal and

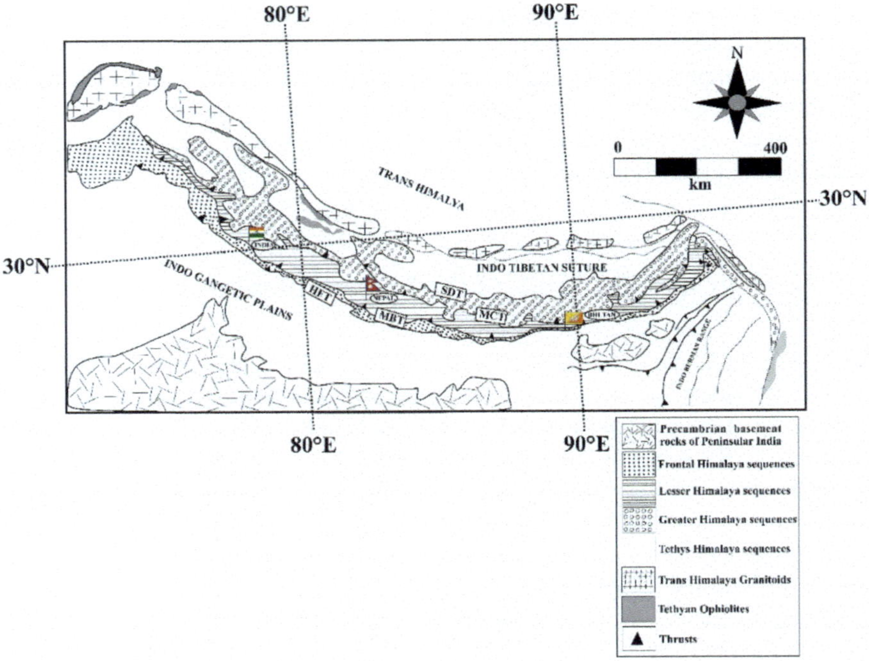

Fig. 4.2 Geologic map of the Himalayas showing the Lesser Himalaya passed through three Nations India, Nepal and Bhutan from North to South (HFT = Himalayan Frontal Thrust; MBT = Main Boundary Thrust; MCT = Main Central Thrust; STD = South Tibetan Detachment) (Modified after Saha (2013))

Munsiari & Vaikrita Thrusts in Kumaun and Garwhal of India (Arita et al. 1982). In the Lesser Himalayas of Nepal, the LMCT is obscured in the eastern parts, so here it is distinguished by the difference in the structural orientation between the MCT zone and the TS II. The MCT zone is characterised by homoclinally northward-dipping foliation, while the TS II has the foliation folded into a dome and basin structure (Paudel and Arita 2000).

In Nepal, the rocks of the Higher Himalayas with the overlying Tethys sediments occur as nappe over the Lesser Himalayas near the Kathmandu area (Paudel and Arita 2000). Kathmandu Nappe, a large synclinorium, contains bedding and foliation parallel to those of the underlying Lesser Himalayan units which reveals that the horizontal UMCT was folded later along with the autochthon during the propagation of the horses (slice of rocks separated from the surrounding rocks and bounded by multiple faults) (Paudel and Arita 2000). On the other hand, the frontal part of the Lesser Himalayas is occupied by the Palpa Klippe, the base of which is a highly sheared tectonic melange zone, about 10 m wide along the Tansen ± Pokhara motor road (Paudel and Arita 2000).

Thus, the presence of several faults in the Lesser Himalayas results in seismicity of different magnitudes which gives rise to landslides of different scales. In addition to it, a higher degree of construction activities, deforestation, mining, farming and other economic activities have made this terrain fragile and aggravated landslide-like phenomena, particularly during the Monsoon.

4.3 Types of Landslides in the Lesser Himalayas

Landslides can be of 32 types based on the variation in movement and diversity of the slope materials as proposed by Hungr et al. (2014) (Table 4.1). In the Himalayan region, landslides are also of different types such as topples, falls, flows, slides and avalanches (Barnard et al. 2001). Debris flows, slides and avalanches dominate the lower elevations; while rock avalanching is the most common type of landslides at relatively higher elevations (>1500 m asl.), as there was a scarcity of unconsolidated sediments. Landslides, particularly debris flows, triggered by monsoons are much more common than earthquake-induced landslides in the parts of the Leser Himalayas. Only two landslides induced by earthquakes have been found among the 30 biggest landslides in the Garhwal Himalaya of Northern India. However, no evidence of the debris flows has been found in case of earthquake-induced landslides. A maximum of ~10 million m^3 of debris was generated during a 6.6 magnitude earthquake. Landslide materials found during the 1999 Garhwal earthquake are relatively less which may be due to lithology comprising resistant, medium-grade metamorphic rocks with steep river gorge. Therefore, the sliding potential is decreased here because of the lack of storage on the steep valley walls. Anthropogenic activities such as slope undercutting due to the construction of roads or buildings have also been found responsible for the landslides affecting 2/3 of all recorded landslides.

The landslides evidenced along the Chamoli to Badrinath section of the Rishikesh—Badrinath National Highway in India are mostly debris flows, debris slides, rock falls and rock slides (Sarkar et al. 2006). Debris flows have only been found along the drainage lines while the rock slides have been witnessed along the discontinuities dipping downslope. Debris flow occurred on the left bank of the Alaknanda River in September 2004 partially blocked the river for a short period and changed its course, as a result of which the roads at the right bank of the river got completely flooded.

The Dumre Besi landslide along the Mugling–Narayanghat Highway within the Mahabharat Range in Nepal is also a significant landslide in the Lesser Himalayas, which was mainly triggered by heavy rainfall (Regmi et al. 2013). This is a complex type of failure involving different types of movement. The landslides spanned over an area of 0.15 Km^2 ranging from Dumre village above the road, to the bank of the Trishuli River below the road and are still moving (Regmi et al. 2013).

Table 4.1 A summarised format of the modified Varnes classification system of landslides as proposed by Hungr et al. (2014)

Type of movement	Rock	Soil
Fall	Rock/ice fall (Detachment, fall, rolling, and bouncing of rock or ice fragments wherein fragments can be broken during impacts)	Boulder/debris/Silt fall (Detachment, fall, rolling and bouncing of soil fragments such as large clasts in soil deposits, or blocks of cohesive (cemented or unsaturated) soil wherein the weakness of the moving particles doesn't result any impact)
Topple	Rock block topple (Rotational movement of rock columns or plates, separated by steeply dipping joints, over the basal discontinuities, starts with low velocity but ends up with overturning at rapid motion)	Gravel/sand/silt topple (Block toppling of columns of cohesive (cemented) soil, separated by vertical joints)
	Rock flexural topple (Bending and slow rotation of a steeply dipping jointed weak and fissile rock mass, strike of which is perpendicular to the fall line of the slope. Due to the absence of basal joints, bending occurs. This motion can be converted into secondary rotational sliding in the hinge zone of the topple)	
Slide	Rock rotational slide (Slow or moderately slow sliding movement of a weak rock mass on a cylindrical or rotational rupture surface with limited internal deformation)	Clay/Silt rotational slide (Slow to rapid Slide movement of the homogeneous and cohesive soil on a rotational rupture surface with little internal deformation. It can be but may be extremely rapid in sensitive or collapsive soils)
	Rock planar slide (Extremely rapid sliding movement of a rock mass on a planar rupture surface with very little or no internal deformation)	Clay/Silt Planar slide (Slow or rapid slide movement of a block of cohesive soil on an inclined planar rupture surface)
	Rock wedge slide (Extremely rapid sliding of a rock mass on a rupture surface formed of two planes with a downslope-oriented intersection without internal deformation)	Gravel/sand/debris slide (Sliding of the colluvium, weathered soil or pyroclastic deposits on a shallow, planar surface parallel with the ground. Many debris slides become flow-like after moving a short distance and transform into extremely rapid debris avalanches)

(continued)

Table 4.1 (continued)

Type of movement	Rock	Soil
	Rock compound slide (Slow or rapid sliding of a rock mass on a rupture surface consisting of several planes, or a surface of uneven curvature, through significant internal distortion of the moving mass.)	Clay/silt compound slide (Sliding of soil mass on a rupture surface consisting of several planes, or a surface of uneven curvature, through significant internal distortion of the moving mass)
	Rock irregular slide (Very sudden and rapid movement of mechanically strong rock mass on an irregular rupture surface consisting of several randomly oriented joints, separated by segments of intact rock)	
Spread	Rock/slope spread	Sand/silt liquefaction spread (Limited and slow movement of the mass of coherent blocks through near-horizontal stretching as a result of intensive deformation of an underlying weak material, or by multiple retrogressive sliding controlled by a weak basal surface)
		Sensitive clay spread (Extremely rapid lateral spreading of a series of soil blocks, floating on a layer of saturated (loose) granular soil, liquefied by earthquake shaking or spontaneous liquefaction)
Flow	Rock/ice avalanche (Extremely rapid flow-like movement of fragmented rocks after getting separated from a large rock slide or rock fall)	Sand/silt/debris dry flow (Slow or rapid flow-like motion of dry, moist or subaqueous, sorted or unsorted granular material, without excess pore pressure)
		Sand/silt/debris flow slide (Very rapid to extremely rapid flow of sorted or unsorted saturated loose sand, debris, loess and silt on moderate slopes involving excess pore pressure or liquefaction of material originating from the landslide source due to multiple retrogressive failures)

(continued)

Table 4.1 (continued)

Type of movement	Rock	Soil
		Sensitive clay flow slide (Very rapid to the extremely rapid flow of liquefied sensitive clay, due to remoulding during a multiple retrogressive slide failure at, or close to the original water content)
		Debris flow (Very rapid to the extremely rapid flow of saturated debris in a steep channel. Strong entrainment of material and water from the flow path)
		Mudflow (Very rapid to extremely rapid surging flow of saturated plastic soil having Plasticity Index > 5%, in a steep channel, involving significantly greater water content relative to the source material)
		Debris flood (Very rapid flow of water, heavily charged with debris, in a steep channel. Peak discharge comparable to that of a water flood
		Debris avalanche (Very rapid to extremely rapid shallow flow of partially or fully saturated debris on a steep slope, without confinement in an established channel)
		Earthflow (Rapid or slower, continuous flow-like movement of plastic, clayey soil, facilitated by a combination of sliding along multiple discrete shear surfaces, and internal shear strains)
		Peat flow The rapid flow of liquefied peat, caused by an undrained failure
Slope deformation	Mountain slope deformation (Large-scale gravitational deformation of steep, high mountain slopes, manifested by scarps, benches, cracks, trenches and bulges with very slow or unmeasurable movement rates)	Soil slope deformation (Deep-seated, slow to extremely slow deformation of valley or hill slopes formed of mostly cohesive soils)

(continued)

Table 4.1 (continued)

Type of movement	Rock	Soil
	Rock slope deformation (Deep-seated slow to very slow deformation of valley or hill slopes with the development of cracks or faults, lacking well-defined rupture surface)	Soil creep (Prolonged movement of the soil layers on a slope as a result of climate-driven cyclical volume changes due to wetting and drying, frost heave)
		Solifluction (Very slow and shallow soil creep involving the active layer in the Alpine or polar permafrost with the formation of lobes)

4.4 Mechanism of Clay Mineralisation in the Lesser Himalayas

Clay mineralisation in the Lesser Himalayas may occur in two ways:

i. Hydrothermal alteration
ii. Weathering due to heavy rainfall.

Hydrothermal alteration is the mineralogical and chemical transformation of the rocks in interaction with the hot fluid. The source of this heat may be magmatic such as the intrusion of igneous bodies or the frictional heat during fault slip (Scholz 1980; Anders and Wiltschko 1994). In addition, the mechanically weak crustal planes may become the pathway through which hot fluid move rapidly towards the surface, preserving the shallow isotherms that cause fluid flow at a later stage. The majority of the heat in the fold and thrust belt of the Lesser Himalayas are probably derived from the tectonic advection of hot rock to the sub-surface rather than fluid flow as also confirmed by the significant presence of several hot springs near MCT (Derry et al. 2009). Mayangdi Khola Thermal Springs and Kodari Thermal Springs are a few of them, the discharge temperature of which is around 40 °C (Bhattarai 1980). Earlier researchers have also pointed out the movement of hydrothermal fluid along the MCT (Fort 1988; Evans et al. 2001), which was derived during the emplacement of the Miocene leucogranite of the Higher Himalayas and its subsequent cooling during 22–20 Ma (Harris et al. 1993). Further, the deformation and partial melting of the hanging wall of the MCT may also have a role in the movement of this hydrothermal fluid which was circulated later in the highly fractured Lesser Himalayan rocks during and after the advancement of the MCT (Fort 1988; Hodges et al. 1996) (Fig. 4.3). Thus, the advancement of MCT has resulted in hydrothermal alteration in the Lesser Himalayan rocks and produced clay minerals like chlorite, illite, etc. Such presence of clay minerals has been reported from various locations along the sliding surface of the large-scale landslides in the Narayangarh–Mugling (NaMu) Highway of Central Nepal and several clay layers of >1 m thickness have been found in the sliding surfaces in this region, which is underlain by the highly fractured and jointed phyllites (Hasegawa et al. 2009). Earlier investigations have also found the presence of fault gouge in the sliding surface but the overlying phyllites don't show any evidence of shearing which indicates the activity of hydrothermal fluid in the sliding surface (Hasegawa et al. 2009). Although the possibility of clay minerals formation by hydrothermal alteration has been pointed out in many parts of the lesser Himalayas in Nepal, still the role of weathering in clay mineralisation can't be ruled out.

In the Dumre Besi landslides, most of the clay minerals were formed by rock weathering and not by hydrothermal alteration (Regmi et al. 2013). The highly fractured overlying rock mass acts as the fluid percolation pathways which make the fluid viable up to the sliding surface of any landslides. In fact, the fractured rocks along the steep slopes are more susceptible to weathering due to pre-existing stresses, which promote the degradation of rock strength. Thus, the rainwater during monsoon can enter through the faults and cause intense chemical weathering to produce clay minerals. Slates found in the Dumre Besi landslide zone have been found chemically less altered because the weathered slate probably has been removed by erosion and running water (Regmi et al. 2013). Most part of the landslide zone is covered by colluvium and fines and the rocks are only found exposed near the thrust zone. The predominance of the weathered rocks near the thrust zone indicates its influence on weathering. Chlorite, smectite, zeolite, and vermiculite have been reported significantly in these weathered rocks among which smectite and vermiculite were the weathering product, while chlorite and zeolite might have been formed by hydrothermal alteration during regional metamorphism.

4.5 Discussion

4.5.1 Mechanism of Landslides: Physics Behind Landslides

A landslide is facilitated by the tangential component of the gravitational force while exceeding the frictional force on the slide zone. The friction coefficient (μ) on the sliding surface of the landslide is (Dieterich 1979; Ruina 1983)

$$\mu = \mu_0 + a \ln\left(\frac{v}{v_0}\right) + b \ln\left(\frac{v_0 \theta}{d_c}\right) \tag{4.1}$$

where μ_0 is the steady-state friction coefficient, a and b are the friction parameters which determine the changes of μ with the variation of v, and d_c is the slip distance for the evolution of frictional contacts. The state variable θ represents the average life span of a set of frictional contacts which evolves following the Aging (or Dieterich) law,

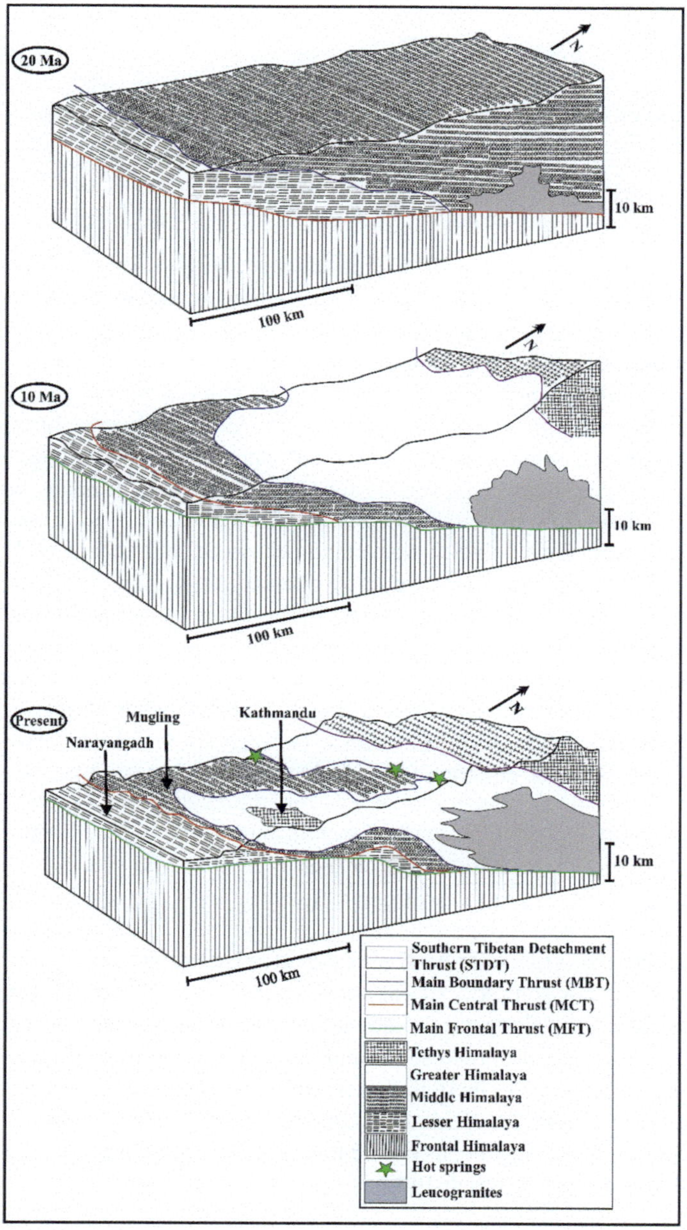

Fig. 4.3 Conceptual schematic diagram revealing the formation and subsequent advancement of several thrusts observed in Himalayan terrain. The diagram also shows the mechanism of surging of hydrothermal fluid and the relative position of the hot springs as the expression of the hydrothermal activities. The relative locations of Narayangadh, and Mugling, where the potential role of clay minerals on landslides has been evidenced, are also displayed in the diagram. (Modified after Hasegawa et al. (2009)

$$\frac{d\theta}{dt} = 1 - \frac{v\theta}{d_c} \tag{4.2}$$

In the case of a steady state, the above equation becomes (Rice Ruina 1983),

$$\mu = \mu_0 + (a - b)\ln\left(\frac{v}{v_0}\right) \tag{4.3}$$

The motion and magnitude of landslides depend on (a–b) value of the materials present in the slide zones. If the slide zone contains materials that have (a–b) > 0, frictional resistance will increase with an increase in slip velocity and stable slide or creep will occur. These materials are called velocity-strengthening (Meller, 2014). On the other hand, slide zone materials having (a–b) values < 0 reveal a reverse response with an increase in slip velocity and facilitate unstable slip. Thus, the velocity-strengthening materials never fail catastrophically (Meller and Kohl 2014). Furthermore, Ikari et al. (2011) found, that rocks or minerals with a low friction coefficient show velocity-strengthening behaviour, whereas those with high friction coefficients show velocity-weakening behaviour (Ikari et al. 2011).

However, the stress equation in the case of landslides is quite different from that used in fault mechanics. It includes elastic stress which indicates the resistance to differential motion due to longitudinal compression at grain contacts, as landslides are not considered as the isolated block on the slope. As the lesser Himalaya is a complex fold and thrust belt, so incipient slip along the landslide surface accelerates or decelerates as a function of the balance between frictional and elastic forces. Hence, the stress balance equation for such a highly faulted region becomes (Segall 2010),

$$\tau_d + \frac{G}{2\pi(1-\nu)} \int_{-\infty}^{+\infty} \frac{d\delta/d\xi}{\xi - x}\, d\xi = \mu\sigma' + \frac{G}{2v_s} v \tag{4.4}$$

τ_d is the driving stress due to gravity, σ' is the effective normal stress, G is the shear modulus, ν is the Poisson ratio, (dδ/dξ) denotes the elastic forcing associated with spatial gradients in slip along the sliding surface at local coordinate x, v_s is the shear-wave speed, μσ' is the frictional stress and ($G/2v_s$) v represents an approximation of radiation damping due to inertial effects, which is significant only when slip rate approaches $v_{dyn} \sim (2\sigma' a v_s)/G$ (Rice 1993).

In the case of the homogeneous elastic material, the critical nucleation length for a landslide to occur is (Dieterich 1979).

$$h^* = \frac{G d_c}{\sigma'(b - a)} \tag{4.5}$$

If the length of the slip surface (L) exceeds the critical nucleation length (h*), the combined effects of frictional and elastic stresses cannot compensate for the driving stress, as a result of which catastrophic failure will take place (Fig. 4.4).

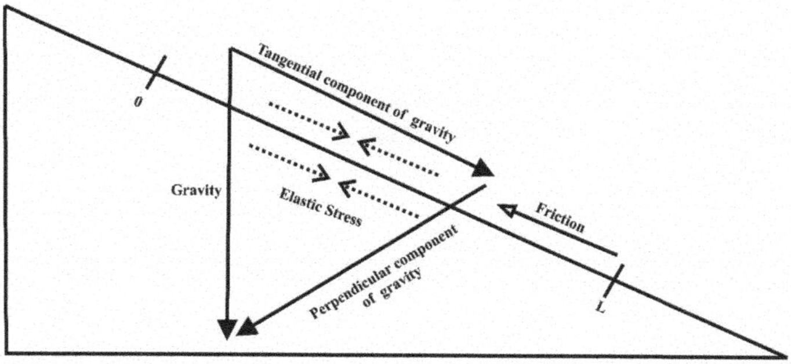

Fig. 4.4 Schematic model of landslide geometry showing the contributions of different forces along the sliding surface between 0 and L (Modified after Handwerger (2016)

Nevertheless, in the case of L < h*, reductions in frictional strength associated with increasing sliding velocity are balanced by elastic stresses along the boundaries of the slipping region having large strain gradients.

4.5.2 *Role of Clay Minerals on Landslides: Chemistry Behind Landslides*

The Lesser Himalayan lithology is dominated by clay minerals, particularly in the landslide-prone zones. These clay minerals have a profound impact on the Dumre Besi landslide and a few other failures in its adjacent areas. The phyllosilicate clay minerals have interlayer spacing and overall negative charge on their surface, which make them highly potential to absorb or adsorb water. Among the clay minerals, smectite has a basal spacing of 10–20 angstrom and those of kaolinite, vermiculite, chlorite and illite have 7, 10–15, 14 and 10 angstroms (Karathanasis 2002). These interlayer spaces have the potential to absorb/adsorb water, so their presence in the sliding zones of landslides evolves fluid pressure and decreases the frictional resistance. So, σ' becomes ($\sigma' - P_f$) and the right-hand function of the Eq. (4.4) exceeds the left-hand function resulting in slides. Moreover, clay minerals may clog in the interstices or pore spaces or the fracture spaces where they are formed, as a result of which, permeability is reduced inhibiting the further flow of water. Thus, the fluid pressure (P_f) is evolved resulting in further reduction of σ', which leads to failure. Hence, in the compressional stress regime around the Kathmandu Nappe of Central Nepal, clay minerals present in the slide planes give additional momentum to landslides.

Furthermore, the magnitude of the slides is highly dependent on the bond strength and internal atomic structure of these clay minerals. Smectite, illite and vermiculite-like clay minerals, with relatively higher interlayer spacing, can absorb a large volume of water. Additionally, the velocity-strengthening behaviour of these clay minerals enhances the shear strength with an increase in sliding velocity which gives rise to short-term but large-scale landslides, as also observed in the weathered rock suits of the Narayangharh–Mugling (NaMu) highway section of the Lesser Himalayas in Nepal, where Dumre Besi landslide took place in 2003. However, chlorite shows radically contrasting behaviour when present in the Lesser Himalayas like stress regime containing a series of small as well as large-scale thrusts. Chlorite consists of the talc-like T-O-T layer (Tetrahedral–Octahedral–Tetrahedral) and brucite-like octahedral sheet which are joined by H bonds (Fig. 5a), and control the mechanical response of chlorite to stress. The behaviour of chlorite with increasing stress can be represented in two stages as follows:

First stage: In response to stress, initially the volume and thickness of the octahedra of the talc-type layer get reduced, but the octahedra of brucite-type layers become tilted and more regular leading to an increase in the volume of the octahedron (Zanazzi et al. 2007) (Fig. 5b). In contrast, tetrahedral distortion increases with P, but the volume slightly decreases without any significant change in thickness, as a result of which the distance between the OH of the brucite-type sheet and the basal O of the tetrahedral sheet decreased. The OH-O distance in chlorite crystal is found around 2.7 Å at 3.2–4.25 GPa pressure but O_2^-–O_2^- comes in contact when the pressure increases up to 8 GPA (Zanazzi et al. 2007). This sort of decrease in bond length increases electrostatic interaction between the acceptor and donor atoms of the constituting hydrogen bonds which leads to an increase in bond strength. Thus, an increase in pressure results in a relatively stronger hydrogen bond enhancing the stiffness of the chlorite structure, which is why, in response to the applied stress to accommodate the strain perpendicular to the c-axis, several atomic scale ripples are generated on the basal plane of the chlorite and allows one plane of atoms to migrate over another without distorting the in-plane bonds (Fig. 5b) (Aslin et al. 2019). Thus, with rising or continued stress, a large number of small-scale ripplocations occur, which can store the strain energy until a threshold.

Second Stage: Further increase in stress on chlorite causes the stored energy to reach its threshold (i.e., after 8 GPa pressure), so the interlayer shortening occurs with a subsequent decrease in OH-O angle from 170° to 155° that results in repulsion between adjacent H^+-H^+ cations and a subsequent decrease in bond strength (Zanazzi et al. 2007; Hofmeister et al. 1999; Welch and Marshall 2001) (Fig. 5c). As a consequence, the small ripples start to migrate with the energy gained and merge into kink bands curving the crystal lattice to overcome the compelling force (Bell et al. 1986) (Fig. 5c).

This is how, the yield strength of chlorite rises proportionally with rising pressure up to the temperature at which it dehydrates, and behaves in a viscoelastic manner. The chlorite crystal retains the strain energy within its crystal thereby preventing slip and inhibiting the occurrences of large-scale landslides. Hence, in the NaMu highway section, more active landslides are evidenced in the areas where the soil dominantly contains illite rather than those containing chlorite and kaolinite with little or no illite content (Neupane 2011). Furthermore, the low frictional strength of chlorite increases its velocity-strengthening behaviour, so the shear strength diminishes with increasing slide velocity and results in slow-motion landslides for longer duration (years to centuries) or creep, as also evidenced frequently in different areas of the Lesser

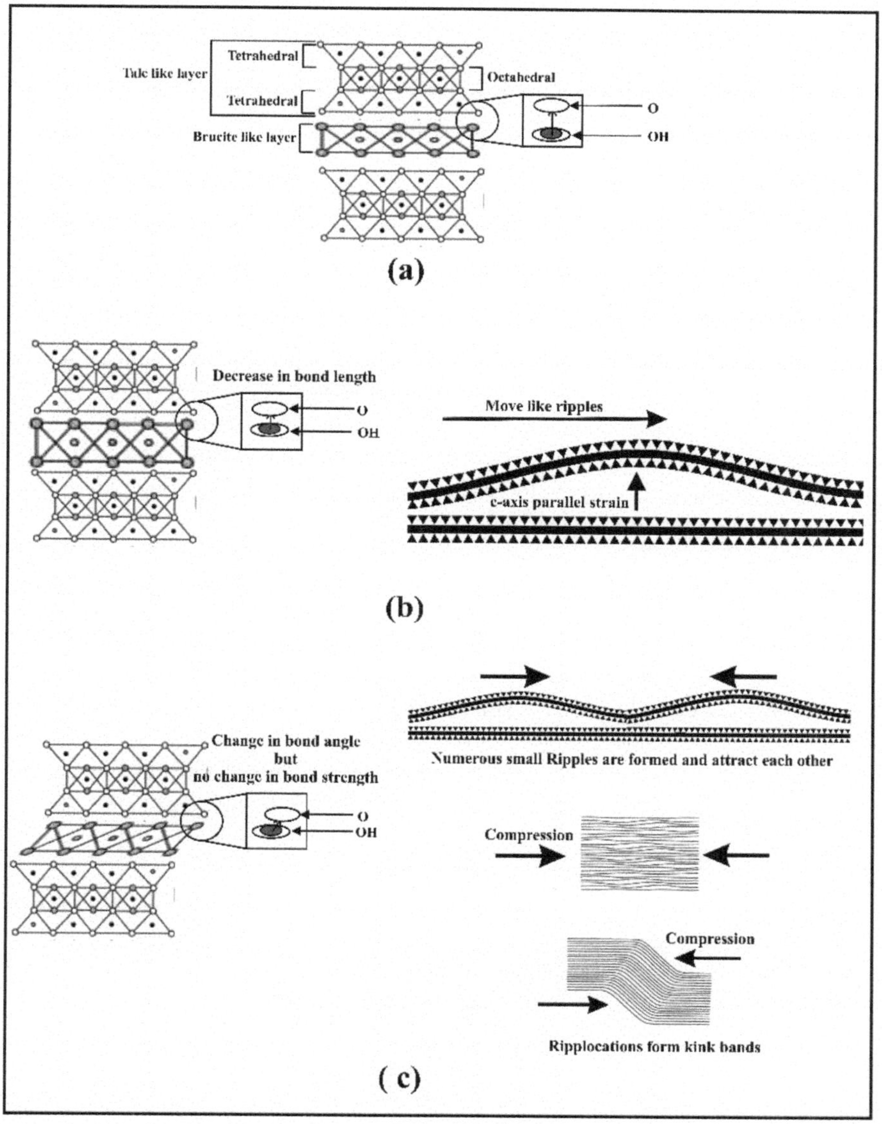

Fig. 4.5 Schematic diagram showing how chlorite behaves under stress when present in the highly stressed Lesser Himalayas. **a** Structure of chlorite at normal conditions showing the talc-like T-O-T layer (Tetrahedral-Octahedral-Tetrahedral) and the brucite-like octahedral sheet are joined by H bonds **b** Structure of the chlorite with increasing stress revealing an increase in the volume of the octahedron resulting the decrease in the distance between the OH of the brucite-type sheet and the basal O of the tetrahedral sheet and subsequent generation of ripples on its basal plane; **c** The structure of chlorite with continued stress showing the interlayer shortening along with a decrease in OH-O angle and subsequent migration of the ripples with the energy gained and formation of kink bands curving the crystal lattice

Himalayas. Thus, the landslides in the lesser Himalayas of Nepal advance through several stages where large-scale failure occurs along the zone of clay mineralisation followed by small-scale slope failures in the same huge landslide body.

4.6 Conclusion

Landslide is one of the deadliest hazards in the tectonically active Himalayan Region, particularly in the Lesser Himalayas. The present study concludes the following:

(i) Advanced degree of weathering due to heavy rainfall in addition to the hydrothermal alteration during the formation of the Main Central Thrust (MCT) has resulted in clay mineralisation in this region, which reduces the frictional strength of the slide planes and triggers landslides. Thus, apart from the stress pattern and fault geometry, the clay minerals present in the slide planes give additional momentum to landslides.

(ii) Unlike other clay minerals, chlorite resists large-scale slip and inhibits the occurrences of rapid landslides due to their potential to retain the strain energy within its crystal. Moreover, the low frictional strength of chlorite increases its velocity-strengthening behaviour, so the shear strength diminishes with increasing slide velocity and results in slow-motion landslides for longer duration (years to centuries) or creep, as also evidenced frequently in different areas of the Lesser Himalayas. So, aside from the conventional physical parameters (such as slope angle, frictional resistance, and effective normal stress) the internal atomic structure of the slope material can take control over the mode of landslides.

Thus, the current article integrates the chemistry and physics behind the landslides in light of the Lesser Himalayas.

Acknowledgements The authors express sincere gratitude to the Director, BSIP, and Director, AcSIR, for their motivation. The authors are also grateful to the editors of this book and the anonymous reviewers for their constructive comments and suggestions.

References

Alimohammadlou Y, Najafi A, Yalcin A (2013) Landslide process and impacts: A proposed classification method. CATENA 104:219–232

Anders M, Wiltschko D (1994) Microfracturing, paleostress and the growth of faults. J Struct Geol 16:795–815

Arita K, Hayashi D, Yoshida M (1982) Geology and structure of the Pokhara-Piuthan area, central Nepal. J Nepal Geol Soc 2:5–29

Aslin J, Mariani E, Dawson K, Barsoum MW (2019) Ripplocations provide a new mechanism for the deformation of phyllosilicates in the lithosphere. Nat Commun 10:686

Barnard PL, Owen LA, Sharma MC, Finkel RC (2001) Natural and human-induced land sliding in the Garhwal Himalaya of northern India. Geomorphology 40:21–35
Bell IA, Wilson CJL, McLaren AC, Etheridge MA (1986) Kinks in mica: Role of dislocations and (001) cleavage. Tectonophysics 127:49–65
Bhattarai DR (1980) Some geothermal springs of Nepal. Tectonophysics 62:7–11
Bilham R, Blume F, Bendick R, Gaur VK (1998) Geodetic constraints on the translation and deformation of India: Implications for future great Himalayan earthquakes. Current Sci, 213–229
Burchfiel BC et al. (1992) The South Tibetan detachment system, Himalayan orogen: Extension contemporaneous with and parallel to shortening in a collisional mountain belt
Célérier J, Harrison TM, Webb AAG, Yin A (2009) The Kumaun and Garwhal Lesser Himalaya, India: Part 1. Structure and Stratigraphy. GSA Bulletin 121:1262–1280
Dahal R (2012) Rainfall-induced landslides in Nepal. Int J Eros Control Eng 5:1–8
DeCelles PG, Robinson DM, Zandt G (2002) Implications of shortening in the Himalayan fold-thrust belt for uplift of the Tibetan Plateau. Tectonics 21:12–11–12–25
Derry LA, Evans MJ, Darling R, France-Lanord C (2009) Hydrothermal heat flow near the Main Central Thrust, central Nepal Himalaya. Earth Planet Sci Lett 286:101–109
Dieterich JH (1979) Modeling of rock friction: 1. Experimental results and constitutive equations. J Geophys Res: Solid Earth 84:2161–2168
Dikshit A, Sarkar R, Pradhan B, Segoni S, Alamri AM (2020) Rainfall induced landslide studies in Indian Himalayan Region: A critical review. Appl Sci 10:2466
Evans MJ, Derry LA, Anderson SP, France-Lanord C (2001) Hydrothermal source of radiogenic Sr to Himalayan rivers. Geology 29:803–806
Le Fort P (1988) Granites in the tectonic evolution of the Himalaya, Karakoram and southern Tibet. Philos Trans R Soc London Ser A, Math Phys Sci 326:281–299
Handwerger AL, Rempel AW, Skarbek RM, Roering JJ, Hilley GE (2016) Rate-weakening friction characterizes both slow sliding and catastrophic failure of landslides. Proc Natl Acad Sci 113:10281–10286
Harris N, Massey J, Inger S (1993) The role of fluids in the formation of High Himalayan leucogranites. Geol Soc, Lond, Spec Publ 74:391–400
Hasegawa S, Dahal R, Yamanaka M, Bhandary N, Inagaki H (2009) Causes of large-scale landslides in the Lesser Himalaya of central Nepal. Environ Geol 57:1423–1434
Hodges K, Parrish R, Searle M (1996) Tectonic evolution of the central Annapurna range. Nepal Himalayas Tecton 15:1264–1291
Hofmeister AM, Cynn H, Burnley PC, Meade C (1999) Vibrational spectra of dense, hydrous magnesium silicates at high pressure: Importance of the hydrogen bond angle. Am Miner 84:454–464
Hungr O, Leroueil S, Picarelli L (2014) The Varnes classification of landslide types, an update. Landslides 11:167–194
Ibarra DE et al (2023) High-elevation Tibetan Plateau before India-Eurasia collision recorded by triple oxygen isotopes. Nat Geosci 16:810–815
Ikari MJ, Marone C, Saffer DM (2011) On the relation between fault strength and frictional stability. Geology 39:83–86
Karathanasis A (2002) CD. Barton. Clay minerals. Encyclopedia of soil science, Marcel Dekker, New York, USA, pp 187–192
Kohn M, Paul S, Corrie S (2010) The lower Lesser Himalayan Sequence: A Paleoproterozoic arc on the northern margin of the Indian plate. Geol Soc Ameri Bull 122:323–335
Mann U, Pradhan B, Prechtel N, Buchroithner M (2012) An automated approach for detection of shallow landslides from LiDAR derived DEM using geomorphological indicators in a tropical forest. Terrigenous Mass MovS: Detect, Model, Early Warn Mitig Using Geoinformation Technol, 1
Martha TR et al (2015) Landslides triggered by the June 2013 extreme rainfall event in parts of Uttarakhand state, India. Landslides 12:135–146

Martin AJ (2017) A review of definitions of the Himalayan Main Central Thrust. Int J Earth Sci 106:2131–2145

Martin AJ, Burgy KD, Kaufman AJ, Gehrels GE (2011) Stratigraphic and tectonic implications of field and isotopic constraints on depositional ages of Proterozoic Lesser Himalayan rocks in central Nepal. Precambr Res 185:1–17

Meller C, Kohl T (2014) The significance of hydrothermal alteration zones for the mechanical behavior of a geothermal reservoir. Geotherm Energy 2:12

Naithani AK, Kumar D, Prasad C (2002) The catastrophic landslide of 16 July 2001 in Phata Byung area, Rudraprayag District, Garhwal Himalaya, India. Current Science 82:921–923

Najman Y et al. (2009) Reconstructing the exhumation history of the Lesser Himalaya, NW India, from a multitechnique provenance study of the foreland basin Siwalik Group. Tectonics 28

Neupane B, Adikari DP (2011) Role of clay minerals in the occurrence of landslides along Narayangarh-Mugling Highway Section, central Nepal. J Nepal Geol Soc 43:301–308

Parkash S (2013) Earthquake related landslides in the Indian Himalaya: experiences from the past and implications for the future. Landslide Sci Pract: Complex Environ 5:327–334

Paudel LP, Arita K (2000) Tectonic and polymetamorphic history of the Lesser Himalaya in central Nepal. J Asian Earth Sci 18:561–584

Regmi AD, Yoshida K, Dhital MR, Devkota K (2013) Effect of rock weathering, clay mineralogy, and geological structures in the formation of large landslide, a case study from Dumre Besei landslide, Lesser Himalaya Nepal. Landslides 10:1–13

Rice JR (1993) Spatio-temporal complexity of slip on a fault. J Geophys Res: Solid Earth 98:9885–9907

Rice and Ruina AL (1983) Stability of steady frictional slipping. J Appl Mech 50:343–349

Ruina A (1983) Slip instability and state variable friction laws. J Geophys Res: Solid Earth 88:10359–10370

Saha D (2013) Lesser Himalayan sequences in Eastern Himalaya and their deformation: Implications for Paleoproterozoic tectonic activity along the northern margin of India. Geosci Front 4:289–304

Sakai H (1985) Geology of the Kali Gandaki supergroup of the Lesser Himalayas in Nepal

Sarkar S, Kanungo D, Patra A (2006) Landslides in the Alaknanda Valley of Garhwal Himalaya, India. Qua J Eng Geol Hydrogeol 39:79–82

Scholz CH (1980) Shear heating and the state of stress on faults. J Geophys Res: Solid Earth 85:6174–6184

Searle MP et al (2008) Defining the Himalayan main central thrust in Nepal. J Geol Soc 165:523–534

Segall P (2010) Earthquake and volcano deformation. Princeton University Press, Princeton, NJ, pp 200–266

Welch M, Marshall W (2001) High-pressure behavior of clinochlore. Am Miner 86:1380–1386

Yagi H, Maruo Y, Saijo K, Nakamura S (1990) The sept 1988 large landslide in the vicinity of MCT, Darbang, Nepal. Landslides 26:45–49

Zanazzi PF, Montagnoli M, Nazzareni S, Comodi P (2007) Structural effects of pressure on monoclinic chlorite: A single-crystal study. Am Miner 92:655–661

Chapter 5
Geotechnical Investigation of Toldi Landslide Along the National Highway-44 (NH-44), Udhampur, J&K

Shifali Chib, Yudhbir Singh, Sumit Johar, and Rajesh Singh Manhas

Abstract The Toldi slide falling in the Samroli area along the NH-44 in the Udhampur district of Jammu and Kashmir has been plagued by frequent failures during insistent rainfall from last more than three years. This slide falls within the Murree Group of rocks comprising interbedded medium- to fine-grained sandstone and mudstone which is considered as the water-sensitive lithology. The current slide on the basis of its areal content is categorized as a medium slide and covers an area of about 51,888 m^2. It is also categorized as a rock slide depending on the kind of movement and the geologic material. The fundamental goal of the present study was to comprehensively understand the underlying failure mechanisms, with the aim of proposing immediate remedial measures to mitigate the losses and disruptions experienced by the population. In order to determine the quality of the rock mass and the slope mass's instability, the current study includes extensive field investigations as well as geotechnical studies in the form of rock mechanics (RMR, SMR and kinematic analysis). The basic rock mass rating inferred that the rock mass falls under fair category; however, the slope mass rating indicated a completely unstable slope with a high likelihood of failure (failure probability of 0.9). Kinematic analysis on the other hand uncovered two prevalent modes of failure at the site: planar and wedge failures. The study also inferred that the unsystematic toe cutting of the water-sensitive lithology is the dominant reason behind this landslide. Other causes include uncontrolled blasting, mechanical excavation, subsequent rainfall, improper drainage, steeper slope angle and shear zones around the slide.

Keywords Toldi slide · Samroli · NH-44 · Udhampur district · Jammu and Kashmir · Murree group · Geotechnical investigations · Kinematic analysis

---

S. Chib (✉) · Y. Singh · S. Johar · R. S. Manhas
Department of Geology, University of Jammu, Jammu 180006, India
e-mail: Shifali.chib@jammuuniversity.ac.in

B. K. Rastogi et al. (eds.), *Natural Hazards and Risk Mitigation*, Springer Transactions in Civil and Environmental Engineering, https://doi.org/10.1007/978-981-97-7658-0_5

5.1 Introduction

Natural disasters occur as a consequence of a variety of complicated variables that are beyond the control of humans, although certain disasters may occur as a result of human activity. Land sliding is recognized as the most common geological event that can occur anywhere in the world which gives rise to a higher mortality rate and ranks as the 7th largest disaster (Alimohammadlou et al. 2013; Sati et al. 2022). A report which was given on worldwide events of landslides (1971–1975) to UNESCO by 'The International Association of Engineering Geology (IAEG)' has recorded that landslides contribute to about 14% of fatalities from natural disasters (Froude and Petley 2018). According to the International Disaster Database (EM-DAT) study, landslide incidents that occurred between 1990 and 2015, account for about 4.9% of all disasters that occurred in Asia (Guha-Sapir et al. 2015). The Himalayan Mountains are recognized as one of the world's most prone regions to landslides. The active tectonic activities in the Himalayan Mountains combined with the precipitation, high elevation and high slope angle made the region more prone to landslides both in the past and in the present. Although rainfall and earthquakes are considered as the two dominant factors for triggering landslides, but rainfall has been recognized as the most significant factor for triggering landslides (Pant and Luirei 1999; Singh, 2006; Rawat and Kothyari 2007; Kothyari et al. 2010, 2012; Singh and Bhat 2010; Singh et al. 2010; Sarkar et al. 2012; Joshi et al. 2023; Kholia et al. 2023; Luirei et al. 2023; Taloor et al. 2023). According to Varnes's (1958) classification of landslides, it is simply described as the outward and downward movements of rock materials in the form of rock, debris, and/or earth under the direct action of gravity. These outward and downward slope movements are further divided on the basis of the type of geologic material and the type of movement (Varnes 1978). These slope movements not only cause blockade of roads but also pose problems in the implementation of work.

Jammu and Kashmir is the northernmost Union Territory of India and geographically 90% of its area falls under the Himalayan range. The Jammu–Srinagar National Highway (NH-44) passes through the Outer Himalayan to the Higher Himalayan tectonic zone which has a very complex geology and immature terrain making the slope more prone to failures. During the monsoon season, this highway frequently remains closed due to several recurrent landslides that cutoff the Kashmir valley from the rest of India. Road clearance and road widening operations are always under progress for smooth connectivity which makes the area more vulnerable. Some experts have noticed that ground vibrations induced by the operation of heavy machinery in the excavation process and the passing of heavy vehicles along the cut slopes of road are one of the most significant triggering factors for slope failure that must be considered (Kundu et al. 2017). Singh et al. (2010, 2016) and Sarkar et al. (2012) in their studies concluded that the increased ground vibrations caused by the passing of heavy and light vehicles in the landslide prone areas can be 0.5 times stronger than the gravity-caused vibrations, and therefore increasing shear stress on the slope and reducing the Factor of Safety (FOS) of the slope material. Because the slopes of the mountain in the study region are being constantly cut with heavy machinery for the road widening operation, therefore, drilling and blasting of the slopes is a common occurrence in the area. This type of constant road widening makes the region more prone to slope failures. As it is well known that the recent activity of landslides in many regions of the Himalayan mountains is also related to the resurgence of the earlier landslides which later resulted in follow-up landslides, therefore, it is necessary to detect and investigate previous landslides that occurred in the past. The findings of such a study can be utilized to evaluate the recent landslides. A substantial amount of work has previously been done in this field, which has significantly improved our understanding of mechanisms, processes, causes, and associated hazards (Singh 2006; Pandey et al. 2014). The reason behind the landslides cannot be directly investigated as it influences all types of geological materials having various types of volumes and shapes. This cause can only be assessed using an amalgamation of several methods to acquire an appropriate conclusion. The present study focuses on field investigations and geotechnical studies of the slope materials so as to characterize the landslide. This will help in understanding the cause and the nature of the landslide.

5.2 General Geology and Structural Set-Up of the Study Area

The present study area (road chainage Ch 78 + 10 km) is a part of the National Highway-44 (NH-44) having latitude $33^\circ 00' 01''$ and longitude $75^\circ 12' 22''$. Geologically, the study area falls within the Murree Group (Miocene age) of the Outer Himalayan tectonic zone of the north-western Himalaya (Fig. 5.1). The sliding area falls within the Upper Murree Formation of the Murree Group comprising of fine to medium-grained sandstones of greyish green colour and unconsolidated interbedded layers of reddish and maroon mudstones/siltstones which are of splintery type. The existence of thrusts, folds and faults complicates the structure of the sliding region. The most prominent structural units around the study area are the Main Boundary Thrust (MBT) in the north which borders the Murree Group from the Siwalik Group and the Murree Thrust in the south which delimits the Murree Group from the Precambrian Souni volcanics. The present sliding area under investigation is sandwiched between these two prominent thrusts thus making it more prone to failures. The area has a rugged and uneven topography and also receives heavy precipitation during the monsoon season.

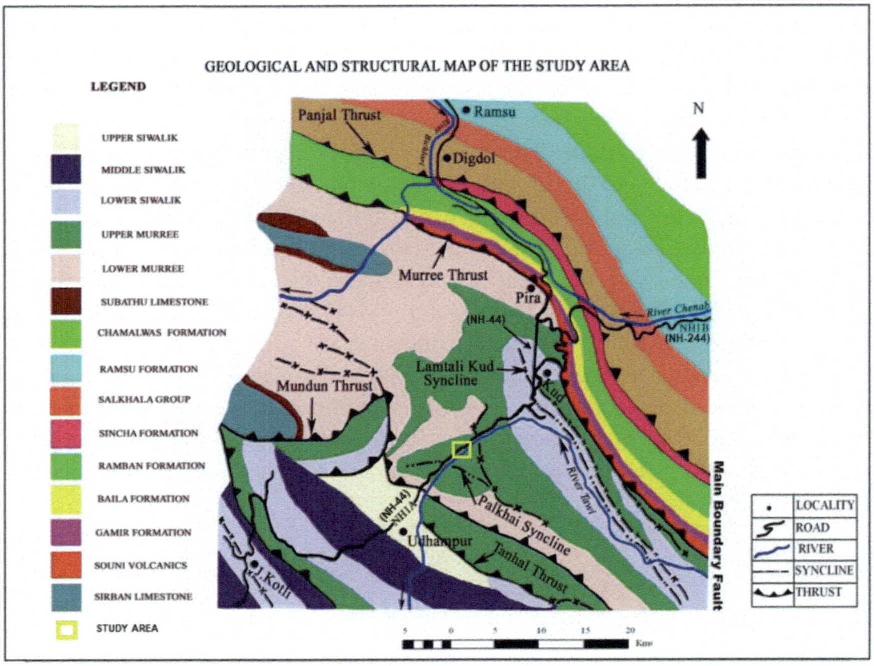

Fig. 5.1 Geological map of the study area (Singh 2006)

5.3 Methodology

The present study is aimed at identifying the type of landslide and dominant causative factor responsible for its initiation through geotechnical study (Fig. 5.2). In this study, considerable emphasis was placed on gathering data on various aspects of joint attributes. Additionally, comprehensive geotechnical data concerning the rock mass was gathered, facilitating the calculation of five key parameters (Uniaxial compressive strength/Point load index, RQD, discontinuity spacing, discontinuity condition, and groundwater condition) essential for establishing the basic Rock Mass Rating (RMR_{basic}). To evaluate the stability of the rock slope, the Slope Mass Rating (SMR) was derived from the RMR_{basic}, incorporating factorial adjustment factors (F1, F2 and F3) based on the slope-joint relationship. Another factor (F4) was introduced, depending on the excavation method employed. Data pertaining to different joint sets were plotted on a stereonet using RocScience Dips 6.008 software, for the identification of dominant joint sets and possible mode of failure and direction.

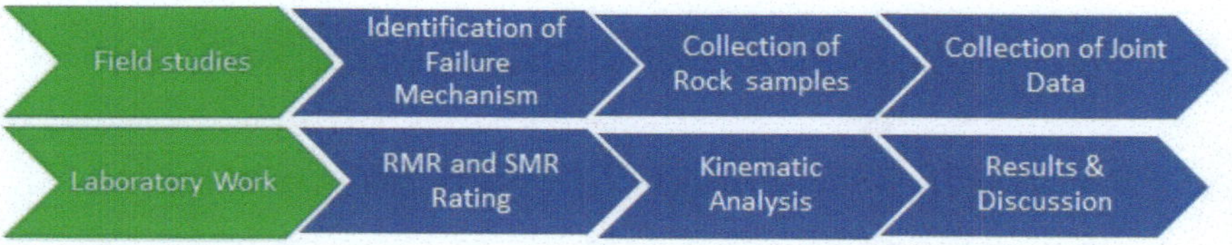

Fig. 5.2 Flowchart of the methodology used for the present study

5.4 Field Observations

The sliding area falls within the Upper Murree Formation of the Outer Himalayan tectonic zone. The rocks exposed at this location are composed of fine to medium-grained sandstones of greyish green colour and unconsolidated interbedded layers of reddish and maroon mudstones. The landslide covers an area of about 51,888 m^2 and on the basis of its areal extent, it is classified as a medium landslide (Van Schalkwyk and Thomas 1991). On the basis of the kind of movement and the kind of the geologic material, this landslide is classified as a rock slide and the sliding area's slope angle ranges from 50° to 60°. The Google Earth images representing the slope, vegetation pattern and local drainage channel are shown in Fig. 5.3.

In the aftermath of the recent sliding incident, extensive transverse and longitudinal cracks have emerged across the slope, with depths ranging from 5 to 6 feet and widths spanning from 1 to 2 ft. Notably, at the crown of the slide, a clear vertical slip measuring 3 m has been recorded. The complete vertical height of the landslide, spanning from crown to run-out distance, is approximately 162 m, while its width measures about 132 m adjacent to the road as shown in Fig. 5.4. An evident shift of 10 ft from its original position has been documented in the constructed culvert. Numerous wide and open cracks are conspicuous both around the crown area and along the slope of the slide. The rock formations in this region exhibit a dip angle between 55° and 78°, displaying four prominent joint sets alongside some random joints (Fig. 5.5a and b). These joint sets present an average spacing ranging from 0.1 to 0.6 m, with aperture dimensions varying between 1.5 and 2 cm. Boulders of various sizes from 7152 to 51,133 kg, are precariously positioned all along the slope and susceptible to falling. Noteworthy 5–6 ft deep cracks (as shown in Fig. 5.5c) have been documented on the slope. The infiltration of water into these wide-open cracks has led to the saturation of slope materials. This saturation, in turn, escalates pore pressure, functioning as a lubricant at the interfaces of bedding planes. The failure appears to have occurred along the mudstone layer, situated around 3–4 m beneath the surface and overlain by sandstone. Furthermore, the presence of slickensides in the area indicates a heightened stress environment as shown in Fig. 5.5d. Prominent sedimentary structures, including ripple marks and graded bedding, are prevalent, with a few sandstone bands revealing rain imprints.

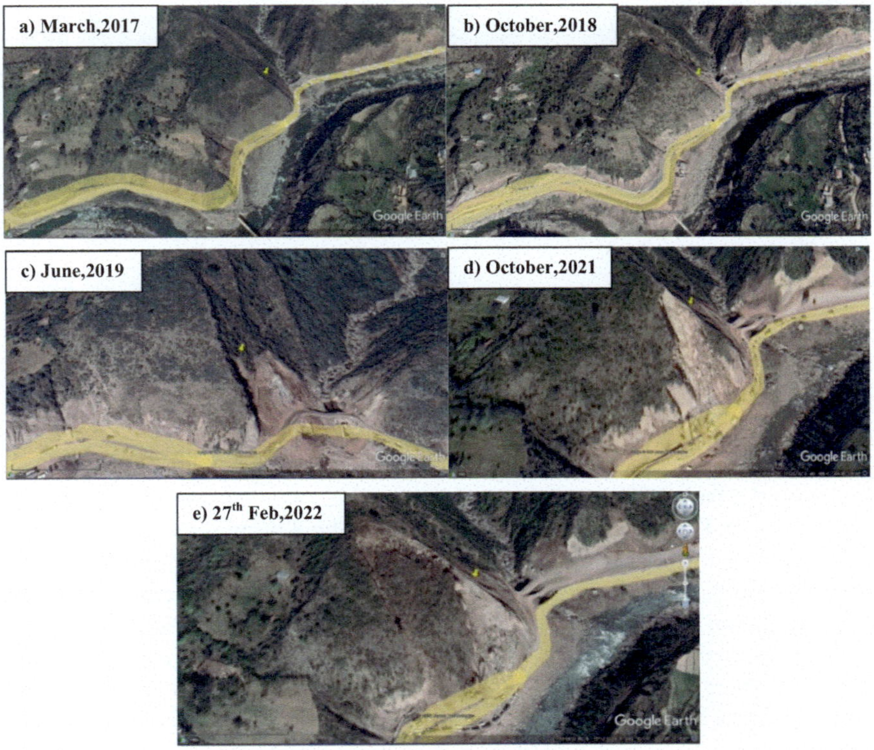

Fig. 5.3 Google Earth images of the sliding area showing vegetation pattern and local drainage channel **a** and **b** fully vegetated area; **c** and **d** initiation of landslide due to unsystematic toe cutting for culvert construction and **e** showing the development of crack and slip plane on 27 February 2022

Fig. 5.4 Field photograph of the sliding area on 27 February 2022

Fig. 5.5 Field photographs showing **a** and **b** highly jointed rock mass **c** wide and open cracks; **d** Polished surfaces and slickensides indicating the presence of fault around the area

5.5 Geotechnical Investigations

For assessing the mechanical properties of the rock in the study area, samples collected were analysed in the laboratory to find rock mass quality. The main objective of the geotechnical investigation is to draw a relationship between these parameters in triggering slope failures. The geotechnical investigation included a comprehensive analysis of the geological and structural aspects of the site. This investigation involved a detailed study of the patterns of movement of slope materials, the type of geological formations present and their interactions. Geotechnical data pertaining to the rock mass was meticulously gathered. This data served as the basis for calculating five important variables: uniaxial compressive strength/point load index, rock quality designation (RQD), discontinuity spacing, discontinuity condition and groundwater condition. These parameters collectively contributed to the computation of the basic Rock Mass Rating (RMR_{basic}) and derivation of the Slope Mass Rating (SMR), which was pivotal for evaluating the inherent stability of the slope.

5.5.1 *Basic Rock Mass Rating (RMR_{basic})*

Bieniawski (1973) developed this geomechanics classification for the very first time in 1973 at the South African Council of Scientific and Industrial Research (CSIR) (Kaiser et al. 1986). He takes into account the following five factors that can be used to assess the fundamental quality of rock masses:

(a) **Uniaxial Compressive Strength (UCS)**: This crucial factor identifies the strength attribute of rock and is used to classify rock types and analyse slope instability (Thurro 1997). Using the Point Load Testing equipment (PLT), the strength properties of the rocks are measured. The point load test is chosen to analyse the field-collected rock samples. The following relation is used to determine the point load strength index:

$$I_L(50) = P / (DW)^{0.75} \sqrt{D_{50}}$$

where P is the peak load at failure in N;

I_L (50) is the point load lump strength index in MPa;
D_{50} is the standard lump size (50 mm);
D is the distance between point loads in millimetres; and
W is the average width of minimum cross-sectional area in millimetres.

(b) **Rock Quality Designation (RQD)**: In order to add another criterion to the quantitative evaluation of rock quality, Deere (1964) created the Rock Quality Designation (RQD) in 1964. But the method of collecting RQD used in this study is indirect. By using a straightforward relationship, Palmstrom (2005) shows how to transform Jv into RQD.

$$RQD = 110 - 2.5(Jv)$$

where Jv is the volumetric joint count or total number of joints per cubic metre.

Palmstrom equation for calculating Jv is given below:

$$Jv = [(1/S_1) + (1/S_2) + (1/S_3) \ldots. + (Nr/5)]$$

where S_1, S_2, S_3 are the joint set spacing in metres; and
Nr is the number of random joints.

(c) **Discontinuity spacing**: The structural discontinuity in the rock has a great influence on the stability of rock slopes. The size of blocks that can slide from a face is determined by the persistence of discontinuities along with spacing.
(d) **Discontinuity condition**: A discontinuity surface's roughness is a measurement of how irregular and wavelike it is inherently in relation to its mean plane. Joint roughness is palpable and can be distinguished in the field.
(e) **Groundwater condition**: This explains how the water pressure affected the underground excavation in particular. The RMR_{basic} for a slope is obtained by adding these five parameters algebraically as shown in Table 5.1.

Results of RMR_{basic}

The results of the rock mass rating inferred that the rock mass falls under 'fair' category as shown in Table 5.1. The cohesion of the rock mass ranges between 0.2 and 0.3 MPa and its angle of internal friction lies between 35° and 45°.

Table 5.1 Table showing the RMR_{basic} ratings and rock mass class

Parameters	Point load index (MPa)	RQD (%)	Discontinuity spacing (m)	Discontinuity condition	Ground water condition	RMR basic	Rock mass class	Rock mass description
Values	3.365	64.35	0.1	Continuous, Slickensides, wall rock separation 1–5 mm, gouge < 5 mm	Completely dry			
Ratings	7	13	8	10	15	53	III	Fair

5.5.2 *Slope Mass Rating (SMR)*

The SMR is a quantification technique used to assess the stability conditions of a rock slope. It is an empirical equation that was first created by Romana (1985) and modified from the Rock Mass Rating idea put out by Bieniawski (1989) by the addition of factorial adjustment variables. Depending on the joint direction, slope direction, joint angle and slope angle, adjustment ratings F_1, F_2 and F_3 for joints are assessed by using Table 5.2 given by Romana in 1985. F_4 is taken to have a value that depends on the excavation method employed here

$$SMR = RMR_{basic} + (F_1 \times F_2 \times F_3) + F_4$$

where F1 shows the relationship between the strikes of the slope face and joints in case of planar failure, whereas, in wedge failure, it shows the relationship between the plunge direction of the joint plane intersections and the direction of the slope face. Its value ranges from 1.00 to 0.15; F2 is the joint dip angle in planar failure, whereas, in wedge failure, it is the plunge amount of the joint plane intersections. Its value also ranges between 1.00 and 0.15; F3 shows the relationship between the dip amount of slope face and joints in planar failure, whereas, in wedge failure, it shows the parallelism between the dip amount of slope face and plunge amount of the joint plane intersections; and F4 is the correction factor for the excavation technique that was determined empirically.

Results of SMR

The results of the slope mass rating inferred that the slope of the sliding area is completely unstable because it falls in the 'Very bad' rock mass description. Big planar as well as large wedges of rock fall from the slope and its failure probability is 0.9. The results of SMR are given in Table 5.3.

Table 5.2 Adjustment factor values for various joint and slope orientations (Romana 1985)

Case of slopes failure		Very favourable	Favourable	Fair	Unfavourable	Very unfavourable
P T W	$\lvert \alpha j - \alpha s \rvert$ $\lvert (\alpha j - \alpha s) - 180 \rvert$ $\lvert (\alpha i - \alpha s) \rvert$	$>30^{\circ}$	$30–20^{\circ}$	$20–10^{\circ}$	$10–5^{\circ}$	$< 5^{\circ}$
P/W/T	F_1	0.15	0.45	0.70	0.85	1.00
P W	$\lvert \beta j \rvert$ $\lvert \beta i \rvert$	$<20^{\circ}$	$20–30^{\circ}$	$30–35^{\circ}$	$35–45^{\circ}$	$>45^{\circ}$
P/W	F_2	0.15	0.40	0.70	0.85	1.00
T	F_2	1.0	1.0	1.0	1.0	1.0
P W	$\lvert \beta j - \beta s \rvert$ $\lvert \beta i - \beta s \rvert$	$>10^{\circ}$	$10–0^{\circ}$	0°	$0–(-10^{\circ})$	$<-10^{\circ}$
T	$\lvert \beta j + \beta s \rvert$	$<110^{\circ}$	$110–120^{0}$	$>120^{\circ}$	–	–
P/W/T	F_3	0	–6	–25	–50	–60

where P—planar failure, T—toppling failure, W—wedge failure, α_s—slope strike, α_j—joint strike, α_i—plunge direction of joint plane intersection, β_s—slope dip, β_j—joint dip, β_i—plunge amount of joint plane intersection

Table 5.3 Results of SMR rating showing its stability condition and probability of failure (Romana 1985)

Class no	V
SMR value	18
Rock mass description	Very bad
Stability	Completely unstable
Type of failures	Big planar failures or soil like or circular or big wedges of rocks
Probability of failure	0.9

5.5.3 Kinematic Analysis

The kinematic analysis in landslide studies is a method used to interpret the potential type of failure of rock that took place due to the unfavourable/critical orientation of structural discontinuities (faults, joints, beddings and foliations). The mode of failure can be a wedge failure, toppling failure or planar failure.

Four major joint sets were recognized in the field and the joint/slope orientations are given in Table 5.4. Based on the data gathered from the different joint sets, kinematic analysis was performed using RocScience Dips 6.008 software to construct pole plot, major joint planes and rose diagram (as shown in Figs. 5.6, 5.7, 5.8 and 5.9) to find the type of failure for Toldi landslide. Stereographic projections show how the primary joint sets are oriented in terms of its dip direction and amount. On a stereonet, the discontinuity orientations are depicted in the form of poles and clusters of poles of discontinuities are recognized by using density contours. The stereonet also be used to determine the direction of a slope failure.

Results of Kinematic Analysis

The contour map created for the research region depicts that the highest density of poles is in the NW direction (Fig. 5.6). Our study shows mainly two modes of failure in the sliding area; the first mode is where the whole unbroken rock mass slipped and formed a slip plane, making it a planar mode of failure. Here J1 and J2 forms a planar mode of failure (Fig. 5.7). A wedge failure is the second type of failure that occurred in the study area Wedge failure occurs between J1 & J2, J1 & J3, J1 & J4 and J2 & J4 (Fig. 5.8). The dominant joint sets J1 (55° due 125°) and J2 (45° due 160°) with respect to slope face (58° due 140°) intersect to form a wedge in the direction 171°. J1 & J3 also forms a wedge with respect to slope face in the direction 200°. Similarly, J1 & J4, J2 & J4 forms a wedge in the direction 122° and 125°, respectively.

Table 5.4 Field data of joint–slope orientation of the study area

Slope orientation	Joints orientation			
	J1	J2	J3	J4
DD/DA	DD/DA	DD/DA	DD/DA	DD/DA
140/58	125/55	160/45	260/35	040/85

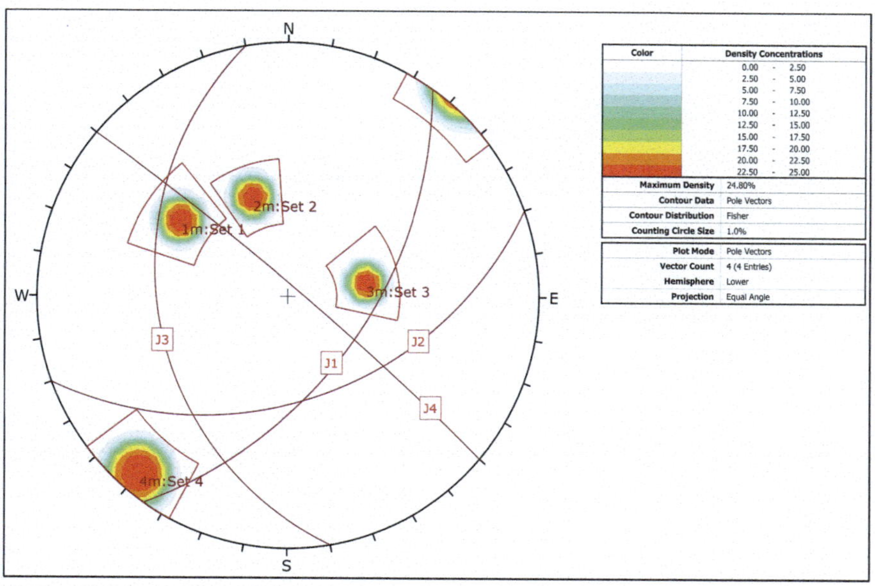

Fig. 5.6 Stereographic projection showing major joint sets and its contour plot

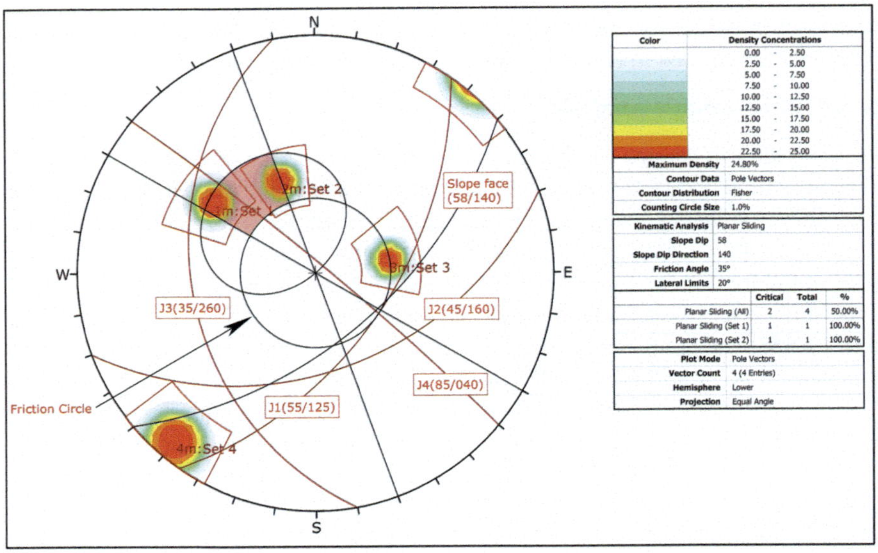

Fig. 5.7 Stereographic projection showing planar mode of failure

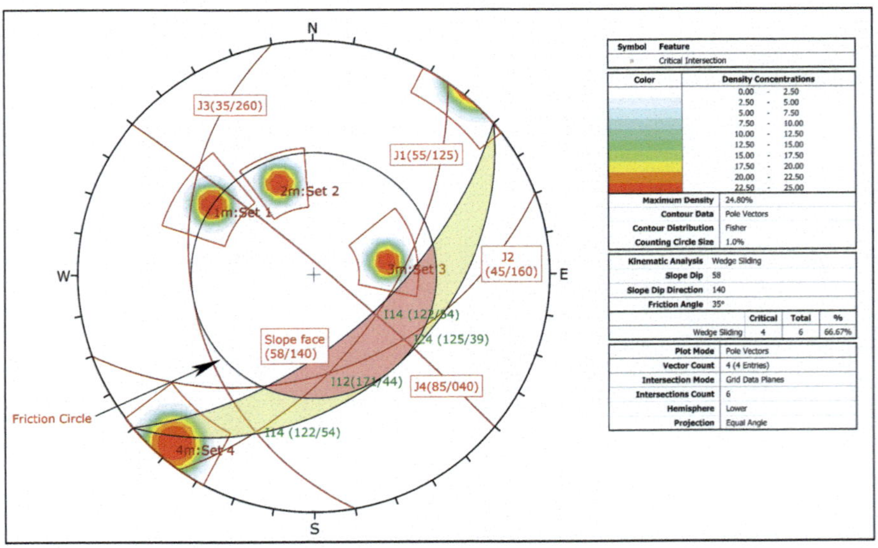

Fig. 5.8 Stereographic projection indicating wedge modes of failure

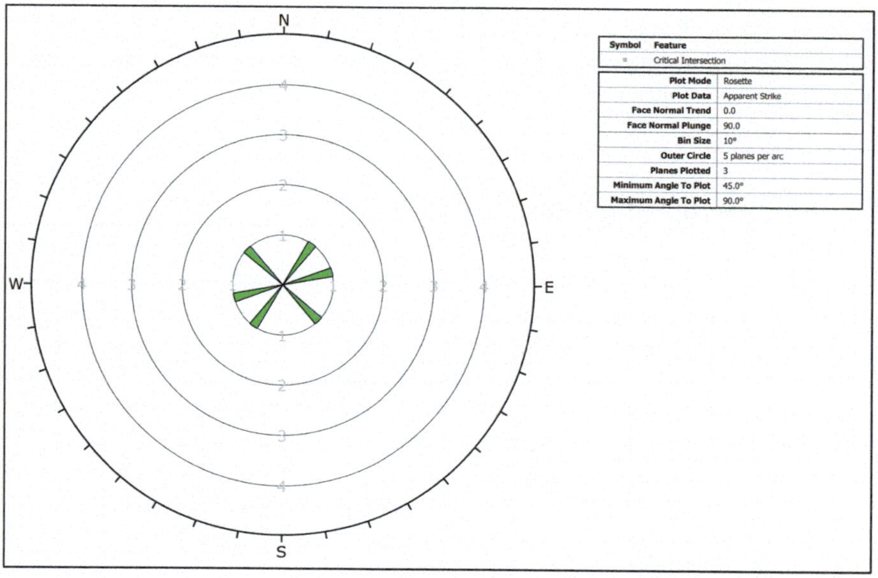

Fig. 5.9 Stereographic projection showing the rosette plot of the study area

5.6 Causes of Failure

Generally, most of the landslides are triggered by heavy precipitation during the monsoon season and occasionally due to intense rainstorms of limited duration (Rawat et al. 2007) which results in surface runoff and rain water percolation into the bedrocks, making these rocks more susceptible to failures. Percolation of water to the bedrock elevates the pore pressure and acts as a lubricant layer at the interfaces of the bedrock and the overlying rocks and as a result whole of the slide is reactivated and moved down the slope. This particular phenomenon is happening in the present sliding area which reactivates landslides in this stretch of road in the past as well as in the present. After rainfall, the lithology of the area also plays a significant role as the area constitutes of interbedded layers of sandstones and mudstones. The layered nature of the lithology in this case made the sliding surfaces for triggering landslides. However, seasonal rains combined with road cut slopes for highway maintenance promotes the reactivation of slope failures along this area in the past and in the present. The steep slopes of the mountains are dipping out hill towards the roadside which aggravates the movement of rock materials. Presence of wide and deep cracks on the slopes and unplanned excavation of materials from the slope so as to widen the road also aggravates the slope failures. Our research also demonstrates that the commencement of a landslide was the result of a combination of various triggering conditions that come into play and cause landslides, i.e. heavy precipitation, sparse vegetation, soil physics, weak lithology, steep slope, unplanned excavation, uncontrolled drilling and blasting, ground vibrations due to heavy traffic and machinery, lack of deflector canals, heavily jointed rocks, wide and open cracks and active tectonics, whereas the main triggering factor recognized was unsystematic toe cutting of the slope.

5.7 Suggested Mitigation Measures

Since the slope under investigation falls under 'completely unstable' category, addressing this situation requires employing some immediate site-specific techniques to enhance overall stability. Methods such as systematic rock bolts, retaining walls, gravity walls and anchor walls can be utilized to reinforce the slope. Implementing a well-designed drainage system is crucial to effectively manage water runoff, thereby preventing excessive saturation of materials and reducing erosion risks. To safeguard against rockfalls, erecting barriers and catchment systems is vital to intercept and redirect falling rocks away from roads. To minimize the potential triggers of instability, it's advisable to avoid blasting activities and the use of heavy machinery, which can generate vibrations.

Acknowledgements We are highly thankful to the Department of Geology, University of Jammu, for providing the infrastructure regarding this research and we also extend our sincere gratitude to the anonymous reviewers for their valuable feedback and constructive suggestions, which significantly contributed to the enhancement of this paper.

References

Alimohammadlou Y, Najafi A, Yalcin A (2013) Landslide process and impacts: A proposed classification method. CATENA 104:219–232

Bieniawski ZT (1989) Engineering rock mass classifications: a complete manual for engineers and geologists in mining, civil, and petroleum engineering. John Wiley & Sons

Bieniawski ZT (1973) Engineering classification of jointed rock masses. Civ Engineering= Siviele IngSwese, 1973(12):335–343

Deere DU (1964) Technical description of rock cores for engineering purpose. Rock Mech Eng Geol 1(1):17–22

Froude MJ, Petley DN (2018) Global fatal landslide occurrence from 2004 to 2016. Nat Hazard 18(8):2161–2181

Guha-Sapir D, Below R, Hoyois P (2015) EM-DAT: International disaster database. Cathol Univ Louvain: Bruss, Belg 27(2015):57–58

Joshi M, Kothyari GC, Kotlia BS (2023) Landslide detection in Kinnaur Valley, NW India using PS-InSAR technique. Phys Geogr:1–15

Kaiser PK, MacKay C, Gale AD (1986) Evaluation of rock classifications at BC Rail tumbler ridge tunnels. Rock Mech Rock Eng 19(4):205–234

Karunakaran C, Rao AR (1979) Geological survey of India. Misc Publ, 41(Part IV):1–66

Kholia N, Kotlia BS, Joshi N, Kandregula RS, Kothyari GC, Dumka RK (2023) Quantitative analysis of Khajjiar and Rewalsar lakes and their surroundings, Himachal Pradesh (India). Remote Sens GIS-Based Approaches J Appl Geophys 211:104976

Kothyari G, Pant P, Joshi M, Luirei K, Malik J (2010) Active faulting and deformation of Quaternary Landform Sub Himalaya, India. Geochrnometria 37:63–71

Kothyari GC, Pant PD, Luirei K (2012) Landslides and neotectonic activities in the Main Boundary Thrust (MBT) zone: Southeastern Kumaun Uttarakhand India. J Geol Soc India 80:101–110

Kundu J, Sarkar K, Tripathy A, Singh TN (2017) Qualitative stability assessment of cut slopes along the National Highway-05 around Jhakri area, Himachal Pradesh, India. J Earth Syst Sci 126(8):1–15

Luirei K, Mehta M, Iqpal A, Nazir S, Kothyari GC (2023) Factors influencing the slope instability of the Mussoorie-Bhitarli area, Garhwal Himalaya. Geosciences J:1–14

Palmstrom A (2005) Measurements of and correlations between block size and rock quality designation (RQD). Tunn Undergr Space Technol 20(4):362–377

Pandey VK, Srinivasan KL, Kulkarni UV (2014) Landslide challenges due to widening of road section between udhampur and chenani along National Highway-44, Jammu and Kashmir, India. Disaster& Dev, 84

Pant PD, Luirei K (1999) Malpa rockfalls of 18 August 1998 in the northeastern Kumaun Himalaya. Geol Soc India 54(4):415–420

Raghuvanshi TK (2019) Plane failure in rock slopes–A review on stability analysis techniques. J King Saud Univ-Sci 31(1):101–109

Rawat RK, Trivedi RK, Soni L (2007) Patalkot landslide, Madhya Pradesh-A case study. J Geol Soc India 69(2):344

Rawat KS, Kothyari GC (2007) The mass movement disaster at the Ampadev locality in the Kumaun Siwaliks, Nainital, and Uttaranchal Himalaya. A Geol Asp, 5:145–151

Romana M (1985) New adjustment ratings for application of Bieniawski classification to slopes. In: Proceedings of the international symposium on role of rock mechanics, Zacatecas, Mexico, pp 49–53

Sarkar K, Singh TN, Verma AK (2012) A numerical simulation of landslide-prone slope in Himalayan region—a case study. Arab J Geosci 5(1):73–81

Sati SP, Sharma S, Kothyari GC, Asim M, Sundriyal YP, Malik K, Joshi A, Dobhal H, Rana N, Juyal N (2022) Mountain highway stability threading on the fragile terrain of upper ganga catchment (Uttarakhand Himalaya), India. J Mt Sci 19(12):3407–3425

Singh Y, Bhat GM (2010) Role of morphometric parameters in landslides along the National Highway-1A between Udhampur and Batote, Jammu and Kashmir, India: A case study. J Himal Geol 31(1):43–50

Singh TN, Verma AK, Sarkar K (2010) Static and dynamic analysis of a landslide. Geomat Nat Haz Risk 1(4):323–338

Singh TN, Singh R, Singh B, Sharma LK, Singh R, Ansari MK (2016) Investigations and stability analyses of Malin village landslide of Pune district, Maharashtra, India. Nat Hazards 81:2019–2030

Singh Y (2006) Geotechnical and structural evaluation of tectonostratigraphic units along the National Highway, NH-1A between Udhampur and Batote Jammu Himalaya. Unpublished (Doctoral dissertation, Ph. D. Thesis University of Jammu, Jammu)

Taloor AK, Kothyari GC, Dumka RK, Alam A, Malik K (2023) Crustal deformation study of Kashmir basin: Insights from PSInSAR based time series analysis. J Appl Geophys 211:104979

Thuro K (1997) Drillability prediction: geological influences in hard rock drill and blast tunnelling. Geol Rundsch 86(2):426–438

Van Schalkwyk A, Thomas MA (1991) Slope failures associated with the floods of September 1987 and February 1988 in Natal and Kwa-Zulu. Republic of South Africa. Geotechnics in the African environment, Balkema, Rotterdam, pp 57–63

Varnes DJ (1958) Landslide types and processes. Landslides Eng Pract 24:20–47

Varnes DJ (1978) Slope movement types and processes. Special Report 176:11–33

Chapter 6
Stability Assessment of Rockfall-Prone Slides in Garhwal Himalayas Using CSMR and Q-Slope Techniques

Neeraj Dahiya, Koushik Pandit, Shantanu Sarkar, and Anindya Pain

Abstract Rockfall is a natural mass-wasting phenomenon in mountainous regions that involves the detachment of rock fragments by sliding, toppling and falling. The fragments sliding and toppling the hill slopes are often weakened by improper road-cutting, and sometimes are subjected to sliding by rainfall or earthquake occurrences. To minimize the risk, slope stability analysis and its stabilization play a crucial role. In the present study, two rockfall-prone slopes near Shivpuri along NH-58 in the Garhwal Himalayas, India, have been studied. A field study was carried out along this route. The rock mass classification techniques such as Continuous Slope Mass Rating (CSMR) and Q-slope were used for slope stability assessment. The stability grade of these slopes evaluated from CSMR has been validated with Q-slope rock mass classification. The results of Q-slope techniques have been plotted on the Q-slope stability chart and stability condition of both the slopes have been observed. The safe slope angle for both the slopes has been determined using the Q-slope technique. These rock mass classification systems are much more effective for quick assessment of slope stability. Hence, the methodology present in this study will be beneficial for a critical stability assessment of slope vis-a-vis for selection of suitable control measures.

Keywords Slope stability · Rockfall · Rock mass classification · Indian Himalayas

N. Dahiya (✉) · K. Pandit · A. Pain
CSIR—Central Building Research Institute (CBRI), Roorkee 247 667, India
e-mail: neerajdahiya413@gmail.com

Academy of Scientific and Innovative Research (AcSIR), Ghaziabad 201 001, India

S. Sarkar
Uttarakhand Landslide Mitigation and Management Center (ULMMC), Dehradun, India

B. K. Rastogi et al. (eds.), *Natural Hazards and Risk Mitigation*, Springer Transactions in Civil and Environmental Engineering, https://doi.org/10.1007/978-981-97-7658-0_6

6.1 Introduction

Rockfall is a common recurring phenomena along the hill-road. In Indian Himalayan roads, rockfall is mainly induced by the factors like steep slopes, the presence of weathered and highly fractured rocks, unfavourable geological structures, heavy rainfall, seismo-tectonic activities and manmade activities. Most of the slopes in the Himalayan belt consist of deformed rocks and frequent joints with unfavourable alignment of structural discontinuities in rock mass, making them particularly prone to rockfall. The construction of hill-roads without appropriate geotechnical and geological investigation lead to an increase in slope failure events (Bhandari 2006; Sarkar et al. 2012). Rockfall events in the hilly area create major obstruction to the developments of any infrastructure and also lead to loss of lives and property. Hence, to minimize the effect of this disaster, the stability of rockfall-prone slopes needs to be pre-analysed if any road-widening is planned. There are many studies by several researchers in Indian Himalayas to evaluate the stability of slopes using empirical and numerical techniques, such as Sarkar et al. 2012; Kanungo et al. 2013; Gupta et al. 2015; Kainthola et al. 2015; Siddique et al. 2017; Neeraj et al. 2018; Ansari et al. 2019; Ansari et al. 2020; Kumar et al. 2020; Pradhan and Siddique 2020; Singh et al. 2020; Sarkar et al. 2021, Dahiya et al. 2022 to name a few. For the present study, two rockfall-prone slopes near Shivpuri along the national highway NH-58 have been selected. The physiomechanical characteristics of rock were investigated in laboratory. Engineering rock mass classification techniques, such as Continuous Slope Mass Rating (CSMR) and Q-slope have been utilized for evaluation of slope stability. A new safe slope angle has also been evaluated using the Q-slope technique for the design of 'stable' cut slope.

6.2 Study Area and Geology

The studied slopes are situated along National Highway NH-58 near Shivpuri in the Garhwal Himalaya. This NH-58 connects Rishikesh with Badrinath and Kedarnath, two well-known shrines of the Char Dham Yatra made by lakhs of pilgrims each year. In this study area, especially during the wet season, slope failures occur frequently. This area is a part of the Garhwal Lesser Himalayas. The Main Boundary Thrust (MBT) line is located 7 km before Shivpuri. Between Shivpuri and Kaudiyala, there is a second fault called the Saknidhar fault. Low-grade Proterozoic metamorphic rocks make up the majority of this belt. The Mussourie Group (Blaini–Krol–Tal Formation) and Jaunsar (Chandpur and Nagthat Formation) rock formations make up the Outer Lesser Himalayas. Berinag Formation is part of the Inner Lesser Himalayas (Valdiya 1980). Several crystalline klippen carried from MCT hanging wall are overlain on the meta-sedimentary sequence of the Lesser Himalayas (Srivastava and Mitra 1994). Sandstone is the predominant rock type in the study area, however limestone, phyllite and slate are also present to some extent.

Fig. 6.1 Field photographs of the studied slopes

6.3 Methodology

6.3.1 *Field and Laboratory Investigations*

Based on the field investigation, two vulnerable sites were chosen. The surface roughness plays a significant role in determining the shear behaviour and fundamental friction angle of the discontinuity plane in the evaluation of rock slope stability. The discontinuities in the rock mass also play very important role in determining the stability of the slope. For the purpose of evaluating stability, all important geometrical, geological and geotechnical information were collected from the sites. For laboratory test, representative rock samples were taken from the slopes. On the collected rock lumps, the Point Load Strength Index test was carried out to determine one of the several input parameters for rock mass characterization, the Uniaxial Compressive Strength (per IS: 8764–1998). The correlation given by Palmstrom (2005) was used to obtain the Rock Quality Designation (RQD) value using a volumetric joint count (Jv). The field photographs of the studied slopes are shown in Fig. 6.1.

6.3.2 *Rock Mass Classification*

Rock mass classification systems are frequently used in the area of engineering geology such as slope stability analysis, tunnelling, mining industry and hydropower projects. In the present study, Rock Mass Rating (RMR) classifications systems has been used for rock mass characterization, whereas the Continuous Slope Mass Rating (CSMR) and the Q-slope classification systems have been used for qualitative slope stability assessment. The following sections provide a brief description of these techniques.

The basic Rock Mass Rating (RMR_{basic}) system was introduced by Bieniawski (1979) that involves only five parameters of RMR, excluding the orientation of discontinuities. The parameters are Uniaxial Compressive Strength of intact rock (UCS), Rock Quality Designation (RQD), spacing of discontinuities, condition of discontinuities and groundwater condition. All of these five parameters were subdivided into five ranges of values and importance ratings were assigned to the different value ranges of the parameters (Bieniawski 1989). A higher rating indicates a good rock mass condition, while a lower rating represent a poor rock mass condition. Ratings of these five parameters are algebraically added to obtain the basic rock mass rating, which ranges between 0 and 100.

Tomas et al. (2007) introduced a new rating system named Continous Slope Mass Rating (CSMR). The CSMR is also calculated using the same equation (Eq. 6.1) as that of Slope Mass Rating (SMR), but the basic difference is in the calculation of the adjustment factors (F_1, F_2 and F_3).

$$\text{CSMR} = \text{RMR}_{\text{basic}} + (\text{F}_1 . \text{F}_2 . \text{F}_3) + \text{F}_4 \quad (6.1)$$

For the qualitative stability assessment of road cut slopes, Barton and Bar (2015) presented the relatively new empirical rock mass classification known as Q-slope. This is a modified version of the Q-system, which was initially created for tunnel excavations (Barton et al. 1974; Barton and Grimstad 2014). It quickly determines the stability of the slopes in the field and making most favourable adjustments to slope angles as rock mass conditions. Equation 6.2 can be used for calculating the Q-slope value (Bar and Barton 2017).

$$\text{Q} - \text{slope} = \frac{RQD}{J_n} \times \left(\frac{J_r}{J_a}\right)O \times \frac{J_{wice}}{SRF_{slope}} \quad (6.2)$$

In the Q-slope stability, using the data chart provided by Bar and Barton (2017), the slope stability may be evaluated by plotting the Q-slope value regarding the average slope angle of that particular slope. Three zones such as stable, unstable and partially or critically stable are identified on this graph. Barton and Bar (2015) developed an (Eq. 6.3) for the safe slope angle (β) that does not need any reinforcement or support for slope heights less than 30 m in order to reduce the danger of slope failure. Later, this equation has been extended to all slope heights (Bar and Barton 2017).

$$\beta = 20 \log_{10}(\text{Q} - \text{slope}) + 65^{\circ} \quad (6.3)$$

6.4 Results and Discussion

In the current study, engineering rock mass classification techniques have been used for slope stability assessment of rockfall-prone slopes. The calculated RMR_{basic} values and quality of rock mass along with ratings of the input parameters for both the studied slopes are listed in Table 6.1. RMR_{basic} results showed that both the slopes have good rock mass quality.

The evaluated CSMR values and stability grade of both the studied slopes along with the type of critical failure, and rating of all input parameters of CSMR are given in Table 6.2. The results reveal that these two slopes are partially stable with a CSMR value of 59 and 56, respectively. It has been observed that these slope are at the verge of failure and any triggering factor may cause a rockfall event.

The stability of these slopes have also been evaluated using the Q-slope method. All of these input parameters were rated for the purpose of evaluating Q-slope values using the common rating system provided by Bar and Barton (2017). For the calculation of SRF_b, Hoek et al. (2002) strength criterion has been used for computing the major principle stress (σ_1'). The highest rating from each of the three SRF slope categories was used to calculate the Q-slope. To determine the stability condition of the slopes, the estimated Q-slope value and average slope angle of all the investigated slopes were plotted on Q-slope stability data chart, as shown in Fig. 6.2. The Q-slope value and estimated the stability condition derived from the Q-slope stability data chart are given in Table 6.3. It has been observed that both the slope falls under the unstable category with Q-slope values of 0.035 and 0.117, respectively.

Bar and Barton (2017) developed an equation to predict the safe slope angle for different probability of failures. Designing the road cut slopes with a safe slope angle in a way that it would be stable in the future is crucial for long life of infrastructures and environmental safety. For both the studied slopes, the safe slope angle (β) having the probability of failure (PoF) as 1% was calculated using Eq. 6.3 and the results are

Table 6.1 RMR_{basic} values along with rating of input parameters of studied slopes

Slope no	UCS	RQD	Joint spacing	Joint condition	Groundwater condition	RMR_{basic}	Rock mass quality
1	12	17	10	16	15	70	Good
2	7	20	10	15	15	67	Good

Table 6.2 Result of CSMR for the studied slopes

Slope no	RMR_{basic}	Type of failure	F_1	F_2	F_3	F_4	CSMR	Stability grade
1	70	P (J_2)	0.65	0.57	−30.5	0	59	Partially stable
2	67	P (J_3)	0.64	0.58	−30.5	0	50	Partially stable

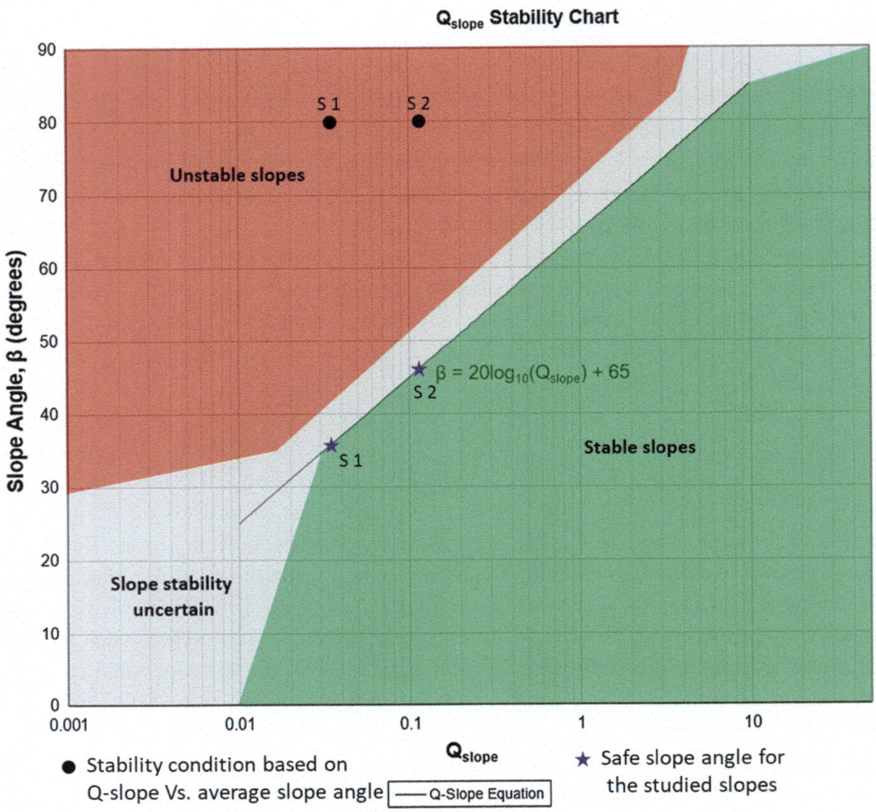

Fig. 6.2 Q-slope stability chart showing stability condition and safe slope angle (β) for the studied slopes

Table 6.3 Q-slope values along with stability condition and corresponding safe slope angle (β) for stability (PoF = 1%)

Slope no	RQD/J_n	J_r/J_a	J_{wice}/SRF_{slope}	Q_{slope}	Average slope angle	Present stability condition from data chart	Safe slope angle (β) for stability (PoF 1%)
1	8.44	0.75	0.005	0.035	80	Unstable	36
2	10.44	2.25	0.005	0.117	80	Unstable	46

given in Table 6.3. The safe slope angles (β) of these two slopes, which were obtained using the Q-slope technique, were found to be decreased significantly from the field average slope angles. Since these slopes are unstable, the lower value of Q-slope may be the cause of this. As illustrated in Fig. 6.2, the evaluated safe slope angle (β) value was plotted against the computed Q-slope value for each slope to determine the stability conditions of the slopes with safe slope angle. The results show that, after reducing the slope angle with a 1% failure probability, both of the studied slopes fall into the stable category of the Q-slope stability chart. Additionally, it has been noted that the values of a derived safe slope angle appear to be unrealistic and impractical in the actual field to follow and implement, particularly for slopes that are unstable or completely unstable. For example, the Q-slope approach advises changing the slope angle from extremely steep to gentle, as demonstrated in these studied slopes. Due to lack of horizontal space, often this degree of slope angle may not be practically achievable in mountainous areas. Since a minor change in slope angle is adequate to improve slope stability, it may be concluded that this type of recommendation for slope angle modification is only suitable for slopes that are partially stable.

6.5 Conclusion

In the present study, slope stability assessment of two rockfall-prone slides has been performed through two rock mass classification techniques i.e., CSMR and Q-slope. It has been analysed that both the slopes have good quality of rock mass according to RMR_b classification, but both of them fall under the partially stable and unstable category when CSMR and Q-slope technique were applied for assessment of slope stability. The Q-slope technique has been found to be effective for evaluating stability grade but in case of unstable slopes, the evaluation of safe slope angle looks to be impractical and unrealistic in field to adopt due to unavailability of space to reduce the slope angle significantly. This kind of suggestion regarding modification in slope angle is only acceptable only for partially stable slopes in which a slight change in slope angle is required to stabilize the slope. Based on the present slope stability analysis, it was observed that both of the studied slopes have fallen under unstable category of stability grade. To stabilize these slopes, suitable remedial measures are required. The methodology presented in the current study may be used for preliminary assessment of slope stability and for elementary design of the support system to improve stability of slopes.

Acknowledgements The authors are thankful to the Director, CSIR–CBRI, Roorkee, for his kind permission to publish the work.

References

Ansari TA, Singh KH, Singh TN, Tripathi JN (2020) Stability analysis of road cut slope near Devprayag in lesser Himalaya, India. J Earth Syst Sci 129:132. https://doi.org/10.1007/s12040-020-01394-0

Ansari TA, Sarma KM, Singh TN (2019) Empirical slope stability assessment along the road corridor NH-7, in the lesser Himalayan. Geotech Geol Eng 37:5391–5407.https://doi.org/10.1007/s10706-019-00988-w

Bar N, Barton N (2017) The Q-slope method for rock slope engineering. Rock Mech Rock Eng 50:3307–3322. https://doi.org/10.1007/s00603-017-1305-0

Barton N, Lien R, Lunde J (1974) Engineering classification of rock masses for the design of tunnel support. Rock Mech 6:189–236. https://doi.org/10.1007/BF01239496

Barton N, Bar N (2015) Introducing the Q-slope method and its intended use within civil and mining engineering projects. In: EUROCK 2015 64th geomechanics colloquium, Salzburg, Austria, pp 157–162

Barton NR, Grimstad E (2014) An illustrated guide to the Q-system following 40 years use in tunnelling. House publisher, Oslo

Bhandari RK (2006) The Indian landslide scenario strategic issues and action points. In: India disaster management congress, New Delhi, pp 29–30

Bieniawski ZT (1979) Thegeomechanics classification in rock engineering applications. In: 4th ISRM congress, Montreux, Switzerland

Bieniawski ZT (1989) Engineering rock mass classifications: a complete manual for engineers and geologists in mining, civil, and petroleum engineering. Wiley–Interscience, pp 40–47

Dahiya N, Pandit K and Sarkar S (2022) A comparison of slope stability assessment techniques using different rock mass classification systems and finite element method (FEM): A case study from the Garhwal Himalayas, India. J Earth Syst Sci 131:242

Gupta V, Bhasin RK, Kaynia AM, Chauhan VK, Saini AS, Tandon RS, Pabst T (2015) Finite element analysis of failed slope by shear strength reduction technique: a case study for Surabhi Resort Landslide, Mussoorie township, Garhwal Himalaya. Geomatics, Nat Hazards Risk. https://doi.org/10.1080/19475705.2015.1102778

Hoek E, Carranza-Torres C, Corkum B (2002) Hoek-Brown criterion—2002 edition. In: Proceedings of NARMS-TAC conference, vol 1. Canada, Toronto, pp 267–273

IS 8764 (1998) Method of determination of point load strength index of rocks (First Revision). Published by Bureau of Indian Standards, 9 Bahadur Shah Zafar Marg, New Delhi–110002

Kainthola A, Singh PK, Singh TN (2015) Stability investigation of road cut slope in basaltic rockmass, Mahabaleshwar, India. Geosci Front 6:837–845. https://doi.org/10.1016/j.gsf.2014.03.002

Kanungo DP, Pain A, Sharma S (2013) Finite element modeling approach to assess the stability of debris and rock slopes: a case study from the Indian Himalayas. Nat Hazard 69:1–24. https://doi.org/10.1007/s11069-013-0680-4

Kumar A, Sharma RK, Mehta BS (2020) Slope stability analysis and mitigation measures for selected landslide sites along NH-205 in Himachal Pradesh, India. J Earth Syst Sci 129:135. https://doi.org/10.1007/s12040-020-01396-y

Neeraj, Pandit K, Sarkar S (2018) Stability assessment of cut slopes along Shivpuri-Kaudiyala road (NH-58), Uttarakhand Himalayas. In: Proceedings of Indian geotechnical conference, 13–15 December, 2018. Indian Institute of Science, Bengaluru, India

Palmstrom A (2005) Measurements of and correlations between block size and Rock Quality Designation (RQD). Tunn Undergr Space Technol 20:362–377. https://doi.org/10.1016/j.tust.2005.01.005

Pradhan SP, Siddique T (2020) Stability assessment of landslide-prone road cut rock slopes in Himalayan terrain: a finite element method based approach. J Rock Mech Geotech Eng 12(1):59–73

Sarkar S, Kanungo DP, Kumar S (2012) Rock mass classification and slope stability assessment of road cut slope in Garhwal Himalaya, India. Geotech Geol Eng 30:827–840. https://doi.org/10.1007/s10706-012-9501-x

Sarkar S, Pandit K, Dahiya N, Chandna P (2021) Quantified landslide hazard assessment based on finite element slope stability analysis for Uttarkashi-Gangnani Highway in Indian Himalayas. Nat Hazards. https://doi.org/10.1007/s11069-021-04518-x

Siddique T, Pradhan SP, Vishal V, Mondal MEA, Singh TN (2017) Stability assessment of Himalayan road cut slopes along National Highway 58, India. Environ Earth Sci 76:759. https://doi.org/10.1007/s12665-017-7091-x
Siddique T, Pradhan SP, Vishal V, Singh TN (2020) Applicability of Q-slope method in the Himalayan road cut rock slopes and its comparison with CSMR. Rock Mech Rock Eng 53:4509–4522. https://doi.org/10.1007/s00603-020-02176-2
Singh HO, Ansari TA, Singh TN, Singh KH (2020) Analytical and numerical stability analysis of road cut slopes in Garhwal Himalaya, India. Geotech Geol Eng 38:4811–4829. https://doi.org/10.1007/s10706-020-01329-y
Srivastava P, Mitra G (1994) Thrust geometries and deep structure of the outer and lesser Himalaya, Kumaun and Garhwal (India): Implications for evolution of the Himalayan fold-and-thrust belt. Tectonics 13(1):89–109. https://doi.org/10.1029/93TC01130
Tomás R, Delgado J, Serón JB (2007) Modification of Slope Mass Rating (SMR) by continuous functions. Int J Rock Mech Min Sci 44:1062–1069
Valdiya KS (1980) The two intracrustal boundary thrusts of the Himalaya. Tectonophysics 66(4):323–348. https://doi.org/10.1016/0040-1951(80)90248-6

Chapter 7
GIS-Based Bivariate Statistical Model for Landslide Susceptibility Assessment Along a Srinagar-Bandipora Highway, Kashmir Himalaya

Iftikhar Hussain Beigh, Mohmad Ashraf Ganaie, Syed Kaiser Bukhari, and Shabir Ahmad

Abstract The Srinagar-Bandipora highway is one of the historical roads of Kashmir Valley and has always remained a supportive lifeline for the people in this region. This road is particularly vulnerable to landslides due to extreme weather conditions, complex mountainous terrain, and anthropological activity. Because of this, there are always frequent traffic disruptions and a severe threat to ongoing traffic and all activities associated with the region's social and economic development. As a result, it is critical to identify and mitigate the region's landslide risk, which causes significant inconvenience and economic and human losses. In the current study, landslide susceptibility mapping (LSM) was performed utilizing a bivariate frequency ratio model (FRM) and a geographical information system (GIS) platform. Besides, the landslide inventory details were prepared from primary and secondary data sources. The landslide susceptibility map (LSM) was generated from landslide inventory data and fifteen landslide causative factors (CFs). Therefore, the frequency ratio model examined the link between landslide CFs and landslides. LSM divided the area into five susceptibility categories, viz., very high (20.17Km^2), high (39.22Km^2), medium (62.49Km^2), low (76.87Km^2), and very low (56.90Km^2). Subsequently, the landslide susceptibility map was then evaluated and compared to the existing landslide inventory data using the area under the curve (AUC) method. Moreover, the LSM created will be helpful to various stakeholders, including planners, designers, engineers, and the local community, for the upcoming building and maintenance projects in the study region.

Keywords Landslide susceptibility · Frequency ratio · AUC · Himalaya · GIS

I. H. Beigh (✉) · M. A. Ganaie · S. K. Bukhari
Department of Civil Engineering, National Institute of Technology Srinagar, Srinagar 190006, India
e-mail: iftikharbeigh@nitsri.ac.in

S. Ahmad
S. P. College Srinagar, Srinagar 190001, India

B. K. Rastogi et al. (eds.), *Natural Hazards and Risk Mitigation*, Springer Transactions in Civil and Environmental Engineering, https://doi.org/10.1007/978-981-97-7658-0_7

7.1 Introduction

In mountainous terrain, landslides are a frequent mass movement on steep slopes. Among other forms, they can resemble mudslides, rockfalls, and debris falls. Landslides can result from human activities or natural causes (Gutiérrez et al. 2014). Earthquakes, extended periods of intense rainfall, certain types of soil, and variations in groundwater levels naturally cause them. On the other hand, slopes are vulnerable to failures due to several anthropogenic factors, especially in hilly areas. These include extensive mining and quarrying, tunneling, high traffic density in landslide-prone locations, and steep slopes cut for road construction or extension (Kamp et al. 2008). Therefore, on the slope hills, any of these circumstances could result in one or more slope failures (Subramanian et al. 2017). However, the Jammu and Kashmir Himalayas, part of the Northwest Himalayas, provide 67% of landslides, the Northeast Himalayas 19%, and the Western Ghats 15%, putting India top in terms of fatal landslides with 14% of the global contribution (Martha et al. 2021). Landslides are undoubtedly the most dreadful environmental hazard in the Himalayan highlands and adjacent slopes. As a result, landslides are one of the most devastating natural disasters, frequently occurring along the highways in mountainous areas with cut slopes.

Further, most landslides occur on natural slopes that have been altered by the development of transportation infrastructure such as dams, bridges, and tunnels (Das et al. 2011; Singh et al. 2013). Natural calamities such as landslides can significantly influence the flow of automobiles on road networks. Natural catastrophes destroy many roads, and the damage is frequently extensive. Contrast this with traffic delays due to accidents or maintenance (Bíl et al. 2015). Increased human land use on steep terrain has heightened natural threats (Kellerer-Pirklbauer 2002). To lessen the risk of landslides, development planning and disaster mitigation measures must comprehensively identify and evaluate the landslide hazard (Ahmad et al. 2017). In addition, understanding the geographical distributions of landslides is essential for mitigating the effects of landslides on human life and property since it identifies areas that may or may not be vulnerable to future landslides (Kirschbaum et al. 2015; Petley 2012). Typically, landslides occur in regions where they have occurred previously (Guzzetti et al. 1999; Malamud et al. 2004). In landslide research, geospatial technology is a cutting-edge tool that can practically access inaccessible locations and identify all types of terrain (Anbalagan et al. 2014). However, ground-based data, thematic maps, and remote sensing data must be combined with geographical and sequential information regarding landslides (Pandey et al. 2008). Some recent works (Ahmed 2015; Cárdenas and Mera 2016; Hadmoko et al. 2017) employ GIS to estimate landslide hazards using various criteria. Besides, using qualitative and quantitative methods, landslide zonation maps are constructed (Kaur et al. 2017). They are also using GIS approaches, and the research has advanced by assessing the spatial probability of landslides. GIS is a fundamental and valuable analytical tool for managing and manipulating geographical data about landslide hazards (Wu and Lan 2019).

Furthermore, as an advanced platform, Google Earth Engine (GEE) provides access to satellite data availability and provides an opportunity for software and algorithms to process satellite data products (Mutanga and Kumar 2019). The GEE platform enables us to analyze and process big satellite data and generate temporal datasets for land use and land cover (Yuan et al. 2005). Also, this platform allows for generating change detection and NDVI maps locally and regionally. These maps benefit land planners, policymakers, etc., who want to make decisions about disaster preparedness, land use, and infrastructure development. Numerous research on landslide hazard mapping have utilized GIS, employing various procedures and approaches, such as bivariate statistical analysis, to obtain accurate and trustworthy results. The frequency ratio method, a type of bivariate statistical analysis, is one of the most often employed techniques for determining landslide susceptibility (Beigh and Bukhari 2024; Korup and Stolle 2014; Shafapour Tehrany et al. 2019). This method for assessing landslide susceptibility is user-friendly and effective due to the simplicity of the underlying concepts and GIS application (Lee and Pradhan 2006; Oh and Pradhan 2011). Like other bivariate methods, the frequency ratio approach generates landslide susceptibility maps and incorporates the sensitivity of landslide failure to certain landslide-related factors (Razavizadeh et al. 2017; Vakhshoori and Zare 2016). Despite introducing other robust machine learning algorithms, researchers and practitioners often apply this approach (Korup and Stolle 2014).

The frequency ratio method can potentially have greater precision than other methods (Beigh and Bukhari 2024, 2023; Ding et al. 2017; Huang et al. 2018; Patil et al. 2022; Ramesh and Anbazhagan 2014; Sharma and Mahajan 2018; Vakhshoori and Zare 2016). Therefore, using the frequency ratio method, a study was conducted on a landslide-prone road between Srinagar and Bandipora in Kashmir Himalaya. The present study aims to create a landslide susceptibility map using the GIS-based frequency ratio method and validate it with a compiled landslide inventory map. The highway is one of North Kashmir Himalaya's most essential and dangerous roads. Besides, landslides obstruct roads and cause traffic congestion when it rains or snows. Furthermore, the literature suggests that it is necessary to construct, analyze, and create a trustworthy susceptible map, as no such work has previously been undertaken in the study region. Moreover, this study could help identify places to avoid construction and developmental projects and help design mitigation practices in landslide-prone areas.

7.2 Study Area and Geological Setting

Srinagar-Bandipora Road is a part of the historically known Srinagar-Bandipora-Gilgit Road of Kashmir Himalaya. The area is bounded between 34°26′47.444″ N and 34°5′28.686″ N longitudes and 74°34′1.361″ E and 74°44′40.741″ E latitudes (Fig. 7.1). The study area is covered with an area of 255.64 km^2. The elevation of the study area ranges from 1522 to 2560 m from mean sea level, and the slope is from

0° to 68° approximately. The study region experiences a temperate to Mediterranean climate. Average minimum and maximum temperature varies from –5 °C to 32 °C.

The Bandipora-Srinagar highway is the only highway for the people of Bandipora and Gurez to connect with Srinagar. The high elevation road with the steep slopes has made the road vulnerable to landslides, particularly debris flow and rock fall. Therefore, only some areas are highly vulnerable to landslides. However, anthropogenic activities without any proper guidance have made roads more vulnerable.

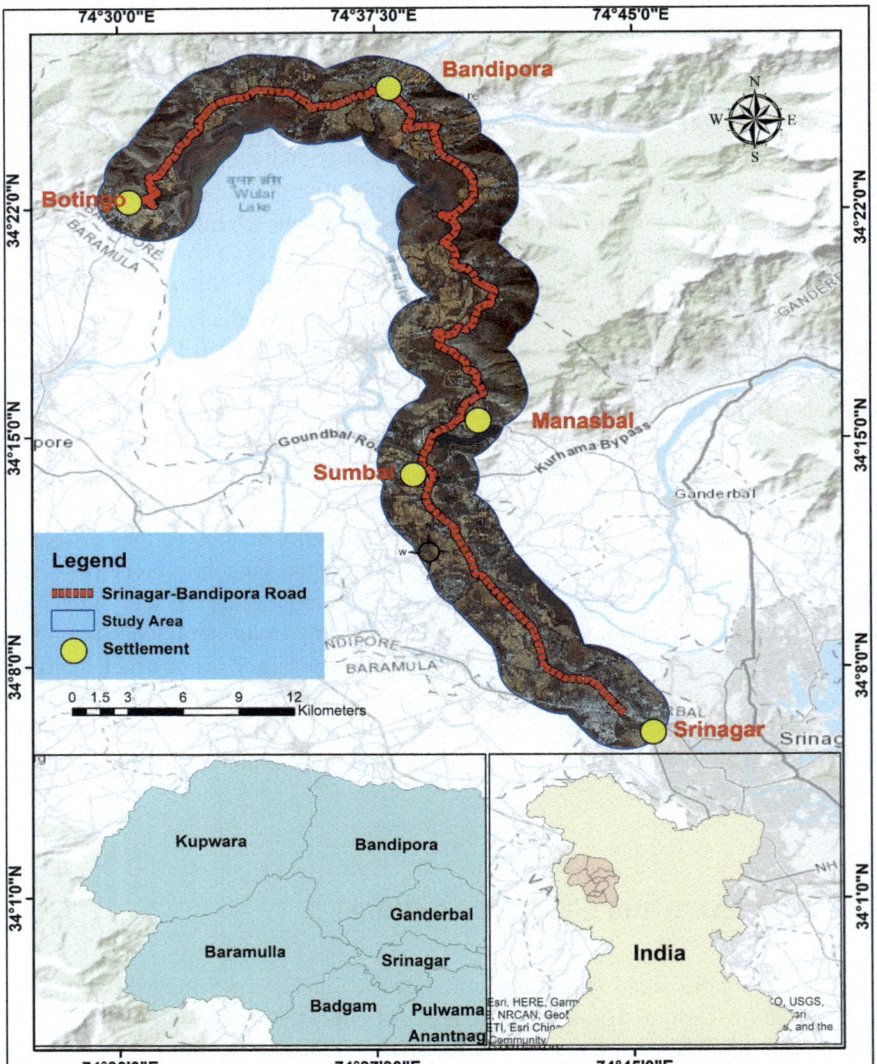

Fig. 7.1 Location map of the study area

The uneven road cuts for road widening along the Bandipora-Srinagar highway have made the roads vulnerable and at high risk. Quarrying is intensive, particularly along the Srinagar—Bandipora Road in the Sadarkoot Sector. The mining runs throughout the year and dislocates many loose stones, pebbles, and slush that enter the lakebed during the monsoon season. Unplanned mining causes lots of destruction during the rainy season. The loose soil and broken rocks induce high-density debris flows, damaging the low-lying village infrastructure. During regular times, the rock falls block the roads and kill laborers and people. These activities have led to environmental degradation and destroyed the region's biodiversity of plants and animals. Thus, the present study has highlighted some serious issues and delineated the landslide-prone areas along the Srinagar-Bandipora highway.

The crisscross pattern of the Srinagar-Bandipora highway is mostly engraved on the elevated terrain. Most of the area along it is under hard rock terrain. The Panjal Traps cover most of the slopes of the highway, followed by Limestones and quartzites. Good exposures of limestones were found near Chewa, Saderkoot Bala, Ajas, and Nadihal. Panjal Traps are well exposed near Ajas, Saderkoot Payeen, and Garoora. Quartzites are mainly located near the Churteng area of S. K. Payeen. The area near Safapora is under the Karewa formations. In addition, the Karewa formations were also found near Nusoo and Wudar Bandipora. The Himalayan rivers traverse the highway near Bazipora, Garoora, Nadihal, and Papacha by streams like Ajas Nar, Aragam Nar, Rampore Nar, and Erin Nar, respectively.

7.3 Database Generation and Organization

High-resolution sentinel imagery was combined with Google Earth imagery, followed by extensive Fieldwork to locate landslides along the present State Highway (SH) and its surroundings. For landslide susceptibility mapping along the Srinagar-Bandipora (SH) in the Kashmir Himalaya, fifteen contributing variables were taken into account, including slope gradient, elevation, slope aspect, topographic wetness index (TWI), slope curvature, stream power index (SPI), proximity to road, proximity to drainage, proximity to lineament, geology, land use, and land cover, normalized difference vegetation index (NDVI), Earthquake Magnitude, Earthquake Peak ground acceleration (PGA) and rainfall. The Jenks natural breaks classification method was also used to classify these maps (Jenks 1967). The drainage map was extracted from DEM, and lineaments were motivated from the NRSC Bhuwan site. The lithological map of Jammu and Kashmir was motivated by the Geological Society of India. Using the Google Earth Engine Platform, the LULC and NDVI were derived from Sentinel and Landsat-8 OLI/TIRS satellite imagery. The Indian Meteorological Department (IMD) gridded datasets for the past 30 years were used to obtain the precipitation (rainfall) data (1990–2020). To establish the integrity of the landslide inventory, we used past Literature, govt. official data, global data archives, and high-resolution imageries followed by detailed field inspections (2019–2021). Earthquake information (1500–2021) was compiled from several sources, including the Global Data Archives (USGS, IRIS, GSI), published articles, and past Literature. The data was collected as point data and was converted into a raster map using the IDW interpolation technique. The details of the datasets used in the study are shown in (Table 7.1). The steps in creating a landslide susceptibility map were sequential and are outlined in Fig. 7.2.

7.3.1 Landslide Inventory Mapping

The first stage in determining a region's landslide vulnerability is observing, locating, and mapping landslides (Zhao et al. 2019). The inventory of historical landslides provides information about ongoing landslides in an area, which validates the landslide susceptibility map. Field surveys or high-resolution aerial images can be employed to complete this task. The current study's landslide occurrences were found and plotted using secondary sources, field investigation, and high-resolution Sentinel and Google Earth imagery. In general, spatial mapping of landslides is essential to investigate the relationship between landslide distribution and predisposing factors. The GPS coordination system was used to find landslides produced as polygons using high-resolution images. These historical landslide locations were shown as point features on the inventory map since they are too small to be highlighted at the current scale (Yilmaz 2009). Besides, landslides have occurred in 90 spots within the study region. The landslide inventory data was divided into two groups for model training (70%) and validation (30%). The inventory map shows that landslides occurred randomly along the study road (Fig. 7.3). Rockfalls and debris slides are the dominant types of landslides along the study road.

7.3.2 Landslide Inducing Factors

For the landslide Susceptibility evaluation of a region, it is crucial to determine what factors contribute to their occurrence. Surface topography limits landslides' density and spatial extent by determining the flow sources and runoff direction. Digital Elevation Model (DEM) data at a resolution of 12.5 m was generated from ALOS-PALSAR-DEM. Parameters like slope gradient, slope aspect, topographic wetness index (TWI), slope curvature, and stream power index (SPI) were retrieved from the dem data using the Spatial Analyst tool in ArcMap10.5 (Fig. 7.4a–e).

Table 7.1 Details of datasets and data sources employed during research work

S. no	Datasets	Description	Scale/ Resolution	Source
1	Digital Elevation Model (DEM)	ALOS PALSAR	12.5 m	Alaska satellite facility https://search.asf.alaska.edu/
2	Slope	Derived from DEM	12.5 m	Alaska satellite facility https://search.asf.alaska.edu/
3	Slope aspect	Derived from DEM	12.5 m	Alaska satellite facility https://search.asf.alaska.edu/
4	LANDSAT-8 OLI/TIRS	Downloaded	30 m	USGS earth explorer https://earthexplorer.usgs.gov
5	Landslide inventory	Acquired	90 points	Field survey, GSI, BRO, Literature, Global landslide database, Google earth, Sentinel-2-B
6	Land use and Land cover (LULC)	Derived from LANDSAT-8 OLI/TIRS (Bands 2, 3 and 4) (Dated-25 and 28 Oct 2020)	30 m	Google earth engine https://code.earthengine.google.com
7	Normalized Difference Vegetation Index (NDVI)	Derived from LANDSAT-8 OLI/TIRS	30 m	Google earth engine https://code.earthengine.google.com
8	Lithology	Digitized from lithological map of Jammu and Kashmir	1:50,000	GSI, 2012 https://bhukosh.gsi.gov.in/Bhukosh/Public
9	Drainage	Derived from DEM	12.5 m	Alaska satellite facility https://search.asf.alaska.edu/
10	Lineaments	Digitized	1:50,000	Bhuvan NRSC https://bhuvan.nrsc.gov.in/
11	Rainfall	India Meteorological Department (IMD)	$0.25^{\circ} \times 0.25^{\circ}$ Grid	http://www.imd.pune.gov.in/

(continued)

Table 7.1 (continued)

S. no	Datasets	Description	Scale/ Resolution	Source
12	Roads	Digitized	Polyline	Google earth imageries
13	Earthquake data (Magnitude) and Earthquake (PGA)	United States Geological Survey (USGS), Geological Survey of India (GSI), Incorporated Research Institutions for Seismology (IRIS)	Point data	https://earthquake.usgs.gov/earthquakes,https://ds.iris.edu/ds/, https://bhukosh.gsi.gov.in/Bhukosh/Public,Publications,Literature

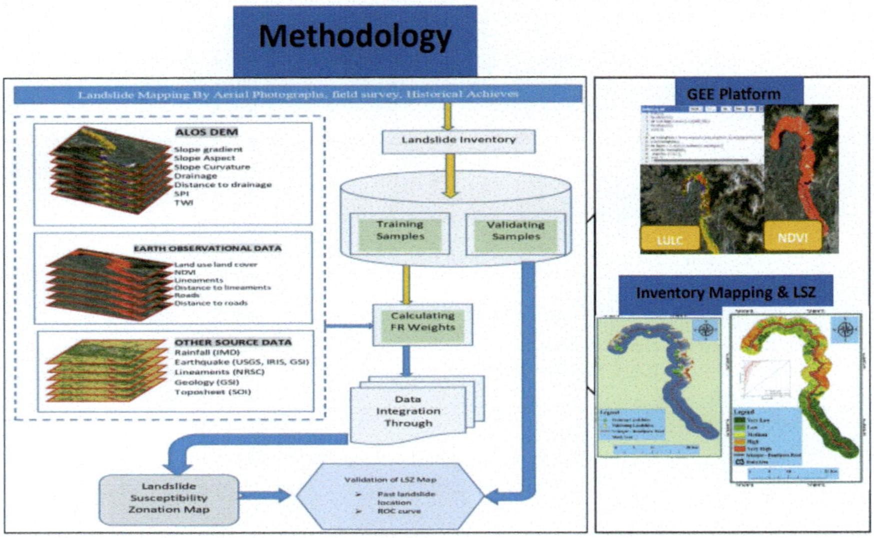

Fig. 7.2 Flow chart of research methodology

7.3.2.1 Slope Gradient

A crucial factor in slope stability analysis and a standard tool for creating landslide susceptibility maps is slope degree, which directly impacts landslide estimation (Gorsevski et al. 2012). In addition, it is crucial to hydrology and geomorphology because it affects how quickly surface and subsurface flows move (Kannan et al. 2013). By applying the Jenks natural breaks classification scheme (Jenks 1967), we were able to reclassify the slope gradient map (Fig. 7.4a) into seven distinct categories: 5 (Gentle Slope), 5–8.5 (Moderate Slope), 8.5–16.5 (Strong Slope), 16.5–24 (Very Strong Slope), 24–35 (Extreme Slope), 35–45 (Steep Slope), and >45 (Very Steep Slope).

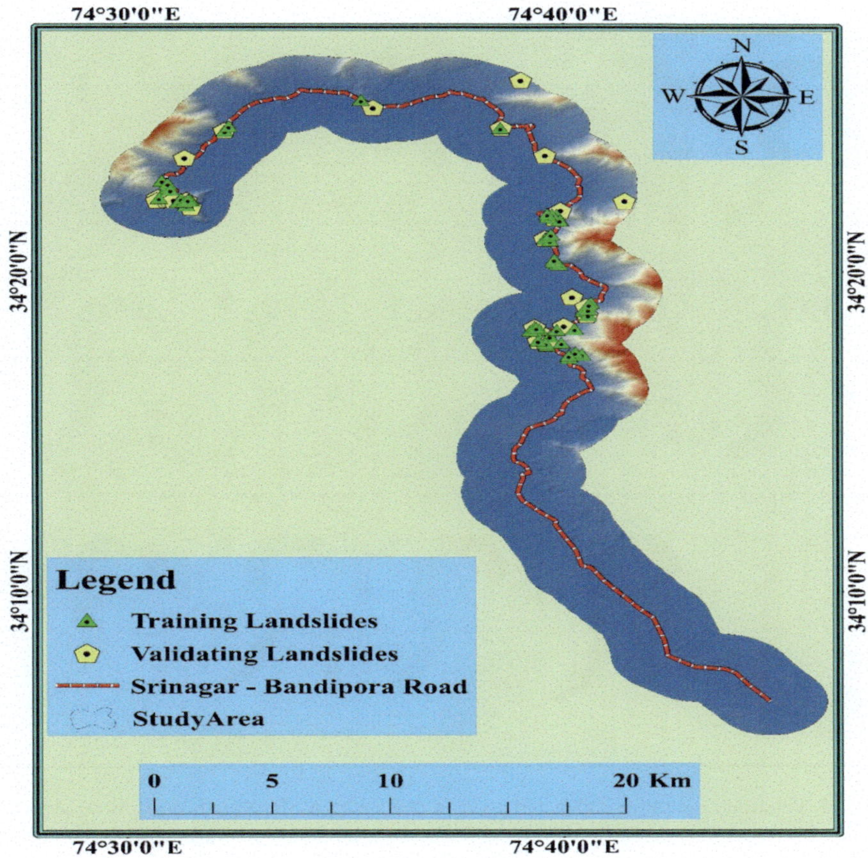

Fig. 7.3 Landslide inventory map of the study area

7.3.2.2 Slope Aspect

The slope's direction is referred to as its aspect. The slope aspect influences the development of landslides by influencing lineaments, rainfalls, wind effects, and sun exposure (Pourghasemi et al. 2012a). There are nine distinct categories for the slope direction, as shown in Fig. 7.4b: Flat (–1), Northwest (292.5–337.5), West (247.5–292.5), Southwest (202.5–247.5), South (157.5–202.5), Southeast (112.5–157.5), East (67.5–112.5), Northeast (22.5–67.5), and North (22.5–337.5). Most of the current study's landslide areas are along slopes to the South, west, southwest, and northeast.

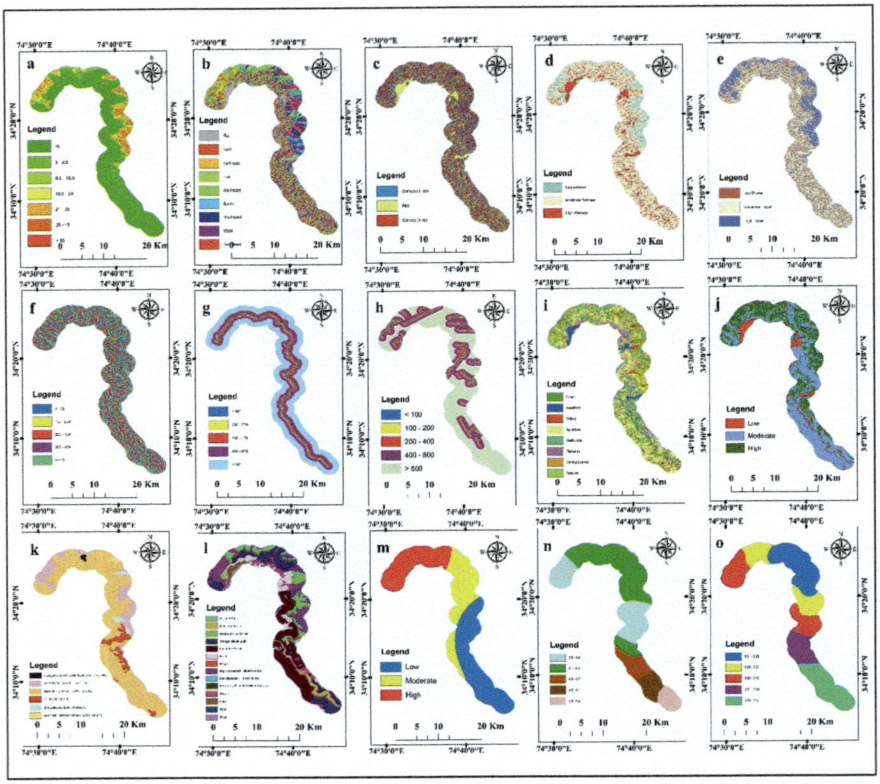

Fig. 7.4 Landslide Causative parameters used in this study: **a** Slope gradient, **b** Slope aspect, **c** Slope Curvature, **d** TWI, **e** SPI, **f** Proximity to Drainage, **g** Proximity to Lineament, **h** Proximity to Road, **i** lulc **j** NDVI, **k** Lithology, **l** Geomorphology, **m** Rainfall, **n** Earthquake (Magnitude), **o** Earthquake (PGA)

7.3.2.3 Slope Curvature

In this study, we focus on the general curvature, the sum of the plane curvature, and the profile curvature. The general curvature, which includes both the plane and the profile curvature, is the quantity we analyze here. It is commonly stated that slope stability is affected by general curvature, the slope's rate of change in aspect or degree. Therefore, the general curvature map can examine the slope morphology and flow characteristics (Nefeslioglu et al. 2008). In the curvature map, positive values indicated a convex shape, negative values indicated a concave shape, and zero indicated a flat shape (Fig. 4c). The most significant historical landslides (28) fall on convex slopes, whereas the least (12) occur on flat slopes.

7.3.2.4 Geomorphology

The geomorphologic map illustrates major geomorphic units, landforms, and underlying geology to help comprehend the processes, materials, structures, and geological controls that influence landslide susceptibility. The study area's geomorphology includes various groups, viz., Piedmont alluvial plain, ridge, moderately dissected hills and valleys, swamp, older flood plain, marsh, younger alluvial plain, lacustrine terrain, low dissected hills and valleys, channel bar, highly dissected hills and valleys, and water bodies such as lakes and rivers, and ponds. The primary cause of landslides in the research area is the presence of high, low, and moderately dissected hills and valleys. Besides, most of the study region is made up of structurally highly dissected hills and valleys. The geomorphology map (Fig. 4l) of the research area was collected from the Geological Society of India (GSI) provided website called BHUKOSH (https://www.bhukosh.gsi.gov.in (accessed on 14 Aug 2022)).

7.3.2.5 TWI

The topographic wetness index (TWI) is another topographical aspect of the runoff model that describes how topography affects the distribution and extent of saturated source areas from which runoff is generated (Pradhan and Kim 2014). TWI denotes the source zone that has become saturated due to surface runoff under the influence of topographic conditions. Therefore, it is positively connected with the quantity of precipitation that percolates into the materials that make up the slope, thereby raising the slope's pore pressure (Yilmaz and Keskin 2009). Following the natural breaks approach, the entire map was divided into three categories: 5 (Low Wetness), 5–10 (Moderate Wetness), and >10 (High Wetness) (Fig. 7.4d).

7.3.2.6 SPI

One of the major preconditioning elements affecting landslide occurrence is erosional processes. The stream power index (SPI), which assesses a stream's capability for erosion, is a factor that adds to the region's stability (Regmi et al. 2014). Due to the steeply sloping terrain and moderately steep channel gradient, higher SPI values suggest a more powerful stream and a greater capacity for erosion. Additionally, lower SPI areas encourage sediment deposition, which facilitates the shift from debris flows to flow patterns with less concentration (Marchi and Dalla Fontana 2005). Therefore, the SPI map of the study area was categorized into three groups based on the natural break method: 0 (Low power), 0–5 (Moderate power), and >5 (High power) (Fig. 7.4e).

7.3.2.7 LuLc

The land use and land cover classification were performed for landslide susceptibility assessment which comprises the different types of land cover within the region. The LuLc classification can help to identify the areas that are more susceptible to landslides based on the features of the land cover types present there. We used Sentinel-2 data sets for the land use–land cover classification. The data were accessed in the Earth Engine platform by filtering the region of interest and images while minimizing seasonal effects, viz., free from cloud clover and snow cover (Maingi and Marsh 2001) were obtained by applying the algorithm ee.Filter.date (start, end). A total of eight classes were identified based on image interpretation, ground truth collection, and other techniques. The delineated classes were forests, horticulture, agriculture, plantation, marshes, pasture, built-up, and water bodies. Random Forest classification is considered the most important and extensively used algorithm for land use/land cover classification using satellite imagery (Li et al. 2016). Using remote sensing data. Quantitative analysis was done to calculate the areal extent of all the eight classes (Fig. 7.4i). Following the natural breaks approach, LuLc has been divided into eight categories, viz., forest waterbody, built-up agriculture, horticulture plantation, marshy/swamps, and pastures with landslide percentage of 15.87, 0, 12.27, 23.81, 1.58, 11.11, 12.7 and 22.22% respectively.

7.3.2.8 NDVI

Normalized Differential Vegetation Index (NDVI) is an index that measures and displays the relative biomass or greenness of the terrain. We incorporated the technique of NDVI, as this technique provides valuable information about the vegetation cover and land use-land cover changes. Using NDVI data with other geospatial data enables us to create more accurate landslide susceptibility maps of the region. So, NDVI studies indirectly contribute to landslide susceptibility mapping by providing information about the vegetation health and land cover changes of the area. We generated the NDVI map (Fig. 7.4i) using the data acquired from Landsat-8 OLI/TIRS satellite imagery through Google Earth Engine platform (Schmid 2017).

The formula for calculating NDVI is as follows:

$$\boldsymbol{NDVI} = \frac{\boldsymbol{NIR} - \boldsymbol{RED}}{\boldsymbol{NIR} + \boldsymbol{RED}} \tag{7.1}$$

where

NIR is the reflectance in the near-infrared band. Red is the reflectance in the red band.

We imported satellite imagery, in the Earth Engine and used the executed JavaScript code to perform NDVI analyses. Following the natural breaks method, the NDVI has been divided into three categories: high, moderate, and low with 23.81, 76.91 and 0% of landslides, respectively.

7.3.2.9 Lithology

The lithology or rock type has a significant impact on the slope stability of the area. Therefore, it regulates the weathering and erosion processes in the region (Citrabhuwana et al. 2016). As a result, while analyzing landslide susceptibility, the research region's lithology must be considered. The lithology map (Fig. 4k) of the study area is generated from the Geological Survey of India (GSI) provided website called BHUKOSH (https://www.bhukosh.gsi.gov.in (accessed on 14 Aug 2022)). The lithological map was digitized using a GIS platform and classified into ten categories. The classified map was then converted into a 20 m x 20 m gridded map. The spatial distribution of numerous rock types in the study region is shown in (Fig. 4k). The study area contains ten lithological types, Micaceous siltstone and arenite with limestone, andesitic and basaltic lava flows, alluvium, moraines, hill wash and scree, golden brown silt, shale, limestone and sandstone, shale, and limestone with calcareous bands. According to the FR analysis, Alluvium, moraines, hill wash and scree class of lithology possess the highest percentage of landslide (47.62%) in the study area. Landslides are influenced by alluvium and limestone deposits because of their substantial capacity to absorb water.

7.3.2.10 Rainfall

Rainfall is the most frequent cause of landslides. Landslides are caused by slopes becoming saturated by heavy rain and flood-like conditions, which increases pore water pressure (Sati 2014). Therefore, rainfall was considered a landslide-triggering parameter, making it one of the critical parameters in landslide susceptibility mapping. The point data of the rainfall was gathered from India Meteorological Department (IMD) grided data for 30 years. Later, this annual Long Period Average (LPA) IMD rainfall dataset was interpolated (IDW) for the research region and reclassified (Fig. 7.4m) into three primary groups (Jenks 1967): Low (878–923 mm), Moderate (924–956 mm), and High (957–984 mm).

7.3.2.11 Earthquake (Magnitude)

The potential for a landslide to be triggered by the Earthquake made it one of the most crucial variables in hazard mapping. For this research, we interpolate earthquake magnitudes recorded worldwide (1500–2022). Using the location of the Earthquake, an interpolated map showing the magnitude of the Earthquake is constructed in five distinct zones. The studied area's earthquake magnitude data ranges in size from 3.5 to 5.8 magnitudes (Fig. 7.4n). As it falls into zones, the research shows that this region is susceptible to earthquakes.

7.3.2.12 Earthquake (PGA)

The peak ground acceleration (PGA) of an earthquake is the highest absolute value of the surface soil's acceleration during an earthquake (Tian et al. 2013; Xu et al. 2010). PGA is a common parameter for describing the magnitude of an earthquake's ground motion (Yang et al. 2022). Previous research has emphasized the need to combine dynamic elements (PGA, MAP) with other "static" components for landslide zonation mapping in areas, even though these factors play an essential role not only in the reactivation of previous landslides but also in the creation of new ones (Owen et al. 2008; Sato et al. 2007; Zahoor et al. 2023). The PGA distribution map for the study region is shown in Fig. 4o. However, to construct the PGA map, seismic data from the motion monitoring station is spatially interpolated. Subsequently, five Susceptible index classes are distinguished within the study area (Fig.7.4o). Moreover, the southeast region is in the severe (>0.35) seismic zone. The northernmost area of the region has an earthquake index zone that is relatively low (0.21–0.25), while the remaining portion of the region is in the moderate (0.35–0.37) zone.

7.3.2.13 Proximity Analysis

The saturation level of the material on the slope is a crucial factor affecting a slope's stability. Another key factor is the distance to rivers. Therefore, the crucial element for landslide susceptibility assessments is how close the slope is to drainage systems (Park et al. 2013). Using the Euclidean distance interpolation method, five distinct buffer categories, including <100, 100–200, 200–300, 300–400, and >400 m, were created to determine the amount to which streams affected slopes (Fig. 7.4f). One of the most important anthropogenic factors influencing landslide occurrences is the proximity of slopes to highways, which can lead to cut slope formations due to road building that alters the natural topography and affects the stability of the slope (Demir et al. 2013). The distance to the road was calculated with the Euclidean distance tool in Arc GIS. Then the resulting map was classified into five categories: <100, 100–200, 200–400, 400–800, and >800 m (Fig. 7.4h). The most visible structural interpretations on the satellite pictures are lineaments that represent faults, fractures, shear zones, etc. As a result, closer lineaments enhance the likelihood of landslides occurring (Kannan et al. 2013). Moreover, most landslides happen at spots with linear features or lineaments (Nagarajanm et al. 1998). Using high-resolution satellite imagery and GIS methods, we mapped the lineaments of our study area. Landslides are likely to occur less frequently away from these planes and more frequently in zones close to them. The distance from lineaments was calculated using the Euclidean distance tool in ArcGIS. To further categorize it, we used the Jenks natural breaks classification system to divide it into the following five groups: <100, 100–200, 200–400, 400–800, and >800 m. (Fig. 7.4g).

7.3.2.14 Methodology Adopted

The frequency ratio method relies on the correlations between the observed landslip distribution and the subclasses of various CFs. Besides, the link between the landslide sites and the factors influencing the event in the study region is obvious. Consequently, to calculate the frequency ratio of each class in a thematic layer, we examined the association between each class and previous landslide incidents. Further, when computing the correlation between total area and landslide-prone zones, we utilize the mean-value "1". Therefore, a correlation value greater than one suggests a strong relationship, whereas a lesser value implies a weaker one. In this study, we divided each CFs into its respective classes, and the frequency ratio is calculated based on each class "i". Further, the normalized frequency ratio (Eq. 7.3) is computed for all classes in the same factor. Furthermore, the landslide susceptibility index (LSI) is calculated by the normalized frequency ratio summation (Eq. 7.4) of each CF (Beigh and Bukhari 2024, 2023; Pradhan and Lee 2010; Wang et al. 2016).

Fr_i (Frequency Ratio values of class i):

$$\mathrm{Fr}_i = \frac{\left(\frac{NL_i}{NL_t}\right)}{\left(\frac{NC_i}{NC_t}\right)} \tag{7.2}$$

Fr_n (Normalized frequency Ratio values):

$$Fr_n = \frac{\mathrm{Fr}_i}{\sum Classi\mathrm{Fr}} \tag{7.3}$$

LHI (Landslide Susceptibility index):

$$\mathrm{LSI} = \sum Fr_n \tag{7.4}$$

where

NL_i = landslide counts in class i
NL_t = Total landslide counts
NC_i = cell point counts of class i
NC_t = Total cell point counts.

7.4 Results and Discussion

All fourteen factors that cause landslides in the research area were added together to make a landslide susceptibility index map with the application of the frequency ratio method. Using Eq. (7.2), the correlation between each class in the fourteen causative parameter maps and the landslide inventory dataset was used to find the frequency ratio (FR) value for each class. The detailed results of the spatial relationship between various landslide influencing factors and landslide locations using the FR model are shown in (Table 7.2). The normalized frequency ratio values range from 0 to 1. The relationship makes weights, which can be used to show how each class affects a factor that causes the risk of a landslide. In the research area, the slope gradient of >45 (Very Steep Slope) locations with the highest value of 7.58 shows that landslides are most likely to occur. Slopes in the south, southwest, west, and northeast are more likely to fail. Landslides of 44.44% occur on the convex slope and the concave slope 36.51%, while only 19.05% on the flat class. The data also shows that 61.91% of landslides are in the Highly Dissected Hills and Valleys geomorphology class. In the research area, there are various types of land use and land cover, but the agriculture class has the highest risk of landslides with 23.81%, followed by the pastures with 22.22%. Landslides have a value of 1.32 (76.19%) in places with a moderate NDVI class, Fr = 0.66 (23.81%) in areas with a high NDVI class, and null in a low NDVI class. When it comes to distance from the lineament, the chance of a landslide increases as the lineament gets closer. At about 150 m away, the FR value is 5.5, which means there is a high chance of a landslide. The possibility of a landslide occurrence often decreases as we move away from a drainage area. At 200–300 m, the FR value was 2.30, which meant a high chance of a landslide. Therefore, the way gully erosion and undercutting change the landscape could affect landslide initiation. The possibility of landslides in road classes with less than 500 m is quite evident than in the classes with >500 m away distance. The SPI with Class high power has the most significant value of FR (2.43), hence the highest possibility of landslides in this class. Similarly, TWI with a low wetness class yielded the highest FR value (1.0), indicating the highest probability of landslides in this class. Landslides were frequently spotted in the research area near and on generally flat valley bottoms with a sizable contributing area, where TWI and SPI expose greater values. The lithological class shale, limestone, and sandstone have the greatest FR (8.36) values. Hence, this type of lithological unit in the study area has the most influence on causing landslides.

The lithology type with highly fractured limestone deposits impacts landslides due to their high water absorption capability. The results demonstrate that the moderate rainfall class has a much higher FR = 1.28 value than the other two classes. Besides, integrating earthquake and landslide maps in the FR model shows that earthquake zones with magnitudes 3.82–4.09 substantially influence landslide occurrences more than other earthquake levels. Likewise, with the earthquake (PGA) parameter, the class range of 0.28–0.30 has the most significant percentage of landslide events

Table 7.2 Spatial relationship between landslide influencing factors and landslide locations using FR model

Influencing factors	Factor units	No of landslide (NLi)	Landslide %	Area (No of pixcel) (Nci)	Area %	FR (FRi)	Normalization (FRn)
Slope	<5 (Gentle slope)	0	0.00	426,721	66.77	0.00	0.00
	5–8.5 (Moderate slope)	3	4.76	63,038	9.86	0.48	0.06
	8.5–16.5 (Strong slope)	9	14.29	43,613	6.82	2.09	0.28
	16.5–24 (Very strong slope)	17	26.98	29,451	4.61	5.86	0.77
	24–35 (Extreme slope)	28	44.44	52,131	8.16	5.45	0.72
	35–45 (Steep slope)	4	6.35	21,475	3.36	1.89	0.25
	>45 (Very steep slope)	2	3.17	2674	0.42	7.59	1.00
Curvature	Concave (–ive)	23	36.51	242,251	37.90	0.96	0.82
	o (Flat)	12	19.05	161,761	25.31	0.50	0.43
	Convex (+ive)	28	44.44	235,091	36.78	1.17	1.00
Aspect	Flat	6	9.52	87,807	13.74	0.69	0.36
	North	4	6.35	58,604	9.17	0.69	0.36
	Northeast	6	9.52	63,472	9.93	0.96	0.50
	East	5	7.94	72,777	11.39	0.70	0.36
	Southeast	4	6.35	75,042	11.74	0.54	0.28
	South	17	26.98	89,311	13.97	1.93	1.00
	Southwest	6	9.52	76,619	11.99	0.79	0.41
	West	10	15.87	66,334	10.38	1.53	0.79
	Northwest	5	7.94	49,137	7.69	1.03	0.53
Geomorphology	Marsh	0	0.00	62,058	9.71	0.00	0.00
	Pond	0	0.00	1802	0.28	0.00	0.00
	Piedmont alluvial plain	10	15.87	78,101	12.22	1.30	0.12

(continued)

Table 7.2 (continued)

Influencing factors	Factor units	No of landslide (NLi)	Landslide %	Area (No of pixcel) (Nci)	Area %	FR (FRi)	Normalization (FRn)
	Ridge	1	1.59	2753	0.43	3.68	0.33
	Moderately dissected hills and valleys	6	9.52	5458	0.85	11.15	1.00
	Swamp	1	1.59	28,272	4.42	0.36	0.03
	Older flood plain	0	0.00	24,844	3.89	0.00	0.00
	Younger alluvial plain	2	3.17	129,290	20.23	0.16	0.01
	Lacustrine terrain	4	6.35	165,994	25.97	0.24	0.02
	Low dissected hills and valleys	0	0.00	1909	0.30	0.00	0.00
	Channel bar	0	0.00	516	0.08	0.00	0.00
	Highly dissected hills and valleys	39	61.90	100,299	15.69	3.94	0.35
	Lake	0	0.00	19,058	2.98	0.00	0.00
	River	0	0.00	18,749	2.93	0.00	0.00
Proximity to drainage	<100	3	4.76	196,604	30.76	0.15	0.07
	100–200	9	14.29	145,455	22.76	0.63	0.27
	200–300	26	41.27	114,475	17.91	2.30	1.00
	300–400	14	22.22	81,291	12.72	1.75	0.76
	>400	11	17.46	101,278	15.85	1.10	0.48
Proximity to lineament	<100	19	30.16	35,013	5.48	5.50	1.00
	100–200	9	14.29	33,673	5.27	2.71	0.49
	200–400	11	17.46	67,565	10.57	1.65	0.30
	400–800	6	9.52	128,304	20.08	0.47	0.09
	>800	18	28.57	374,548	58.61	0.49	0.09
Proximity to roads	<100	18	46.15	36,877	5.77	8.00	1
	100–200	7	17.95	32,308	5.06	3.55	0.4

(continued)

Table 7.2 (continued)

Influencing factors	Factor units	No of landslide (NLi)	Landslide %	Area (No of pixcel) (Nci)	Area %	FR (FRi)	Normalization (FRn)
	200–400	7	17.95	64,183	10.04	1.79	0.2
	400–800	7	17.95	127,061	19.88	0.90	0.1
	>800	0	0.00	378,674	59.25	0.00	0
Rainfall	Low	19	30.16	264,733	41.42	0.73	0.57
	Moderate	25	39.68	197,036	30.83	1.29	1.00
	High	19	30.16	177,334	27.75	1.09	0.84
NDVI	Low	0	0.00	40,912	6.40	0.00	0.00
	Moderate	48	76.19	367,900	57.57	1.32	1.00
	High	15	23.81	230,291	36.03	0.66	0.50
Lithology	Micaceous siltstone and arenite with limestone	0	0.00	3251	0.51	0.00	0.00
	Andesitic and basaltic lava flows	23	36.51	94,940	14.86	2.46	0.29
	Alluvium, moraines, hillwash and scree	30	47.62	483,181	75.60	0.63	0.08
	Golden brown silt	0	0.00	39,970	6.25	0.00	0.00
	Shale, limestone, and sandstone	10	15.87	12,132	1.90	8.36	1.00
	Shale and limestone with calcareous bands	0	0.00	5629	0.88	0.00	0.00
LuLc	Forest	10	15.87	103,246	16.15	0.98	0.30
	Waterbody	0	0.00	33,819	5.29	0.00	0.00
	Built-up	8	12.70	24,523	3.84	3.31	1.00
	Agriculture	15	23.81	207,546	32.47	0.73	0.22
	Horticulture	1	1.59	56,947	8.91	0.18	0.05
	Plantation	7	11.11	88,099	13.78	0.81	0.24
	Marshy/Swamps	8	12.70	77,953	12.20	1.04	0.31

(continued)

Table 7.2 (continued)

Influencing factors	Factor units	No of landslide (NLi)	Landslide %	Area (No of pixcel) (Nci)	Area %	FR (FRi)	Normalization (FRn)
	Pastures	14	22.22	46,970	7.35	3.02	0.91
TWI	Low wetness	25	39.68	56,816	8.89	4.46	1.00
	Moderate wetness	38	60.32	491,319	76.88	0.78	0.18
	High wetness	0	0.00	90,968	14.23	0.00	0.00
SPI	Low power	6	9.52	182,354	28.53	0.33	0.14
	Moderate power	19	30.16	298,453	46.70	0.65	0.27
	High power	38	60.32	158,296	24.77	2.44	1.00
Earthquake data Mag	3.82–4.09	41	65.08	189,520	29.65	2.19	1.00
	4.09–4.25	22	34.92	284,684	44.54	0.78	0.36
	4.25–4.36	0	0.00	63,896	10.00	0.00	0.00
	4.36–4.57	0	0.00	51,775	8.10	0.00	0.00
	4.57–4.99	0	0.00	49,228	7.70	0.00	0.00
Earthquake data PGA	0.21–0.25	21	33.33	163,191	25.53	1.31	0.79
	0.26–0.27	19	30.16	117,561	18.39	1.64	1.00
	0.28–0.30	23	36.51	142,033	22.22	1.64	1.00
	0.31–0.34	0	0.00	82,445	12.90	0.00	0.00
	0.35–0.37	0	0.00	133,873	20.95	0.00	0.00

than other classes. All CFs are related to landslides, resulting in the varied potential for landslide occurrences. The LSI values on the FR model-generated landslide susceptible (LSM) range from 1.93 to 13.93, as depicted in Fig. 7.5.

The LSM generated by the frequency ratio model in this work was classified using the Jenks natural breaks method (Jenks 1967) into five classes: very high, high, medium, low, and very low within ArcGIS 10.5 (Fig. 7.5). Further, landslide classes with very low, low, medium, high, and very high hazard correspond to 22.26, 30.07, 24.45, and 15.34%, and 7.89%, respectively (Table 7.3). The training landslide frequency and area distribution in various landslide susceptibility classes are shown in Fig. 7.6 and Table 7.3. The current study validated the LSM by utilizing 30% of the total landslides (test samples). Therefore, the FR model-generated LSM yields an overall AUC accuracy of 94.70%. Moreover, a detailed field investigation in the study region was conducted to validate the findings (Figs. 7.7 and 7.8).

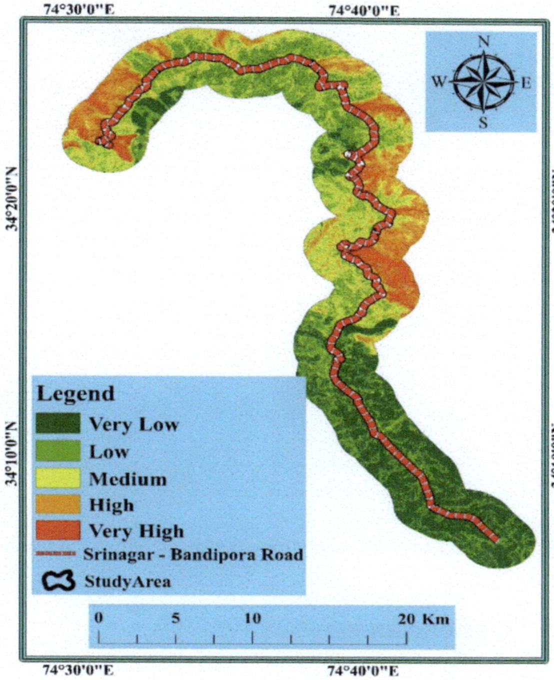

Fig. 7.5 Landslide susceptibility zonation map of the study area using frequency ratio model

Table 7.3 The percentage distribution of landslides in various landslide susceptible zones

Sr. no	Susceptibility zones	Landslide (%)	No of pixels within the zone	Area of zone (%)
1	Very low	1.59	142,246	22.26
2	Low	1.59	192,169	30.07
3	Medium	3.17	156,232	24.45
4	High	14.29	98,043	15.34
5	Very high	79.37	50,413	7.89

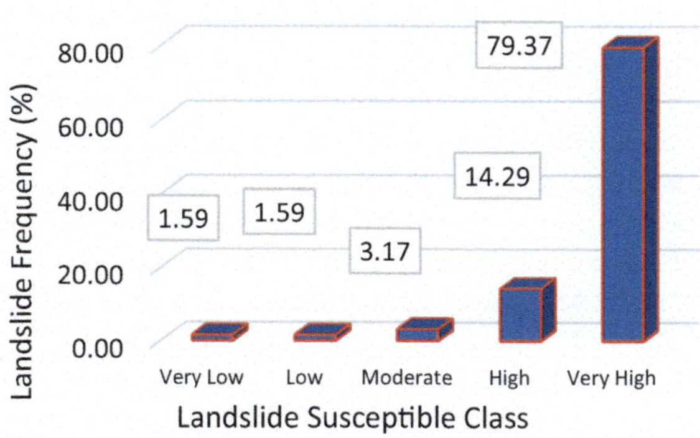

Fig. 7.6 Landslide frequency in various landslide susceptible classes

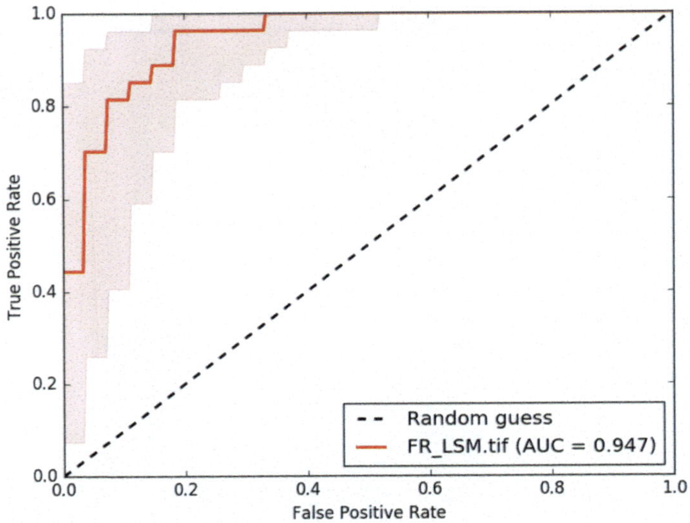

Fig. 7.7 ROC Curve for validation assessment of derived landslide susceptible zonation map

7.5 Conclusions

The study concludes by addressing the lack of information on the risk of landslides along the Srinagar-Bandipora State Highway in North Kashmir Himalaya. Based on the available information, the following conclusions may be derived from the present study:

- The study offers important insights into the region's landslide susceptibility by integrating fifteen contributing elements and using the FR model for landslip susceptibility mapping. The CFs with the most significant impact on the model were slope, lithology, land use/cover, proximity to roads/Lineaments, and geomorphology. The study shows that the FR model can effectively accomplish landslide susceptibility Assessment.
- The Landslide Susceptibility Map (LSM) for the current study was divided into five groups: very high, high, medium, low, and very low. Further, the high and very high susceptible classes account for 23.23% of the total study area. Further, LSM was confirmed using landslide occurrences, and the verification result revealed 94.80% accuracy. The validated results indicated excellent agreement between landslide occurrence and the produced LSM of the area.
- Furthermore, there is a high risk of landslides in Safapora, Malgam, Garoora, Nusso, Mangnipora, and, a medium risk in Watlub, Gund-Saderkoot, and Nadihal.
- The LSM generated by the Frequency Ratio (FR) model can be an invaluable resource for planners and engineers. It can help identify places where preventive steps should be taken to lessen the likelihood of future landslides. This knowledge

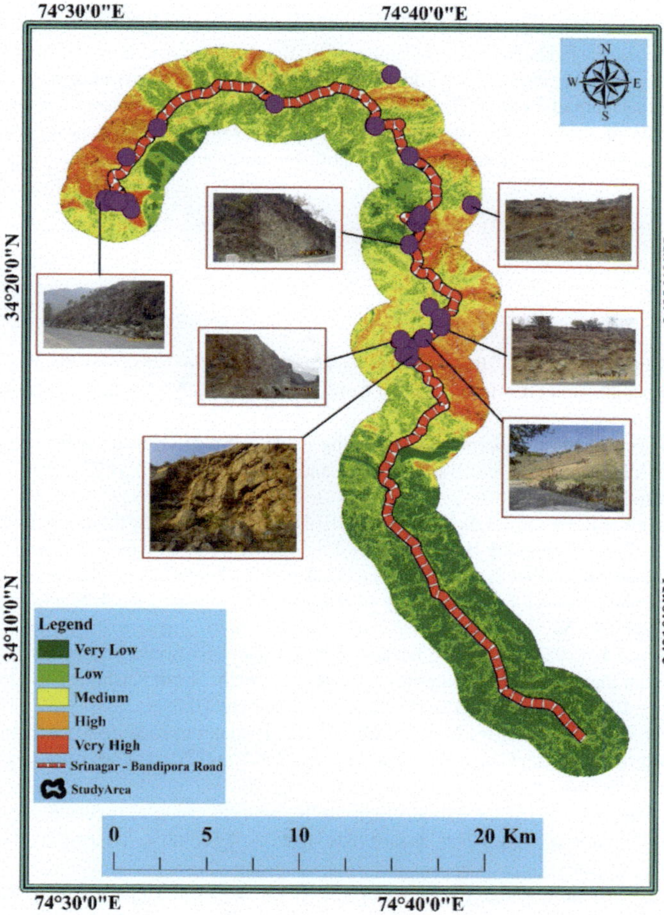

Fig. 7.8 Photo plates of fieldwork with FR-generated LSZ map

can be used to reduce the impact of natural disasters, contributing to the safety and well-being of local communities.

- Overall, this study improves our understanding of landslide susceptibility in the north Kashmir Himalaya and emphasizes the significance of taking practical steps to reduce landslide hazards. Moreover, the study's findings can improve decision-making processes and design effective solutions to promote the region's resilience and safety.

Acknowledgements The authors thank the Border Roads Organization (BRO) Srinagar, J&K, India for providing necessary landslide data. The authors also thank the local administration for letting the fieldwork go smoothly. We also thank Abdul Ahad Thakur, Senior Technician Engineering Geology Lab, and Mushtaq Ahmad Tantray, Geotechnical Lab Assistant, Department of Civil Engineering National Institute of Technology, for their help during fieldwork and lab assistance. Finally, the authors thank anonymous reviewers for their insightful remarks and suggestions.

References

Ahmad RA, Singh RP, Adris A (2017) Seismic hazard assessment of Syria using seismicity, DEM, slope, active faults and GIS. Remote Sens Appl Soc Environ 6:59–70. https://doi.org/10.1016/j.rsase.2017.04.003

Ahmed B (2015) Landslide susceptibility mapping using multi-criteria evaluation techniques in Chittagong Metropolitan Area, Bangladesh. Landslides 12(6):1077–1095. https://doi.org/10.1007/s10346-014-0521-x

Anbalagan R, Kumar R, Parida S, Lakshmanan K (2014) GIS based post-earthquake landslide hazard zonation mapping of Lachung Basin, Sikkim. Int J Emerg Technol Adv Eng 4(1):431–441

Beigh IH, Bukhari SK (2024) Assessment of GIS-based landslide susceptibility using bivariate and multivariate approach—A case study of Kashmir Himalayas. J Rock Mech Tunn Technol (JRMTT) 30(1):39–56

Beigh IH, Bukhari K (2023) Landslide hazard zonation using Bivariate Frequency ratio method along National highway-1 from Baramulla-Uri Road stretch, North Kashmir Himalayas, India. Disaster Adv 16(6):8–17. https://doi.org/10.25303/1606da08017

Bíl M, Vodák R, Kubeček J, Bílová M, Sedoník J (2015) Evaluating road network damage caused by natural disasters in the Czech Republic between 1997 and 2010. Transp Res Part A Policy Pract 80:90–103. https://doi.org/10.1016/j.tra.2015.07.006

Cárdenas NY, Mera EE (2016) Landslide susceptibility analysis using remote sensing and GIS in the western Ecuadorian Andes. Nat Hazards 81(3):1829–1859. https://doi.org/10.1007/s11069-016-2157-8

Citrabhuwana BN, Kusumayudha SB, Purwanto (2016) Geology and slope stability analysis using markland method on road segment of Piyungan–Patuk, Sleman and Gunungkidul regencies, Yogyakarta special region Indonesia. Int J Econ Environ Geol 7(1):42–52

Das I, Stein A, Kerle N, Dadhwal VK (2011) Probabilistic landslide hazard assessment using homogeneous susceptible units (HSU) along a national highway corridor in the northern Himalayas, India. Landslides 8:293–308

Demir G, Aytekin M, Akgun A, ˙Ikizler SB, Tatar O (2013) A comparison of landslide susceptibility mapping of the eastern part of the North Anatolian Fault Zone (Turkey) by likelihood-frequency ratio and analytic hierarchy process methods. Nat Hazards 65(3):1481–1506

Ding Q, Chen W, Hong H (2017) Application of frequency ratio, weights of evidence and evidential belief function models in landslide susceptibility mapping. Geocarto Int 32(6):619–639. https://doi.org/10.1080/10106049.2016.1165294

Gorsevski PV, Donevska KR, Mitrovski CD, Frizado JP (2012) Integrating multi-criteria evaluation techniques with geographic information systems for landfill site selection: A case study using ordered weighted average. Waste Manage 32(2):287–296

Gutiérrez F, Parise M, De Waele J, Jourde H (2014) A review on natural and human-induced geohazards and impacts in karst. Earth Sci Rev 138:61–88

Guzzetti F, Carrara A, Cardinali M, Reichenbach P (1999) Landslide hazard evaluation: A review of current techniques and their application in a multi-scale study, Central Italy. Geomorphology 31:181–216

Hadmoko DS, Lavigne F, Samodra G (2017) Application of a semiquantitative and GIS-based statistical model to landslide susceptibility zonation in Kayangan Catchment, Java, Indonesia. Nat Hazards 87:437–468. https://doi.org/10.1007/s11069-017-2772-z

Huang F, Yao C, Liu W, Li Y, Liu X (2018) Landslide susceptibility assessment in the Nantian area of China: a comparison of frequency ratio model and support vector machine. Geomat Nat Haz Risk 9(1):919–938. https://doi.org/10.1080/19475705.2018.1482963

Jenks GF (1967) The data model concept in statistical mapping. Int Year Book Cartogr 7:186–190

Kamp U, Growley BJ, Khattak GA (2008) GIS-based landslide susceptibility mapping for the 2005 Kashmir earthquake region. Geomorphol 101:631–642

Kannan M, Saranathan E, Anabalagan R (2013) Landslide vulnerability mapping using frequency ratio model: A geospatial approach in Bodi-Bodimettu Ghat section, Theni district, Tamil Nadu, India. Arabian J Geosci 6(8):2901–2913

Kaur H, Gupta S, Parkash S (2017) Comparative evaluation of various approaches for landslide hazard zoning: A critical review in Indian perspectives. Spat Inf Res 25:389–398

Kellerer-Pirklbauer A (2002) The influence of landuse on the stability of slopes with examples from the European Alps. MittEn Naturwissenschaftlichen VerEs Für Steiermark 132:43–62

Kirschbaum D, Stanley T, Zhou Y (2015) Spatial and temporal analysis of a global landslide catalog.Geomorphology, 249:4–15. https://doi.org/10.1016/j.geomorph.2015.03.016

Korup O, Stolle A (2014) Landslide prediction from machine learning. Geol Today 30(1):26–33. https://doi.org/10.1111/gto.12034

Lee S, Pradhan B (2006) Probabilistic landslide hazards and risk mapping on Penang Island, Malaysia. J Earth Syst Sci 115(6):661–672

Maingi JK, Marsh SE (2001) Assessment of environmental impacts of river basin development on the riverine forests of eastern Kenya using multi-temporal satellite data. Int J Remote Sens 22(14):2701–2729. https://doi.org/10.1080/01431160010031298

Malamud BD, Turcotte DL, Guzzetti F, Reichenbach P (2004) Landslide inventories and their statistical properties. Earth Surf Proc Land 29:687–711. https://doi.org/10.1002/esp.1064

Marchi L, Dalla Fontana G (2005) GIS morphometric indicators for the analysis of sediment dynamics in mountain basins. Environ Geol 48:218–228. https://doi.org/10.1007/s00254-005-1292-4

Martha TR, Roy P, Jain N, Khanna K, Mrinalni K, Kumar KV, Rao PVN (2021) Geospatial landslide inventory of India—an insight into occurrence and exposure on a national scale. Landslides:1–17

Mutanga O, Kumar L (2019) Google earth engine applications. Remote Sens, 11, Article No. 591. https://doi.org/10.3390/rs11050591

Nagarajanm R, Mukherjeem A, Roy A, Khire MV (1998) Temporal remote sensing data and GIS application in landslide hazard zonation of part of Western Ghat, India. Int J Remote Sens 19:573–585

Nefeslioglu HA, Gokceoglu C, Sonmez H (2008) An assessment on the use of logistic regression and artificial neural networks with different sampling strategies for the preparation of landslide susceptibility maps. Eng Geol 97:171–191. https://doi.org/10.1016/j.enggeo.2008.01.004

Oh HJ, Pradhan P (2011) Application of a neuro fuzzy model to landslide susceptibility mapping for shallow landslides in a tropical Hilly area. Comput & Geosci 37(9):12641276

Owen LA, Kamp U, Khattak GA, Harp EL, Keefer DK, Bauer MA (2008) Landslides triggered by the 8 October 2005 Kashmir earthquake. Geomorphology, 94:1–9

Pandey A, Dabral PP, Chowdary VM, Yadav NK (2008) Landslide hazard zonation using remote sensing and GIS: a case study of Dikrong river basin, Arunachal Pradesh, India. Environ Geol 54(7):1517–1529. https://doi.org/10.1007/s00254-007-0933-1

Park S, Choi C, Kim B, Kim J (2013) Landslide susceptibility mapping using frequency ratio, analytic hierarchy process, logistic regression, and artificial neural network methods at the Inje area, Korea. Environ Earth Sci 68:1443–1464

Patil A, Panhalkar S, Shinde S (2022) Frequency ratio approach for landslide susceptibility mapping of Phonda Ghat of Maharashtra. In: Singh RB, Kumar M, Tripathi DK (eds) Remote sensing and geographic information systems for policy decision support. Advances in geographical and environmental sciences. https://doi.org/10.1007/978-981-16-7731-1_2

Petley D (2012) Global patterns of loss of life from landslides. Geology 40:927–930. https://doi.org/10.1130/G33217.1

Pourghasemi HR, Pradhan B, Gokceoglu C (2012a) Application of fuzzy logic and analytical hierarchy process (AHP) to landslide susceptibility mapping at Haraz watershed, Iran. Nat Hazards 63:965–996

Pradhan AMS, Kim YT (2014) Relative effect method of landslide susceptibility zonation in weathered granite soil: A case study in Deokjeok-ri Creek, South Korea. Nat Hazards 72(2):1189–1217

Pradhan B, Lee S (2010) Landslide susceptibility assessment and factor effect analysis: backpropagation artificial neural networks and their comparison with frequency ratio and bivariate logistic regression modeling. Environ Modell Softw 25(6):747–759. https://doi.org/10.1016/j.envsoft.2009.10.016

Ramesh V, Anbazhagan S (2014) Landslide susceptibility mapping along Kolli hills Ghat road section (India) using frequency ratio, relative effect and fuzzy logic models. Environ Earth Sci 73(12):8009–8021

Razavizadeh S, Solaimani K, Massironi M, Kavian A (2017) Mapping landslide susceptibility with frequency ratio, statistical index, and weights of evidence models: a case study in northern Iran. Environ Earth Sci 76(14). https://doi.org/10.1007/s12665-017-6839-7

Regmi NR, Giardino JR, McDonald EV, Vitek JD (2014) A comparison of logistic regression-based models of susceptibility to landslides in western Colorado, USA. Landslides 11:247–262. https://doi.org/10.1007/s10346-012-0380-2

Sati VP (2014) Towards sustainable livelihoods and ecosystems in mountain regions. Springer International Publishing, pp 137–150

Sato HP, Hasegawa H, Fujiwara S, Tobita M, Koarai M, Une H, Iwahashi J (2007) Interpretation of landslide distribution triggered by the 2005 Northern Pakistan earthquake using SPOT 5 imagery. Landslides 4:113–122

Schmid JN (2017) Using google earth engine for landsat NDVI time series analysis to indicate the present status of forest stands. https://doi.org/10.13140/RG.2.2.34134.14402/6

Shafapour Tehrany M, Kumar L, Neamah Jebur M, Shabani F (2019) Evaluating the application of the statistical index method in flood susceptibility mapping and its comparison with frequency ratio and logistic regression methods. Geomat, Nat Haz and Risk 10(1):79–101. https://doi.org/10.1080/19475705.2018.1506509

Sharma S, Mahajan AK (2018) A comparative assessment of information value, frequency ratio and analytical hierarchy process models for landslide susceptibility mapping of a Himalayan watershed, India. Bull Eng Geol Environ. https://doi.org/10.1007/s10064-018-1259-9

Singh R, Umrao RK, Singh TN (2013) Stability evaluation of road cut slopes in the Lesser Himalaya of Uttarakhand, India: conventional and numerical approaches. Bull Eng Geol Environ 73(3):845–857. https://doi.org/10.1007/s10064-013-0532-1

Subramanian SS, Ishikawa T, Tokoro T (2017) Stability assessment approach for soil slopes in seasonal cold regions. Eng Geol 2017(221):154–169

Tian YY, Xu C, Hong HY, Zhou Q, Wang D (2019) Mapping earthquake-triggered landslide susceptibility by use of artificial neural network (ANN) models: An example of the 2013 Minxian (China) mw 5.9 event. Geomatics, Nat Hazards Risk 10:1–25. https://doi.org/10.1080/19475705.2018.1487471

Vakhshoori V, Zare M (2016) Landslide susceptibility mapping by comparing weight of evidence, fuzzy logic, and frequency ratio methods. Geomat Nat Haz Risk 7(5):1731–1752. https://doi.org/10.1080/19475705.2016.1144655

Wang Q, Li W, Yan S, Wu Y, Pei Y (2016) GIS based frequency ratio and index of entropy models to landslide susceptibility mapping (Daguan, China). Environ Earth Sci 75(9):1–16. https://doi.org/10.1007/s12665-016-5580-y

Wu Y, Lan H (2019) Landslide analyst—a landslide propagation model considering block size heterogeneity. Landslides 16(6):1107–1120. https://doi.org/10.1007/s10346-019-01154-2

Li X, Chen W, Cheng X, Wang L(2016) A comparison of machine learning algorithms for mapping of complex surface-mined and agricultural landscapes using ziyuan-3 stereo satellite imagery. Remote Sens 8. https://doi.org/10.3390/rs8060514

Xu C, Xu X, Yu G (2013) Landslides triggered by slipping-fault-generated earthquake on a plateau: An example of the 14 april 2010, ms 7.1, Yushu, China earthquake. Landslides, 10:421–431. https://doi.org/10.1007/s10346-012-0340-x

Yang Z, Pang B, Dong W, Li D (2023) Spatial pattern and intensity mapping of coseismic landslides triggered by the 2022 Luding Earthquake in China. Geomat, Nat Hazards Risk 15(5):1323

Yilmaz I, Keskin I (2009) GIS based statistical and physical approaches to landslide susceptibility mapping (Sebinkarahisar, Turkey). Bull Eng Geol Environ 68:459–471

Yilmaz I (2009) Landslide susceptibility mapping using frequency ratio, logistic regression, artificial neural networks and their comparison: a case study from Kat landslides (Tokat-Turkey). Comput Geosci 35:1125–1138

Yuan F, Sawaya KE, Loeffelholz B, Bauer ME (2005) Land cover classification and change analysis of the Twin Cities (Minnesota) Metropolitan Area by multitemporal Landsat remote sensing. Remote Sens Environ 98:317–328

Zahoor F, Ansari A, Rao KS, Satyam N (2023) Seismic hazard assessment of Kashmir region using logic tree approach: Focus on sensitivity of PSHA results Towards declustering procedures and GMPEs. Pure Appl Geophys 180(3):789–827. https://doi.org/10.1007/s00024-023-03239-5

Zhao Y, Wang R, Jiang Y, Liu H, Wei Z (2019) GIS-based logistic regression for rainfall-induced landslide susceptibility mapping under different grid sizes in Yueqing South-eastern China. Eng Geol 259:105147

Chapter 8
Existing and Potential Changes in Himalayan Glaciers: In Climate Change Perspective

Manish Mehta

Abstract The Himalayas contain the largest glacier system outside the Polar regions, serving as a vital water tower for the rivers that drain this massive mountain range and its extensive adjoining alluvial fan. The recent retreat of Himalayan Glaciers has had a significant impact on at least half a billion people. According to the updated glacier inventory data from the Geological Survey of India (GSI; Raina V K and Srivastava D (2008), Glacier Atlas of India. GSI Publication 315) there are more than 9,575 glaciers in the Indian-administered part of the Himalayas, covering an area of approximately 37,000 km^2. Meanwhile, ICIMOD identifies a staggering 540,000 individual glaciers in the Hindu Kush-Himalayan Region, spanning an area of 60,000 km^2, with an estimated ice reserve of 6,000 km^3. These glaciers release meltwater seasonally into the tributaries of the Indus, Ganges, and Brahmaputra Rivers. Around 300 million people depend upon water from these rivers to sustain agricultural and economic activities on the Indian subcontinent's plains. Regions where water supply heavily depends on melting snow and ice are projected to experience hydrological disruptions due to recent warming. Beyond being a constant source of water, this glacier system serves as a climate change indicator in the region, influencing river discharge, debris production, and transportation, triggering avalanches, flash floods, and causing siltation in lower valleys. Additionally, it directly and indirectly affects the mountain ecology that sustains life. Historically, Himalayan Glaciers have exhibited substantial fluctuations, a pattern that continues into the present day. The ongoing shrinkage of glaciers due to climate change is expected to bring about a major impact on water resources, precipitating dramatic changes in the environment. This paper provides a brief discussion of some critical aspects of glacier shrinkage and its potential impacts.

Keywords Himalaya · Glacier · Status of Himalayan glaciers · Review of Himalayan glaciers

---

M. Mehta (✉)
Wadia Institute of Himalayan Geology, Dehradun, India
e-mail: msmehta75@gmail.com

B. K. Rastogi et al. (eds.), *Natural Hazards and Risk Mitigation*, Springer Transactions in Civil and Environmental Engineering, https://doi.org/10.1007/978-981-97-7658-0_8

8.1 Introduction

A glacier is a mass of ice that forms through the recrystallization of snow, slowly moving downward or forward in all directions under the influence of gravity, primarily guided by the topography of the underlying bedrock. As it progresses down a valley, its leading edge forms a towering ice wall known as the snout, representing the glacier's terminal point. Similar to surface streams, glaciers rely on accumulated snow for sustenance. Over time, as the seasons progress, the layers of snow accumulation can deepen, eventually giving birth to a glacier. The processes of accumulation and ablation (including melting and evaporation) are intricately linked to hydrological and climatological factors. Snow accumulates during winter and melts away during the summer months. Due to their dynamic and delicate nature, glaciers are highly responsive to their surroundings, constantly altering both their size and shape. They expand during colder climates and contract in warmer conditions. The interplay between growth (expansion) and loss (contraction) is referred to as glacier mass balance, a crucial component in comprehending the intricate connections among climate, mass balance, and glacier behavior.

Glaciers are frozen masses that cover approximately 10% of the Earth's surface area out of which 0.3% are in Asia (Haeberli et al. 1989). These glaciers serve as reservoirs for numerous rivers that originate from them, playing a pivotal role in sustaining a significant portion of the Earth's landmass. These icy formations are not only sources of perennial water, but they also offer valuable insights into regional climate change. They aid in the assessment of debris production and transportation, fluctuations in river discharge, and the identification of hazards such as avalanches, glacier lake outburst floods (GLOFs), flash floods, and siltation. Additionally, they have both direct and indirect impacts on the preservation of mountain ecology. Furthermore, glaciers contribute to the planning and management of river valley projects, as well as the development of mini/micro-hydroelectric power generation and irrigation systems. They also play a role in climate and weather prediction, which is particularly significant given the rapid decline in snow cover and glacier ice volume worldwide.

Glaciers and their environment hold essential keys to our understanding of current, past, and future environmental conditions. Over the Quaternary period, covering nearly two million years, a one third of the Earth's surface was ice-covered. The dramatic expansion and contraction of glaciers in response to climate shifts have sculpted breath taking landscapes by eroding rocks, transporting sediments, and depositing debris far from its source.

Throughout Earth's approximately 4.6 billion-year history, at least four major ice ages have occurred, with the present time being an interglacial period within the Pleistocene glaciation. Since the last glacial period ended 12,000 years ago, minor local and regional advances and retreats have taken place. In historical times, glaciers expanded during the Little Ice Age (LIA), a cooling period spanning roughly from 1650 to 1850 (Rowan 2017; Mehta et al. 2021a, b). Following this, until around 1940, glaciers globally receded due to warming. However, from 1950 to 1980, glacier retreat slowed and, in many cases, reversed due to a slight global cooling phase (Aniya and Wakao 1997; Rowan 2017). In recent decades, glacier retreat has accelerated significantly and become widespread, endangering many glaciers globally. This acceleration has intensified since the 1980s, resulting in a rapid recession.

The Himalayan Glacier systems have a crucial role in the global climate. The connection between Eurasian snow cover in winter and the Indian monsoon in the subsequent summer is well established (Huguchi and Ohata 1996). In the Himalayas, a substantial population resides in valleys, and 50–70% of the annual discharge from Western Himalayan Rivers and slightly less from Eastern Himalayan Rivers comes from snow and glaciers (Vohra 1981). Despite the vast amount of snow cover and the number of glaciers in the Himalayas, their influence on the hydrological cycle and environment remains inadequately understood. The lack of glacier-related information poses a significant challenge to sustainable development in mountainous regions and adjacent plains, particularly concerning issues like hydro-power, water supply, and environmental quality, all of which directly impact national growth. The fluctuations of glaciers, changes in runoff, annual balance, sediment transport, and their roles in shaping the subcontinent's climate bear immense significance.

The Earth's ecosystem, including its glacier system, faces immense pressure due to population growth and the demands of industrial and technological advancements. Thus, comprehending this crucial natural resource and its interconnectedness with other ecosystem factors becomes paramount. It's imperative that developmental initiatives maintain the equilibrium established by nature over geological epochs and prevent irreversible processes/impacts that could hinder long-term progress. Consequently, the scientific examination of Himalayan Glaciers becomes of utmost significance. Such study is vital for effective water resource management, hydropower generation, climate, and weather forecasting, as well as for understanding the interconnected ecological system. This understanding aligns with national priorities, encompassing aspects like national security, human well-being, livelihood, and sustainable resource development.

8.2 Glaciers of the Indian Himalaya

The Himalaya is widely acknowledged as hosting one of the world's most significant glacier systems, with its glaciers playing a crucial role in contributing freshwater to the primary river systems of the Indian subcontinent. Encompassing extensive areas, the Himalayan region contains valuable water resources in the form of ice and snow cover. The glaciated expanse of the Indian Himalayan ranges has been approximated at 37,000 km^2.

Recent glacier inventory data, as provided by the Geological Survey of India (GSI), reveals the existence of over 9575 glaciers in the part of the Himalaya administered by India (Raina and Srivastava 2008). This territory includes regions such as Jammu and Kashmir, Himachal Pradesh, Uttaranchal, Sikkim, and Arunachal Pradesh (Fig. 8.1 and Table 8.1). Another assessment by the Space Application Centre (SAC) in 2016 identified 32,392 glaciers across the larger Hindu Kush-Himalayan (HKH) region, which encompasses approximately 71,182 km^2. In concurrence, the International Centre for Integrated Mountain Development (ICIMOD) reported 37,955 glaciers in the HKH region, covering a combined area of around 25,438 km^2. Distribution of these glaciers is uneven across the vast expanse of the Himalayan ranges, with a higher concentration observed in the northwestern part of the Indian Himalaya in comparison to the northeastern counterpart. This non-uniform distribution is a consequence of distinct climatic conditions and variations in altitudes. These characteristics are defined by fluctuations in the regional firn line limit, which spans altitudes ranging from 4500 to 5700 m as documented by various studies (Mullar et al. 1977; Sharma and Owen 1996; Dobhal et al. 2013; Mehta et al. 2021a, b, 2023; Rana et al. 2023).

Notably, the records detailing the fluctuations of these glaciers are only 150 years old. Research conducted by Mayewski and Jeschke (1979) evaluated 122 glaciers within the Himalaya and Karakoram regions, concluding that the majority of these glaciers are in a state of retreat. Similarly, Scherler et al. (2011) documented that over 65% of monsoon-influenced glaciers in their study exhibited retreat. However,

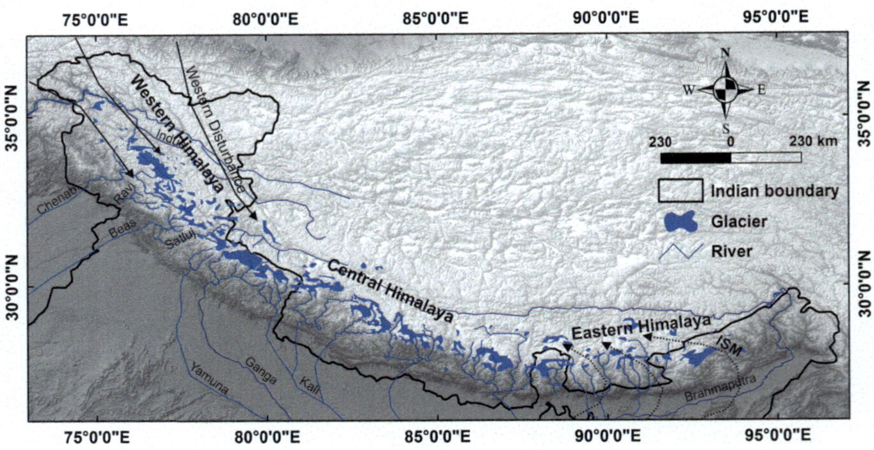

Fig. 8.1 Distribution of glaciers and river system of Himalaya

Table 8.1 Distribution of glaciers of India after Raina and Srivastava (2008)

States of India	Number of glaciers	Area (km^2)
Jammu and Kashmir	5262	29,163
Himachal Pradesh	2735	4516
Uttarakhand	968	2857
Sikkim	449	706
Arunachal Pradesh	161	223
Total	9575	37,465

glaciers covered with substantial debris and characterized by stagnant, low-gradient terminus areas tended to have stable fronts. These debris-covered glaciers are prevalent in the rugged central Himalaya but are scarce in less rugged Tibetan Plateau landscapes, where retreat rates are higher. In contrast, in the Karakoram region of the northwestern Himalaya, influenced by westerlies, over 50% of the observed glaciers are either advancing or maintaining stability. The researchers underscored the absence of a uniform response among Himalayan Glaciers to climate change. They highlighted the significance of debris cover in comprehending glacier retreat a factor that has hitherto been overlooked in predictions regarding future water availability. Extensive research focusing on the recession of Himalayan Glaciers has been undertaken by Indian scientists since the 1970s. Various studies, including those by Vohra (1981), Gautam and Mukherjee (1989), Mehta et al. (2011, 2012, 2021a, b, 2023), Pratap et al. (2013, 2015), Dobhal et al. (2013), Shukla et al. (2020), Garg et al. (2018), Kumar et al. (2017, 2021a, b), and others, collectively illustrate that glaciers are retreating at an average pace of 18–20 m per year. Moreover, it has been observed that the rate of recession for both small glaciers (less than 5 km in length) and larger glaciers (greater than 10 km in length) is relatively similar. This similarity indicates that the prospects for small glaciers in the future are not optimistic. However, it's worth noting that the majority of glaciers in the region are small in size, rather than large.

Glacial meltwater has sustained civilization for centuries in the mountains and nearby plains. However, glaciers can turn highly destructive when they abruptly release vast quantities of meltwater and sediments, causing catastrophic floods and debris flows. Notable instances include the Kedarnath and Rishiganga disasters, and significant Himalayan events resulting in substantial devastation in Uttarakhand (Mehta et al. 2017, 2021b). India's glaciers contribute about $8500 \times 106\ m^3$ of water annually to Himalayan-origin rivers (Raina 2009), translating to 5.35 million acre-feet (m.a.f) from the Ganga Basin. This is augmented by snowmelt during April and June (Raina 2009). Given the increasing power demand, developing run-of-river hydro projects in high altitudes, primarily driven by meltwater, has become crucial. A lack of glaciological studies has created critical data gaps in comprehending hydrological concerns (Table 8.2).

Table 8.2 Glacier Inventory of Indian Himalaya (IH) and Hindu Kush-Himalaya (HKH) prepared by different agencies

Organizations and year	No of glacier	Area (Km^2)
Raina and Srivastave (2008); Raina (2009)	9575 (IH)	37,466
SAC (2016)	32,392 (HKH)	71,182
ICIMOD (2011)	37,955 (HKH)	25,438

8.3 Topography and Atmospheric Status

The Himalayan topography serves as a distinct platform for snow, ice, and glacier accumulation, forming an extensive glacier system. The region's characteristics, including low latitudes, high altitudes, and complex atmospheric conditions, differentiate it from counterparts like the Alps. Variations in altitude within these areas contribute to diverse precipitation and temperature patterns (Liu and Chen 2000; Thompson et al. 2000; Dobhal et al. 2013), significantly influencing the hydro-meteorological features of various river systems.

Situated between monsoonal and westerly atmospheric influences, the Indian Himalayan Glaciers significantly impact regional precipitation and hydrology (Immerzeel et al. 2010; Bookhagen and Burbank 2010). In the Western Himalayas, glaciers receive more winter accumulation due to westerly disturbances, while those in the Eastern Himalayas accumulate more during the summer monsoon (Mehta et al. 2021a, b; Choudhury et al. 2022). Central Himalayan Glaciers receive contributions from both westerly disturbances and the monsoon (Mehta et al. 2012, 2014). Thus, based on precipitation patterns, the Indian Himalayan Glaciers can be categorized into three sub-divisions: Western Himalaya (influenced by westerly disturbances), Eastern Himalaya (monsoon dominated), and Central Himalaya (affected by both westerly disturbances and the monsoon) (Fig. 8.1).

In the Western Himalayan region, roughly 80% of the average annual winter precipitation occurs, contrasting with 20% during summer, leading to arid to semi-arid conditions. Conversely, the Eastern region experiences around 20% of precipitation in winter and 80% in summer, resulting in humid to per-humid conditions (Bahadur 1998). The central Himalayan Glaciers exhibit characteristics of both monsoonal and westerly influences, classifying them as summer accumulation-type glaciers similar to those found in the eastern Himalayas.

The extensive altitudinal range (3800–7000 m asl) of the glaciers spanning approximately 3000 m, along with diverse climate and topography, combines polar, temperate, and tropical conditions. This encompasses polar conditions characterized by sub-freezing temperatures, temperate conditions coinciding with the melting point of ice, and tropical conditions marked by high temperatures and significant precipitation. Beyond direct glacier precipitation, notable contributions originate from snow avalanches descending from lateral slopes onto these glaciers. Avalanche snow transforms into glacier ice within 2–4 years, whereas direct snow accumulation on the glacier surface requires 7–8 years for this transformation. Such environmental factors yield intricate behavior in Himalayan Glaciers. The notably elevated altitudes in the cold environment act as a significant barrier against the intrusion of moisture-laden warm air currents from the Indian and Pacific oceans, as well as against the incursion of cold air masses from the Atlantic and Arctic regions (Bahadur 1992).

8.4 Salient Features of Himalayan Glaciers

Glaciers exhibit a diverse array of morphological and morphometric attributes, underpinned by variations in their thermal, hydrological, and dynamic properties. This variability is especially pronounced in mountainous terrains like the Himalayas. In this region, glacier typologies have been established, with particular emphasis on the mountain valley type, as delineated by seminal works such as Ahlmann (1948), Flint (1972), and Charlsworth (1948). These glaciers manifest in a broad spectrum of sizes and shapes, ranging from diminutive niche glaciers to extensive behemoths like the Siachen Glacier, sprawling across an impressive expanse of 72 km ranking as the second-largest non-polar glacier. The configuration of a glacier is fundamentally influenced by the topographical characteristics and orientation of the valley it occupies. Himalayan Glaciers, as a general trend, adopt a funnel-like structure. This entails a capacious accumulation region at the upper reaches, which then tapers into a narrower terminus area downstream. The accumulation zone, situated at the glacier's head, can manifest either as a solitary cirque basin or as an aggregation of such basins (Fig. 8.2). The classification of Himalayan Glaciers was proposed by Mason (1930), who categorized them into two principal groups: longitudinal and transverse glaciers. Longitudinal glaciers align parallel to the mountain range, and they tend to colonize more expansive valleys while displaying greater dimensions. In contrast, transverse glaciers form perpendicular to the mountain slope.

Fig. 8.2 A panoramic view of a typical Himalayan Glacier (Durung-Drung Glacier Western Himalaya)

Table 8.3 Some salient features of Himalayan Glaciers (After ICIMOD 2011)

Sl. no	Features	Western Himalaya	Central Himalaya	Eastern Himalaya
1	Source/ Basin	Indus	Ganga	Brahmaputra
2	Climate	Arid to semi-arid	Semi-arid and semi-humid	Humid to per-humid
3	Latitudes	30–37 °N	27–31 °N	27–31 °N
4	No of glaciers	18,495	7963	11,497
5	Glacier area	21,192 km^2	9011 km^2	14,019 km^2
6	Ice volume	2696 km^3	793 km^3	1302 km^3
7	Snow line (m)	4700–5000 m	5100–5300 m	5200–5600 m
8	Maxi length	74 km	30 km	10.9 km
9	Snout altitude	3490–4200 m	3900–4200 m	4000–4500 m
10	Glacier surface	Clean with debris-covered lower part	Partially debris covered	Thick debris covered

A distinctive hallmark of Himalayan Glaciers is their propensity to be enveloped in debris. It is noteworthy that debris coverage is more pronounced in the Eastern and Central Himalayan sectors, whereas it is relatively subdued in the Western sector. This array of characteristics encompasses glaciers that span the gamut from modest cirque and niche variants, which span an area of 0.1–4 km^2, to substantial valley glaciers that stretch from 5 to 216 km^2. This holistic characterization encapsulates the comprehensive physical attributes and dynamic processes inherent to glaciers within the Himalayan context. Some of the salient features of Himalayan Glaciers are given in (Table 8.3).

8.5 Significance of Himalayan Glaciers

8.5.1 Potential Water Resource

The Himalayan region harbors significant water resources in the form of snow, ice, and glaciers, which are gradually released into perennial rivers, thereby catering to the needs of downstream populations. The hydrological significance of these meltwater sources, encompassing snow, glaciers, permafrost, and glacial lakes, is paramount in sustaining the intricate river networks in the Himalayas. The aggregate ice volume in the Himalayan expanse approximates 4791 km^3, with diverse glacier systems detailed in Table 8.1 (ICIMOD 2011). Within this region, 22 major rivers emanate (depicted in

Fig. 8.1), of which the Indus and Brahmaputra, two of the Indian subcontinent's major river systems, originate in Tibet. Elevated precipitation at higher altitudes, coupled with snow and ice melt contributions, leads to higher specific water yields from mountainous watersheds than the entire river basin. Consequently, glacier-fed waters hold pivotal significance within mountain hydrology, encompassing snowmelt, glacier ice melt runoff, and monsoon precipitation.

The seasonal dynamics of water contribution from glaciers and snowpack to Himalayan Rivers exhibit variation. The lowest flows transpire from November to March, while peak runoff surfaces in July and August (Fig. 8.3), corresponding to heightened temperatures, maximal glacier melt, and monsoonal rainfall. These intricate processes are often analyzed at mesoscale catchment outlets, yielding insights into snow and glacier contributions.

Noteworthy studies have quantified these contributions in distinct regions. For instance, within the Chenab Basin, it was established that snow and glacier melt contribute to 49% of the Chenab's flow at Akhnoor, and 80% of the Sutlej River's annual flow arises between May and September (Singh et al. 1997). Bhakra Dam receives 59% of its runoff from snow and glacier melt, while the upper Indus contributes 35% of discharge from just 17% of glacierized areas (Collins and Hasnain 1994; Singh and Jain 2002). At Gangotri Glacier, the largest in the Garhwal Himalaya, discharge figures were 565×10^6 cubic meters and 479×10^6 cubic meters in 1999 and 2000, respectively (Kumar et al. 2021a, b).

Similarly, the Dokriani Glacier study displayed an annual discharge of 52.3×10^6 cubic meters in 1998, 42.7×10^6 cubic meters in 1999, and 56.1×10^6 cubic meters in 2000, with 60% of the yearly discharge occurring solely in July and August (Thayyen et al. 2007). The dynamics differ between the western and eastern Himalayas, with

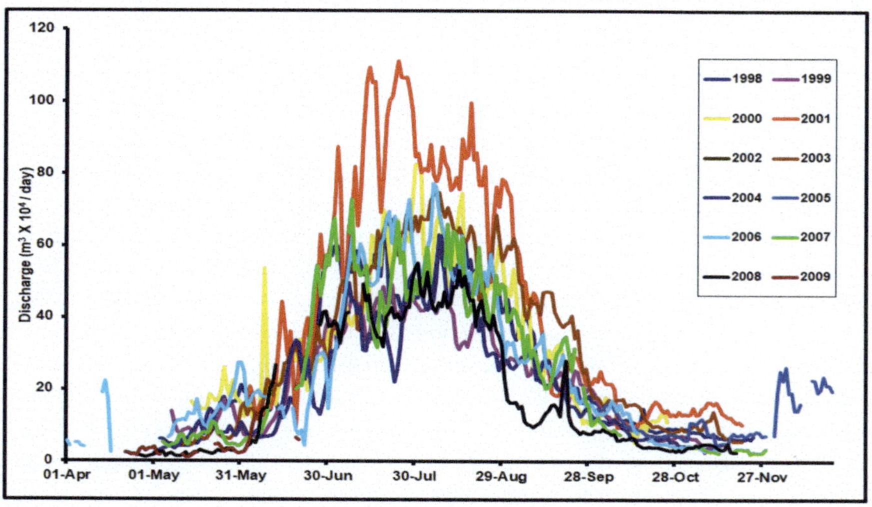

Fig. 8.3 Contribution of discharge from glacierized area from 1998 to 2009. An example from Dokriani Glacier, Bhagiarthi river basin, Garhwal Himalaya (WIHG)

Table 8.4 Significance of hydrological potential of snow and glacier melt contribution from Indian Himalayan Region (IHR)

Rivers	Site	Average snow and glacier melt contribution to annual flows (%)	References
Chenab	Akhnoor	49	Singh et al. (1997)
Ganga	Devprayag	28	Singh et al. (1993)
Ganga	Rishikesh	40	Mauraya et al. (2010)
Satluj (Indian Part)	Bhakra Dam	68	Singh and Jain (2002)
Beas	Pandoh Dam	35	Kumar et al. (2007)
Dhauliganga	Tapovan	77	Arora et al. (2010)
Din gad River (Dokriani) Glacier)	Snout	57–86	Thayyen et al. (2007)
Bhagirathi	Bhojwasa	~97	Singh et al. (2008)

predominant runoff originating from snow and glacier melt in the former, and the opposite in the latter.

In essence, the Himalayan region's frozen water reserves significantly shape the hydrology of its river systems, with meltwater contributions from snow, glaciers, and other sources being crucial for maintaining river flow and meeting the water needs of downstream communities. The snow/ice melt contribution from different glacier basins is shown in Table 8.4.

8.6 Climate Change and Its Impact on the Glacier Regime

In recent times, discernible fluctuations in air temperature have become apparent across the studied region. Within the central and eastern Himalayan realm, Shrestha et al. (1999) documented an upward trajectory in temperature trends, spanning from 0.06 to 0.12 °C per annum, encompassing the majority of the Nepal Himalaya. Expanding upon this, Bhutiyani et al. (2007) highlighted a pronounced escalation in atmospheric temperature within the northwestern Himalayan expanse over the past century. Analyzing the temporal span from 1988 to 2008, Shekhar et al. (2010) observed a notable amplification in both the highest and lowest temperature values, manifesting as an approximate augmentation of 1 °C and 3.4 °C, respectively, across the Himalayan domain. Concurrently, this thermal upswing coincided with diminishing trends in precipitation levels, in tandem with a decline in cumulative snowfall during seasonal intervals (Shekhar et al. 2010; Dimri and Kumar 2008). Diverging from this pattern, the Karakoram Range experienced an opposing trajectory, characterized by a decrease in maximum and minimum temperatures by 1.6 °C and 3.0 °C, respectively (Shekhar et al. 2010), concomitant with an elevation in winter precipitation rates (Fowler and Archer 2006).

The patterns of glacier accumulation and ablation exhibit distinct variations in both temporal and spatial dimensions across the designated area. To elucidate, in the eastern sector, the summer season typically manifests a juxtaposition of maximal accumulation and intensified melting processes. This phenomenon is marked by the accumulation peak at elevated altitudes and the concomitant occurrence of melt processes at lower elevations. Conversely, the western reaches demonstrate a prevailing trend of summertime melting and wintertime accumulation. Notably, the glacier termini in the western region tend to extend to lower altitudes, approximately around 2,500 m, whereas in the eastern and central sectors, this extension occurs at notably higher elevations, ranging from 3,500 to 4,500 m. This variance is chiefly attributed to the comparatively lower temperatures prevalent at the higher latitudes of the more westerly mountain ranges (Armstrong 2011).

Research conducted over the past century has established a significant correlation between the phenomenon of climate change and the process of glacier retreat, as documented by studies such as those by Haeberli et al. (2005) and Kaser et al. (2006). This connection is particularly pronounced in the eastern and central sectors of the Himalaya, as demonstrated by the investigations carried out by Kumar et al. (2021a, b), Mehta et al. (2021a, b), and Mehta et al. (2023), which have consistently identified escalated rates of glacier retreat in these regions. Conversely, in the northwestern Himalayan realm, there has been an intriguing observation of amplified expansion and thickening of larger glaciers, particularly within the Karakoram range, starting from the 1990s. This intriguing phenomenon has been accompanied by a notably elevated occurrence of glacier surges, as expounded upon by Hewitt (2005), Quincey et al. (2011), Bolch et al. (2012), and Bhambri et al. (2022). Moreover, studies that have harnessed the power of multi-sensor remote sensing analysis and data derived from the Gravity Recovery and Climate Experiment (GRACE) have also contributed to our understanding of these glacier dynamics. These investigations have reported anomalous positive mass balance outcomes in the Karakoram region for the same period under scrutiny, as discussed in studies by Bishop et al. (2008) and Jacob et al. (2012).

The alteration in global climate patterns coupled with the ongoing reduction in glacier size is exerting a direct influence on the Earth's landscape and posing a substantial threat to water supplies on a global scale. The sustained regression of glaciers around the world throughout the twentieth century remains an ongoing phenomenon. Temporal sequences of cumulative changes in glacier length and volume serve as empirical indicators of climate transformations at the planetary level. The aforementioned elucidation incontrovertibly underscores the pivotal role of climate change in steering the growth and contraction of glaciers, consequently warranting their designation as a quantifiable indicator for assessing climatic conditions across both spatial and temporal dimensions. This interplay between climate dynamics and glacier behavior stands as a salient concern of international significance. Alterations in the prevailing climate patterns are unmistakably mirrored in the mass and length alterations of glaciers. The oscillation between glacial advancement and retreat yields noteworthy repercussions that reverberate across the glacier's surroundings, exerting discernible impacts on its immediate environment and the broader physiographic context. Interpolated gridded data (30.755 N and 79.250 E) derived from CRU-TS 4 for 117 years (1901–2018) show a continuous increase in temperature of the accumulation periods (January to April and November to December) since 1901, with the insignificant change in the ablation periods (May–October) (Fig. 8.4).

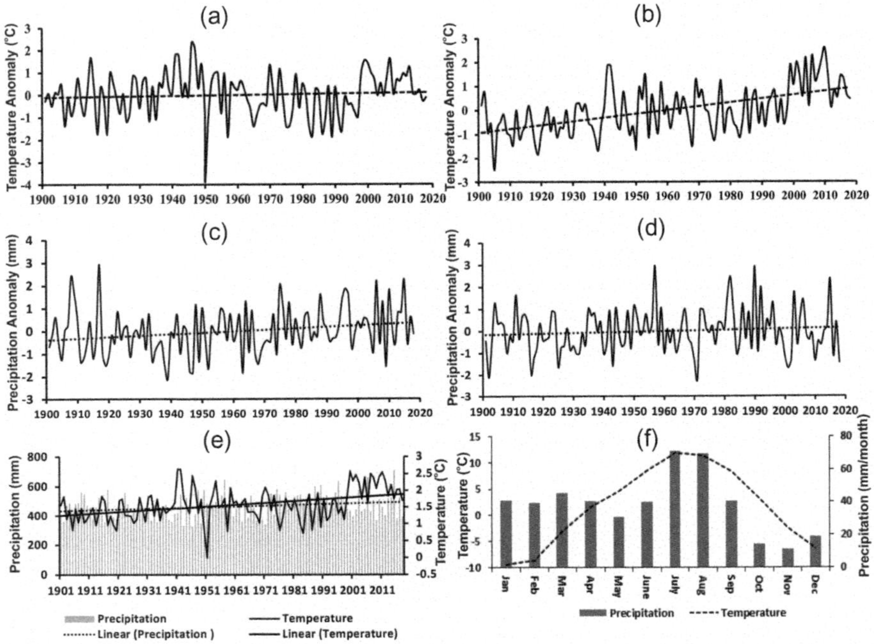

Fig. 8.4 Annual and seasonal variability in the climatic conditions of the study area derived from CRU-TS 4 data (30.755N and 79.250E) for the period 1901–2018. **a** Ablation periods temperature anomaly showed almost similar trend throughout the period 1901–2018 (Summer average 7.7 °C; SD 0.54, between 1901 and 2018). **b** The accumulation period temperature anomaly (1901–2018) depicts the gradual increase in the temperature trends (Winter average –4.5 °C; SD 0.6; between 1901 and 2018). **c** Ablation periods precipitation anomaly showed a slightly increased trend with an average precipitation of 292 ± 60 mm during the period between 1901 and 2018. **d** The accumulation period precipitation anomaly showed not much changes, and average precipitation between 1901 and 2018 was 170 ± 46 mm. **e** Annual average temperature and total annual precipitation of the region data suggested that the maximum average annual temperature was recorded in 1942 (2.6 °C) and the minimum annual average temperature was recorded in 1951 (–0.07 °C), the annual average precipitation between 1901 and 2018 was about 462 mm, while the maximum precipitation occurred in 2015 (709 mm) and minimum precipitation occurred in 1946 (323 mm). **f** The Zanskar area's monthly mean precipitation and temperature data between 1901 and 2018. The figure shows that the minimum (–10 °C) and maximum (18 °C) temperatures occurred in January and July, respectively (*Figure modified after* Mehta et al., 2021a, b)

8.7 Status of Himalayan Glaciers

Most widely reported, mountain glaciers are currently experiencing a period of recession, a trend that began in the mid-nineteenth century with the end of the Little Ice Age (Kamp et al. 2011; Bloch et al. 2012). The rate of recession continued through the twentieth century and has accelerated over the past three decades (Barry 2006; Kolbert 2006; WGMS 2008; Kamp et al. 2011; Bolch et al. 2012). There is clear evidence that the retreat of glaciers in many locations of the world has accelerated in recent decades (Zemp et al. 2008; Cogley 2011). However, glacier systems of the Himalaya have not responded uniformly to recent climate warming; some glaciers of western Karakoram Himalaya are advancing (Hewitt 2005; Gardelle et al. 2012; Bhambri et al. 2012), while some glaciers of central and eastern Himalayas are retreating (Bolch et al. 2012 cited Kääb et al. 2012).

The Himalayan Mountain Range contains thousands of glaciers of widely varying properties, which are spread over nearly 37 000 km^2 and 2400 km of east–west range (Raina and Srivastava 2008). This large geographic extent, complex topography, and variable climatic conditions across the Himalayas have resulted in various sets of glacier properties. The primary climatic factor from west to east is that of a decreasing influence by the mid-latitude westerlies and an increasing influence by the Indian monsoon (Bookhagen and Burbank 2010). Thus, the distribution of glaciers in the Himalaya is uneven, with a higher concentration of glaciers in the NW than in the NE of the mountain range. Glaciers located in the central Himalaya are influenced by the westerlies and the Indian summer monsoon (ISM) (Armstrong 2011). In general, ~70% of the glaciers' ablation zone is covered by debris with thickness ranging from millimeters to tens of centimeters (Rain and Srivastava 2008; Scherler et al. 2011; Kamp et al. 2011). Due to the thick debris covered and with a low gradient of the termini, many glaciers typically have a stable front position. As a consequence of this complex climate system, glacial geometry, properties, and geology, the recession rates of the glaciers are variable (Scherler et al. 2011).

The glacier surface change measured for more than 80 glaciers in central western Himalaya suggested that most of the central Himalayan Glaciers have been retreating since the end of the nineteen century (Bhambri et al. 2011). Bhambri et al. (2011) reported 82 glaciers in the Alaknanda and Bhagirathi River Basin of central Himalaya using remote sensing data and archival records for the period 1968–2006 indicate that there has been a 4.2% reduction in the glacierized area. The study also indicates that around 6% reduction was associated with the Alaknanda Basin, whereas around 3% in the Bhagirathi Basin. Similar observations were made in Chenab, Parbati, and Baspa Basin (466 glaciers), which show an overall 21% deglaciation from 1962 to 2001 (Kulkarni et al. 2005). Other satellite-based studies carried out in the Gori Ganga Basin, Millam Glaciers between 1954 and 2006 showed that glaciers receded 1328 m with an average of ~25 m a^{-1} (Raj 2011).

For instance, of the nearly 10,000 glaciers in the Indian Himalaya, only eleven have been studied in detail for mass balance and little more than a hundred glaciers are being monitored for terminus fluctuations (Dobhal et al. 2008, 2013, 2021; Mehta et al. 2011, 2013, 2014, 2021a, b, 2023; Bhambri and Bolch 2009; Shukla and Quadir 2016; Kumar et al., 2017, 2021a, b; Rana et al. 2023). Many debris-covered glaciers are retreating relatively slowly, e.g., Chorabari (6 m a^{-1}), Dunagiri (3 m a^{-1}), Shankulpa (6.8 m a^{-1}), Bhagirathi Kharak (1.5 m a^{-1}), Pensilungpa (6 m a^{-1}) but many others are retreating at faster rates of up to 25 m a^{-1} (Vohra, 1981; Swaroop et al. 2001; Nainwal et al., 2008; Raina and Srivastava 2008; Dobhal et al. 2013; Kumar et al. 2017; Mehta et al. 2011, 2023). Studies carried out by numerous authors during the last hundred years provide a link between climate change and glacier recession (Haeberli et al. 2005; Kaser et al. 2006; Bolch et al. 2012; Yao et al. 2012; Shukla et al. 2017,). Enhanced retreat rates have been observed in central and adjacent parts of the Himalaya. In the course of the last 30 years, glaciers lost ~5.5% of volume in Nepal, and similar results have also been observed in India, Bhutan, and China (Bajracharya et al. 2007).

These variations in frontal recession may or may not reflect changes in mass, the critical parameter for long-term glacier health. The lack of data on volume change and ice thickness studies in the Himalaya has hampered the proper evaluation of glacier mass change in the region. Field-based retreat trends of some Himalayan Glaciers are given in Table 8.5.

8.8 Mass Balance of the Himalayan Glacier

In-situ studies conducted across the Himalayas indicate a declining trend and fluctuating rates of mass loss, as demonstrated in Table 8.6. The average annual specific balance of Himalayan Glaciers falls within a wide range from –0.03 to –1.43 m w.e. a^{-1}. This variability in specific balance can be attributed to factors such as glacier geometry (including size, shape, aspect, hypsometry, and length) and variations in debris cover, whether thick, patchy, or clean. Despite this variability, assessments combining in-situ and geodetic mass balance data from the entire Hindu Kush-Karakoram-Himalayan (HKH) region indicate a diverse range of behaviors, as highlighted in studies by Cogley (2017), Kääb et al. (2012), Gardelle et al. (2012), Bolch et al. (2012), and Dobhal et al. (2013).

Previous research indicates that the distinct climatic conditions of the High Mountain of Asia (HMA) result in varied responses among glaciers (Kääb et al. 2012; Gardelle et al. 2012; Maussion et al. 2014). Brun et al. (2017) reported an average specific balance of -0.18 ± 0.04 m w.e. a^{-1}, across the entire HMA during 2000–2016, with greater mass loss along the Himalayan range (-0.41 ± 0.20 m w.e. a-1). Particularly negative specific balances were observed in the Langtang (–2.0 m w.e. a^{-1}), Everest (–1.5 m w.e. a^{-1}), and Kanchenjunga (–1 m w.e. a^{-1}) regions.

However, Gardelle et al. (2012) suggested a positive specific mass balance (0.11 ± 0.22 m w.e. a^{-1}) for central Karakoram Glaciers between 1999 and 2008. Bhushan et al. (2018) estimated the mass budget of ten glaciers in the Zanskar Basin of the western Himalaya, reporting a regional glaciers' mean specific balance of -0.38 ± 0.09 m w.e. a^{-1} for the period 1999–2014, with individual glaciers ranging between

Table 8.5 Retreat trends of glacier snouts of the Himalayan glacier. Fields were surveyed and monitored in different time periods

Name glacier	Observation period	No of year	Total recession (m)	Average rate (m/y)	References
Kolai (J&K)	1912–1961	49	800	16.3	Vohra (1981)
Pensilugpa (Ladakh)	1971–2019	48	270	5.6	Mehta et al. (2021a, b)
Parkachik (Ladakh)	1971–2021	50	210	4	Rana et al. (2023)
Durung-Drung (Ladakh)	1971–2019	48	624	13	Mehta et al. (2023)
Chhota Shigri (HP)	1987–1989	3	54	18.5	Kumar and Dobhal (1994a, b)
Gepang Gath Glacier	1989–2017	28	846	30	Kumar et al. (2021a)
Bandarpunch (U.K.)	1960–1999	39	995	25.51	Shukla et al. (2000)
Jaundar Bamak (U.K.)	1975–2010	35	1399	40	Mehta et al. (2012)
Jhajju Bamak (U.K.)	1962–2010	48	700	13.46	Mehta et al. (2012)
Gangotri Glacier (U.K)	1935–2004	69	1519.13	22.02	Vohra (1981), Kumar et al. (2008)
Dokriani Glacier (U.K.)	1962–2007	45	751.35	16.7	Dobhal and Mehta (2010)
Chorabari Glacier (U.K.)	1962–2010	48	327	6.8	Dobhal, et al. (2013)
Satopanth (U.K.)	1962–2006	44	1163.65	26.45	Nainwal et al. (2008)
Tipra Glacier (U.K.)	1962–2008	46	663	14.41	Mehta et al. (2012)
Dunagiri (U.K.)	1992–1997	5	15	3	Swaroop et al. (2001)
Uttari Nanda Devi	1980–2017	37	797	22	Kumar et al. (2021b)
Changbang	1980–2017	37	620	17	Kumar et al. (2021b)
Ramni	1980–2017	37	1138	32	Kumar et al. (2021b)
Bethartoli	1980–2017	37	358	10	Kumar et al. (2021b)
Trishul	1980–2017	37	777	21	Kumar et al. (2021b)

(continued)

Table 8.5 (continued)

Name glacier	Observation period	No of year	Total recession (m)	Average rate (m/y)	References
Dakshini Nanda Devi	1980–2017	37	603	16	Kumar et al. (2021b)
Raunthi	1980–2017	37	565	16	Kumar et al. (2021b)
Pindar Glacier (U.K)	1845–1966	121	2840	23.4	Vohra (1981)
Milam Glacier (U.K.)	1968–2017	49	1565	32	Mal et al. (2019)
Shankulpa (U.K.)	1881–1957	76	518	6.8	Vohra (1981)
Lumding (Nepal)	1960–2001	41	1353	33	Bajracharya et al. (2007)
Chhule (Nepal)	1960–2001	41	772	18.82	Bajracharya et al. (2007)
Bhote Koshi (Nepal)	1960–2001	41	645	15.73	Bajracharya et al. 2007
Khumbu (Nepal)	1960–2001	41	842	20.53	Bajracharya et al. (2007)
Imja (Nepal)	1960–2001	41	2340	57.07	Bajracharya et al. (2007)
Raphsthreng Glacier (Bhutan)	1984–1998	14	490	35	Mool et al. (2001)
Luggye Glacier (Bhutan)	1988–1993	5	800	160	Mool et al. (2001)

-0.13 ± 0.07 and -0.66 ± 0.09 m w.e. a^{-1}. Moreover, Bhushan et al. (2018) found the mean specific balance of the Pinsilungpa Glacier to be -0.53 m w.e. a^{-1} between 1999 and 2014. Whereas, (Mehta et al., 2021a, b), estimated the average net annual mass balance of Pensilungpa Glacier was ~– 3.67×10^6 m^3 w.e. a^{-1} with the –0.36 m w.e. a^{-1} specific balance between 2016 and 2019. Overall, glaciers across the Himalayan ranges are experiencing variable rates of wastage, with the annual mean net mass balance ranging between -0.07×106 and -8.6×106 m^3 w.e. a^{-1} (Table 8.6).

8.9 Summary and Conclusions

The Himalaya have the greatest concentration of glaciers outside the polar region. Meltwater from these glaciers is the storehouse of rivers originating from the Himalaya, which control the lifeline of the large part of the Indian landmass.

Table 8.6 Annual specific balance and mass loss of the Himalayan glaciers in different regions of India (*Modified after* Mehta et al. 2021a, b)

Sn	Glacier name	Period of observation	Glacier area (km^2)	Specific balance (m w.e. a^{-1})	Net mass bal. (10^6 m^3 w.e. a^{-1})	Region	References
1	Neh Nar	1975–1984	1.25	–0.54	–0.67	Kashmir	Raina and Srivastava (2008), Raina (2009)
2	Kolahoi	1983–1984	11.9	–0.27	–3.2	Kashmir	Kaul (1986)
3	Ruling	1979–1981	1.1	–0.11	–0.12	Kashmir	Srivastava et al. (2001)
4	Pensilungpa	2016–2019	10.62	–0.36	–3.96	Kashmir	Mehta et al. (2021a, b)
5	Nardu	2000–2003	4.5	–0.39	–1.8	Himachal	Koul and Ganjoo (2010)
6	Gara	1974–1982	5.19	–0.37	–1.94	Himachal	Raina (2009)
7	Gor Gorang	1976–1984	2.02	–0.43	–0.87	Himachal	Raina (2009)
8	Shaune Gorang	1981–1991	4.98	–0.4	–2.0	Himachal	Raina (2009)
9	Chhota Shigri	2002–2014	15.5	–0.56	–8.6	Himachal	Wagnon et al. (2007), Azam et al. (2012, 2016)
10	Hamtha	2000–2012	3.2	–1.43	–4.6	Himachal	Mishra et al. (2014)
11	Dunagiri	1984–1990	4.39	–0.60	–2.66	Uttarakhand	Raina (2009)
12	Dokriani	1992–2013	7.00	–0.32	–2.25	Uttarakhand	Dobhal et al. (2021)
13	Chorabari	2003–2010	6.60	–0.73	–4.4	Uttarakhand	Dobhal et al. (2013)
14	Gangu La	2012–2014	0.3	–1.38	–0.39	Bhutan	Tshering and Fujita (2016)
15	AX010	1995–1999	0.6	–0.69	–0.41	Nepal	Fujita et al. (2001)
16	Mera	2007–2015	5.1	–0.03	–0.15	Nepal	Wagnon et al. (2013), Sherpa et al. (2017)
17	Pokalade	2009–2015	0.1	–0.69	–0.07	Nepal	Wagnon et al. (2013), Sherpa et al. (2017)

(continued)

Table 8.6 (continued)

Sn	Glacier name	Period of observation	Glacier area (km^2)	Specific balance (m w.e. a^{-1})	Net mass bal. (10^6 m^3 w.e. a^{-1})	Region	References
18	Yala	2011–2012	1.6	–0.89	−1.4	Nepal	Baral et al. (2014)
19	Changme Khangpu	1979–1983	5.6	–0.27	−1.5	Sikkim	Raina (2009)

Glaciers are the product of climate and climate change and therefore are considered very sensitive to climate. During the Pleistocene (about 2 million years), the glaciers occupied about 30% of the total area of the earth as against 10% at the present (https://www.britannica.com/science/Pleistocene-Epoch/Pleistocene-events-and-environments). According to an estimate, there are about ten thousand glaciers *in* the Indian part of the Himalaya (Raina and Srivastava 2008). These glaciers are the perennial source of water to the Indus, Ganga, and Brahmaputra Rivers which are the life-sustaining fluvial systems of the large part of the Indian landmass. The Himalaya cover one of the most dynamic and complex mountain ranges in the world due to tectonic activity, and they are vulnerable to global warming and increasing human activities. Recent studies on Himalayan glaciers show that most of the glaciers are shrinking. The continuous shrinking of glaciers and snow cover due to climate change (warming) and potential impacts on the ecological and socioeconomic landscape is a matter of wide discussion.

Fluctuations in glaciers, especially retreat and thinning, may give rise to catastrophic hazards, like avalanches, rockfall, and landslides. Excessive melt waters, often in combination with liquid precipitation, may trigger flash floods or debris flows. Glacial surges are another hazard in the western Himalayan mountains. In the eastern and central Himalayas, where glacier surfaces are heavily covered with debris, the melting of glaciers due to climate change has resulted in the formation of glacial lakes in exposed areas behind end moraines, raising significant alarm. Many of these high-altitude lakes pose potential hazards. The moraine dams enclosing them are relatively fragile and could rupture suddenly, unleashing large quantities of water and debris in a glacial lake outburst. Such events can trigger devastating floods downstream, causing severe damage to life, property, forests, agricultural lands, and infrastructure. The retreat of glaciers is expected to have profound socioeconomic repercussions in mountainous regions, although the precise local effects are uncertain and will vary. Areas losing substantial glacier coverage will undergo changes in hydrology. Initially, melting glaciers will increase water runoff, but as glacier ice diminishes, runoff will decline. Additionally, due to the slow revegetation of high-altitude terrain, deglaciated areas will face erosion and reduced stability, emphasizing the need to safeguard roads, communication networks, and other structures. Concerns for regions reliant on tourism include uncertain snow cover during peak winter sports seasons, natural hazards like rock and ice falls, and the loss of scenic beauty.

In summary, it can be suggested that current inferences on the health of glaciers in the face of global warming remain uncertain. Glacial retreat primarily stems from rising temperatures, influenced by precipitation levels. Besides climate change, factors like topography, glacier size, and type also impact retreat rates. A comprehensive regional study is needed to establish a clearer link between glacier recession and recent climate shifts. However, given the current climate trajectory, Himalayan Glaciers are expected to persist in their degeneration, losing ice volume over time.

Acknowledgements The author is grateful to the Director of the Wadia Institute of Himalayan Geology, Dehradun, for their encouragement and for providing the necessary facilities. I deeply acknowledge Dr. Vinit Kumar for his assistance in carrying out this work.

References

Ahlmann HW (1948) Glaciological research on the North Atlantic coast. Rece Geol Soci Sr:1–83

Aniya M, Wakao Y (1997) Glacier variations of Heilo Patagonico Norte, Chile between 1945–46 and 1995–96. Bull Glacier Res 15:11–18

Arora M, Rathore D, Singh R, Kumar R, Kumar A (2010) Estimation of melt contribution to total streamflow in River Bhagirathi and River DhauliGanga at Loharinag Pala and Tapovan Vishnugad Project Sites. J Water Res Prot 2:636–643. https://doi.org/10.4236/jwarp.2010.27073

Armstrong RL (2011) The glaciers of the Hindu Kush-Himalayan region – a summary of the science regarding glacier melt/retreat in the Himalayan, Hindu Kush, Karakoram, Pamir, and Tien Shan mountain ranges. ICIMOD, Kathmandu, Nepal, pp 17

Azam MF, Wagnon P, Ramanathan A, Vincent C, Sharma P et al (2012) From balance to imbalance: a shift in the dynamic behaviour of Chhota Shigri Glacier (Western Himalaya, India). J Glaciol 58:315–324

Azam MF, Ramanathan AL, Wagnon P, Vincent C, Linda A et al (2016) Meteorological conditions, seasonal and annual mass balances of Chhota Shigri Glacier, western Himalaya, India. Ann Glaciol 57:328–338

Bahadur J (1992) Snow and Glacier and their contribution to India's water resource. Water Sci Educ Ser, (1):72

Bahadur J (1998) Himalayan eco-hydrology—an emerging topic. J Indian Assoc Hydrol, Roorkee

Bajracharya SR, Mool PK, Shrestha BR (2007) Impact of climate change on Himalayan glaciers and glacial lakes: Case studies on GLOF and associated hazards in Nepal and Bhutan. International Centre for Integrated Mountain Development (ICIMOD), Publication 119, Kathmandu

Baral P, Kyastha RB, Immerzeel WW, Pradhananga NS, Bhattaria C, Shahi S, Galos S, Springer C, Joshi SP, Mool PK (2014) Preliminary results of mass-balance observations of Yala Glacier and analysis of temperature and precipitation gradients in Langtang Valley, Nepal. Ann Glaciology. 55:9–14

Barry R (2006) The status of research on glaciers and global glacier recession: A review. Prog Phys Geogr 30:285–306

Bhambri R, Bolch T (2009) Glacier mapping: a review with special reference to the Indian Himalayas. Prog Phys Geogr 33(5):672–704

Bhambri R, Bolch T, Chaujar RK, Kulshreshtha SC (2011) Glacier changes in the Garhwal Himalayas, India 1968–2006 based on remote sensing. J Glaciol 57(203):543–556

Bhambri R, Bolch T, Chaujar RK (2012) Frontal recession of Gangotri Glacier, Garhwal Himalayas from 1965 to 2006, measured through high resolution remote sensing data. Curr Sci 102(3):489–494

Bhambri R, Hewitt K, Haritashya UK, Chand P, Kumar A (2022) Characteristics of surge-type tributary glaciers, Karakoram. Geomorphology 403:108161

Bishop MP, Bush AB, Collier E, Copland L, Haritashya UK, John SF, Swenson SC, Wahr J (2008) Advancing glaciers and positive mass anomaly in the Karakoram Himalaya, Pakistan. In: Paper presented at the American Geophysical Union fall meeting, 15–19 December 2008, San Francisco, USA

Bhutiyani MR, Vishwas SK, Pawar NJ (2007) Longterm trends in maximum, minimum and mean annual air temperatures across the Northwestern Himalaya during the twentieth century. Clim Change 85:159–177

Bolch T, Kulkarni A, Kääb A, Huggel C, Paul F, Cogley G, Frey H, Kargel JS, Fujita K, Scheel M, Bajracharya S, Stoffel M (2012) The state and fate of Himalayan glaciers. Science 336(6079):310–314. https://doi.org/10.1126/science.1215828

Bhushan S, Syed TH, Arendt AA, Kulkarni AV, Sinha D (2018) Assessing controls on mass budget and surface velocity variations of glaciers in Western Himalaya. Sci Rep 8(8885)

Bookhagen B, Burbank DW (2010) Towards a complete Himalayan hydrologic budget: The spatiotemporal distribution of snow melt and rainfall and their impact on river discharge. J Geophys Res 115:F03019

Brun F, Berthier E, Wagnon P, Kääb A, Treichler D (2017) A spatially resolved estimate of high mountain Asia glacier mass balance from 2000 to 2016. Nat Geosci 10:468–474

Choudhury S, Phukan S, Duarah BP, Goswami DC, Mehta M (2022) Glacier inventory of the Subansiri River Basin in the Brahmaputra catchment using Landsat satellite data: A case study of the Daisaphu Glacier changes. Geol J 57(12):4939–4954

Cogley JG (2011) Present and future states of Himalaya and Karakoram glaciers. Ann. of Glaciology 52(59):69–73

Cogley JG (2017) Present and future states of Himalaya and Karakoram glaciers. Ann Glaciol 52:69–73

Collins DN, Hasnain SI (1994) Runoff and sediment transfer from the glacierised basins at Himalayan scale, Alpine glacier project working paper, 15 University of Manchester, U.K, p 14

Cruz RV, Harasawa H, Lal M, Others (2007) Asian climate change 2007: Impact adaptation and vulnerability. In: Contribution of working group II to 4Th Assessment report of the IPCC, Cambridge University press Cambridge UK, pp 469–506

Charlsworth HW (1948) Quternary era. Edward Arnold, London

Dimri AP, Kumar A (2008) Climatic variability of weather parameters over the western Himalayas: a case study. In: Proceedings of the national snow science workshop, 11–12 January (2008), Chandigarh, India. Snow and Avalanche Study Establishment

Dobhal DP, Gergan JT, Thayyen RJ (2008) Mass balance studies of Dokriani glacier from 1992 to 2000, Garhwal Himalaya, India. Bull Glaciol Res, Jpn Soc Snow Ice 25:9–17

Dobhal DP, Mehta M, Srivastava D (2013) Influence of debris cover on terminus retreatand mass changes of Chorabari Glacier, Garhwal region, central Himalaya, India. J Glaciol 59(217):961–971

Dobhal DP, Pratap B, Bhambri R, Mehta M (2021) Mass balance and morphological changes of Dokriani Glacier (1992–2013), Garhwal Himalaya, India. Quat Sci Advance 4:1–9

Dobhal DP, Mehta M (2010) Surface morphology, elevation change and terminus retreat of Dokriani Glacier, Garhwal Himalaya: Implication for climate change. Himalayan Geology 31:71–78

Flint RF (1972) Glacial and Pleistocene geology. John Wiley and Sons, New York, p 555

Fowler HJ, Archer DR (2006) Conflicting signals of climate change in the upper Indus basin. Jounral Clim 19(17):4276–4293

Fujita K, Kadota T, Rana B, Kayastha RB, Ageta Y (2001) Shrinkage of Glacier AX010 in Shorong region, Nepal Himalayas in the 1990s. Bull Glaciol Res 18:51–54

Garg PK, Shukla A, Tiwari RK, Jasrotia AS (2017) Assessing the status of glaciers in part of the Chandra basin, Himachal Himalaya: A multiparametric approach. Geomorphology 284:99–114. https://doi.org/10.1016/j.geomorph.2016.10.022

Garg S, Shukla A, Mehta M, Kumar V, Samuel SA, Bartarya S, Shukla UK (2018) Field evidences showing rapid frontal degeneration of the Kangriz glacier, Suru river valley, Jammu and Kashmir. J Mt Sci 15:1199–1208
Gardelle J, Berthier E, Arnaud Y (2012) Slight mass gain of Karakoram glaciers in the early twenty-first century. Nat Geosci 5:322–325
Gautam CK, Mukherjee BP (1989) Mass balance vis-à-vis snout position of Tipra Bank Glacier, district Chamoli, U.P. Proceeding Natl Meet Himal Glaciol:141–148
Haeberli W, Mullar P, Alean P, Bosch H (1989) Glacier changes following the little Ice age-A survey of the international data basis and its perspectives. In: Oerlemans J (ed) Glacier fluctuations and climate change. Academic Publishers, Dordrecht, etc., Kulwer, pp 77–101
Haeberli W, Zemp M, Hoelzle M, Frauenfelder R, Hoelzle M and Kaab A (2005) Fluctuations of glaciers 1995– 2000 vol 8, Intl Comm on snow and ice, Intl Assoc of Hydrol Sci. UNESCO, Paris. (http://www.geo.unizh.ch/wgms)
Hewitt K (2005) The Karakoram anomaly? Glacier expansion and the 'elevation effect, Karakoram Himalaya. Mt Res Dev 25:332–340
Immerzeel WW, Beek LPHV, Bierkens MFP (2010) Climate change will affect the Asian water towers. Science 328:1382–1385
ICIMOD (2011) The status of glaciers in the Hindu Kush-Himalaya region (eds. Bajracharya and Shrestha), Kathmandu, pp 127
Jacob T, Wahr J, Pfeffer WT, Swenson S (2012) Recent contributions of glaciers and ice caps to sea level rise. Nature 482:514–518
Kaser G, Cogley JG, Dyurgerov MB, Meier MF, Ohmura A (2006) Mass balance of glaciers and ice caps: Consensus estimates for 1961–2004. Geophys Res Lett 33:1–5
Kamp U, Byrne M, Bolch T (2011) Glacier Fluctuation between 1975 and 2008 in the Greater Himalaya range of Zaskar, Southern Ladakh. J Mt Sci 8:374–389
Kääb A, Berthier E, Nuth C, Gardelle J, Arnaud Y (2012) Contrasting patterns of early twenty-first-century glacier mass change in the Himalayas. Nature 488:495–498
Kaul MN (1986) Mass balance of Liddar glaciers. Trans Inst Indian Geogr 8:95–111
Koul MN, Ganjoo RK (2010) Impact of inter-and intra-annual variation in weather parameters on mass balance and equilibrium line altitude of Naradu Glacier (Himachal Pradesh), NW Himalaya, India. Climatic Change 99(1–2)
Kolbert E (2006) Field notes from a Catastrophe: Man, nature, and climate change. Bloomsbury, New York
Kumar S, Dobhal DP (1994a) Snout fluctuation study of Chhota-Shigri glacier, Lahaul and Spiti district, Himachal Pradesh. J Geol Soc India 44:581–585
Kumar K, Dumka RK, Miral MS, Satyal GS, Pant M (2008) Estimation of retreat rate of Gangotri glacier using rapid static and kinematic GPS survey. Curr Sci 94(2):258–262
Kumar V, Shukla T, Mehta M, Dobhal DP, Bisht MPS, Nautiyal SK (2021a) Glacier changes and associated climate drivers for the last three decades, Nanda Devi region, Central Himalaya, India. Quat Int 575–576(20):213–226
Kumar V, Mehta M, Shukla T (2021b) Spatially resolved estimates of glacial retreat and lake changes from Gepang Gath Glacier, Chandra Basin, Western Himalaya, India. J Geol Soc India 97:520–526
Kumar V, Mehta M, Mishra A, Trivedi A (2017) Temporal fluctuations and frontal area change of Bangni and Dunagiri glaciers from 1962 to 2013, Dhauliganga Basin, central Himalaya, India. Geomorphology 284:88–98
Kumar S, Dobhal DP (1994b) Snout fluctuation study of Chhota-Shigri glacier, Lahaul and Spiti District, Himachal Pradesh. Jour Geol Soc India 44:581–585
Kumar V, Singh P, Singh V (2007) Snow and glacier melt contribution in the Beas River at Pandoh Dam, Himachal Pradesh, India. Hydrol Sci J 52:376–388
Kulkarni AV, Rathore BP, Mahajan S, Mathur P (2005) Alarming retreat of Parbati Glacier, Beas basin, Himachal Pradesh. Current Sci 88(11):1844–1850

Liu X, Chen B (2000) Climatic warming in the Tibetan Plateau during recent decades. Int J Climatology 20:1729–1742

Mal S, Mehta M, Singh RB, Schickhoff U, Bisht MPS (2019) Recession and morphological changes of the debris-covered Milam Glacier in Gori Ganga valley, Central Himalaya, India, derived from satellite data. Front Environ Sci 7:1–17

Mason K (1930) The Glaciers of the Karakoram and neighbourhood. Rec Geol Surv India, 63:214–278

Maussion F, Scherer D, Mölg T, Collier E, Curio J et al (2014) Precipitation seasonality and variability over the Tibetan Plateau as resolved by the High Asia reanalysis. J Clim 27:1910–1927

Mayeswki PA, Jeschke PA (1979) Himalayan and trans-Himalayan glacier fluctuations since ad 1812. Arct Alp Res 11:267–287

Mehta M, Dobhal DP, Pratap B, Verma A, Kumar A, Srivastava D (2013) Glacier changes in Upper Tons River basin, Garhwal Himalaya, Uttarakhand, India. Z Für Geomorphol. https://doi.org/10.1127/0372-8854/2012/0095

Mehta M, Kumar V, Kunmar P, Sain K (2023) Response of the thick and thin debris-covered glaciers between 1971 and 2019 in Ladakh Himalaya, India—A case study from Pensilungpa and Durung-Drung glaciers. Sustainability 15:2–21

Mehta M, Dobhal DP, Kesarwani K, Pratap B, Kumar A, Verma A (2014) Monitoring of glacier changes and response time in Chorabari Glacier, Central Himalaya, Garhwal, India. Current Sci 107(2):281–289

Mehta M, Kumar V, Garg S, Shukla A (2021a) Little Ice Age glacier extent and temporal changes in annual mass balance (2016–2019) of Pensilungpa Glacier, Zanskar Himalaya. RegNal Environ Chang 21(38):21–38

Mehta M, Dobhal DP, Pratap B, Verma A, Kumar A, Srivastava D (2012) Glacier changes in Upper Tons River basin, Garhwal Himalaya, Uttarakhand, India. Z Geomorphol 57(2):225–244

Mehta M, Dobhal DP, Bhist MPS (2011) Change of Tipra Glacier in the Garhwal Himalaya, India, between 1962 and 2008. Prog Phys Geogr 35(6):721–738

Mehta M, Shukla T, Bhambri R, Gupta AK, Dobhal DP (2017) Terrain changes, caused by the 15–17 June 2013 heavy rainfall in the Garhwal Himalaya, India: A case study of Alaknanda and Mandakini basins. Geomorphology 284:53–71

Mehta M, Kumar V, Sain K, Tiwari SK, Kumar A, Verma A (2021b) Causes and consequences of Rishiganga flash flood, Nanda Devi Biosphere Reserve, central Himalaya, India. Curr Sci 121(11):1483–1487

Mishra R, Kumar A, Singh D (2014) Long term monitoring of mass balance of Hamtah Glacier, Lahaul and Spiti district, Himachal Pradesh. Geol Surv India 147(pt. 8):230–231

Mool PK, Bajracharya SR, Joshi SP (2001) Inventory of glacier, glacial lakes and glacial lake outburst floods monitoring and early warning systems in the Hindu Kush-Himalayan Region – Nepal, (Book chapter) pp 365

Mullar F, Caflish T, Mullar G (1977) Introduction of compilation and Assemblage of data for a World Glacier Inventory, Zurich, TTS for WGI Swiss Federal Institute of Technology, Zurich

Nainwal HC, Negi BDS, Chaudhary M, Sajwan KS, Gaurav A (2008) Temporal changes in rate of recession: Evidence from Satopanth and Bhagirath Kharak Glacier, Uttarakhand, using total station survey. Curr Sci 94:653–660

Pratap B, Dobhal DP, Bhambri R, Mehta M (2013) Near-surface temperature lapse rate in Dokriani Glacier catchment, Garhwal Himalaya, India. Himal Geol, 34 (2):183–186

Pratap B, Dobhal DP, Mehta M, Bhambri R (2015) Influence of debris cover and altitude on glacier surface melting: a case study on Dokriani Glacier, central Himalaya, India. Ann Glaciol 56(70):9–16

Quincey DJ, Braun M, Glasser NF, Bishop MP, Hewitt K, Luckman A (2011) Karakoram glacier surge dynamics. Geophys Res Lett 38:L18504. https://doi.org/10.1029/2011GL049004

Raj KBG (2011) Recession and reconstruction of Milam Glacier, Kumaon Himalaya, observed with satellite imagery. Curr Sci 100(9):1420–1425

Rana AS, Kunmar P, Mehta M, Kumar V (2023) Glacier retreat, dynamics and bed overdeepenings of Parkachik Glacier, Ladakh Himalaya, India. Ann Glaciol. https://doi.org/10.1017/aog.2023.50

Raina VK (2009) Himalayan glaciers. A state-of-art review of glacial studies, glacial retreat and climate change. Ministry of Environment and Forests, India, 56. Available at http://go.nature.com/pLgJ6D

Raina VK, Srivastava D (2008) Glacier Atlas of India. GSI Publication 315

Rowan AV (2017) The 'Little Ice Age' in the Himalaya: A review of glacier advance driven by Northern Hemisphere temperature change. The Holocene 27:292–308. https://doi.org/10.1177/0959683616658530

SAC (2016) Monitoring snow and glaciers of Himalayan Region, Space Applications Centre, ISRO, Ahmedabad, India, 413 p. ISBN: 978-93-82760-24-5

Scherler D, Bookhagen B, Strecker MR (2011) Spatially variable response of Himalayan glaciers to climate change affected by debris cover. Nat Geosci 4:156–159

Singh P, Kumar N (1997) Effect of orography on precipitation in the western Himalayan region. J Hydrol 199:183–206

Sharma MC, Owen LA (1996) Quaternary glacial history of NW Garhwal, Central Himalaya. Quatern Sci Rev 15:335–365

Singh P, Haritashya UK, Kumar N (2008) Modelling and estimation of different components of streamflow for Gangotri Glacier basin, Himalayas". Hydrol Sci J 53:309–322

Singh P, Jain SK (2002) Snow and glacier melt in the Satluj river at Bhakra Dam in the Western Himalayan region. Hydrol Sci J 47:93–106

Singh P, Jain SK, Kumar N, Singh UK (1993) Snow and glacier contribution in the Ganga river at Devprayag. CS(AR) 132, NIH, Roorkee, pp 1993–1994

Singh P, Jain SK, Kumar N (1997) Estimation of snow and glacier melt runoff contribution in the Chenab River at Akhnoor. Mt Res Dev 17:49–56

Shekhar MS, Chand H, Kumar S, Srinivasan K, Ganju A (2010) Climate change studies in the western Himalaya. Ann Glaciol 51(54):105–112

Sherpa SF, Wagnon P, Brun F, Berthier E, Vincent C, Lejeune Y, Arnadu Y, Kayastha RB, Sinisalo A (2017) Contrasted surface mass balances of debris-free glaciers observed between the southern and the inner parts of the Everest region (2007–15). J Glaciol 63:637–651

Shrestha AB, Wake CP, Mayewski PA, Dibb JE (1999) Maximum temperature trends in the Himalaya and its vicinity: An analysis based on temperature records from Nepal for the period 1971–94. J Clim 12:2775–2767

Shukla SP, Siddiqui MA (2001) Recession of snout front of Milam glacier, Goriganga valley, Pithoragarh district, Uttar Pradesh. In: Symposium on snow, ice and glaciers-A Himalayan perspective Lucknow 9–11 March 1999, pp 71–75

Shukla A, Qadir J (2016) Differential response of glaciers with varying debris cover extent: evidence from changing glacier parameters. Int J Remote Sens 37(11):2453–2479

Shukla A, Garg S, Mehta M, Kumar V, Shukla UK (2020) Temporal inventory of glaciers in the Suru sub-basin, western Himalaya: Impacts of the regional climate variability. Earth Syst Sci Data 12:1245–1265

Shukla A, Ali I, Hasan N, Romshoo SA (2017) Dimensional changes in the Kolahoi Glacier from 1857 to 2014. Environ Assess Monit 189(1):1–18. https://doi.org/10.1007/s10661-016-5703-7

Shulka SP, Siddiqui MA, Hampaiash P, Swaroop S (2000) Secular movement studies of selected glaciers of Yamuna and Ghagra Basins, Dhradun and Tehri, Uttarakashi and Pithoragarh Districts, U.P. Geol Surv India Rec 133:161–165

Srivastava D, Sangewar CV, Kaul MK, Jamwal KS (2001) Mass balance of Rulung Glacier – a Trans-Himalayan glacier, Indus basin, Ladakh. In: Proceeding Symposium Snow, Ice and Glaciers, March 1999. Geol Surv India, Special Publication, 53:41–46

Swaroop S, Oberoi LK, Srivastava D, Gautam CK (2001) Recent fluctuation in snout front of Dunagiri and Chaurabari glaciers, Dhauliganga and Mandakini—Alaknanda basins, Chamoli district, Uttar Pradesh. In: Symposium on snow, ice and glaciers-A Himalayan perspective Lucknow 9–11 March 1999, pp 77–81

Thayyen RJ, Gergan JT, Dobhal DP (2007) Role of glaciers and snow covered on Headwater River Hydrology in Monsoon region-Micro scale study of Din Gad catchment (Dokriani Glacier) Garhwal. Curr Sci, 92(3):376–382

Thompson LG, Yao T, Mosley-Thompson E, Davis ME, Henderson K, Lin PN (2000) A high-resolution millennial record of the south Asian Monsoon from Himalayan Ice Cores. Science 289:1916–1919

Tshering P, Fujita K (2016) First in situ record of decadal glacier mass balance (2003–2014) from the Bhutan Himalaya. Ann Glaciol 57(71):289–294

Vohra CP (1981) Himalayan glaciers. In: Lall JS, Maddie AD (eds) Himalayan aspects of change. Oxford University Press, Delhi, pp 138–151

WGMS (World Glacier Monitoring Service) (2008) Global Glacier Changes: Facts and Figures. WGMS/UNEP

Wagnon P, Linda A, Arnaud Y, Kumar R, Sharma P et al (2007) Four years of mass balance on Chhota Shigri Glacier, Himachal Pradesh, India, a new benchmark glacier in the western Himalaya. J Glaciol 53:603–611

Wagnon P, Vincent Y, Arnaud E et al (2013) Seasonal and annual mass balances of Mera and Pokalde glaciers (Nepal Himalaya) since 2007. The Cryosphere, Copernicus 7(6):1769–1786

Yao T, Thompson L, Yang W, Yu W, Gao Y, Guo X, Pu J (2012) Different glacier status with atmospheric circulations in Tibetan Plateau and surroundings. Nat Clim Chang 2(9):663–667

Zemp M, Roer I, Kääb A, Hoelzle M, Paul F, Haeberli W (2008) Global glacier changes: facts and figures. World Glacier Monitoring Service, United Nations Environment Programme, Geneva, p 80

Chapter 9
Transition of Earthquake Hazards into Disasters

Afroz Ahmad Shah, M. Gazali Rachman, and Muhsana Mahoor

Abstract The global burden of earthquake hazards has intensified over the decades, particularly in developing countries with inadequate safety measures. Despite two centuries of progress in earthquake science, including mapping fault systems and establishing seismic building codes, many high-risk nations have failed to implement these regulations effectively. This paper advocates for a paradigm shift in addressing earthquake hazards by proposing a dedicated UN forum to enhance global mitigation efforts. Simultaneously, it calls for developing comprehensive nationwide preparedness frameworks aimed at reducing vulnerability and saving lives.

The global burden of earthquake hazards, which have increasingly turned into disasters, has grown over the decades, particularly in countries with inadequate earthquake safety measures. This risk is especially pronounced in much of the developing world. For over two centuries, advances in earthquake science have led to mapping fault systems and establishing seismic building codes, warning systems, and other protective measures. However, in many high-risk countries—such as Pakistan, Afghanistan, India, Nepal, Bangladesh, Indonesia, Haiti, and Morocco—these regulations remain under-implemented despite the known dangers. Recent disasters, such as the 2023 earthquakes in Syria and Turkey, which claimed more than 50,000 lives, and the moderate but deadly earthquake in Morocco that killed over 2,000, underscore the urgent need for a more robust response. Both well-mapped and lesser-known fault systems continue to pose significant threats, especially in the Global South, where the impacts of such disasters disproportionately affect millions of people. This paper

A. A. Shah (✉) · M. G. Rachman
Department of Earth Sciences, Faculty of Science, University of Brunei Darussalam, Jalan Tungku Link Gadong BE, Bandar Seri Begawan 1410, Brunei Darussalam
e-mail: afroz.shah@ubd.edu.bn

M. G. Rachman
e-mail: muhammad.gazali@upnyk.ac.id

M. G. Rachman
Geological Engineering Department, Faculty of Technology Mineral, Universitas Pembangunan Nasional "Veteran" Yogyakarta, Depok, Indonesia

M. Mahoor
Department of Earth Sciences, Faculty of Science, University of Kashmir, Srinagar, India

B. K. Rastogi et al. (eds.), *Natural Hazards and Risk Mitigation*, Springer Transactions in Civil and Environmental Engineering, https://doi.org/10.1007/978-981-97-7658-0_9

argues for a paradigm shift in addressing earthquake hazards, drawing inspiration from the climate change discourse at the United Nations. Establishing a dedicated UN forum for earthquake hazards could catalyze global efforts to enhance mitigation and adaptation strategies, paving the way for a safer future. Additionally, the development of comprehensive, nationwide earthquake preparedness frameworks is critical. Addressing long-standing gaps in mitigation and adaptation could significantly improve countries' vulnerability profiles and save countless lives.

9.1 Introduction

Earthquakes occur due to tectonic plate interactions, a process that has been ongoing since the Earth's formation, making these events a fundamental indicator of the planet's dynamic nature (Fig. 9.1). Damaging earthquakes have captured human attention from early history, likely due to their significant impact on life and property. This concern has driven the development of science and technology to address earthquake-related challenges (Bilham 2009). Over the past 200 years, earthquake scientists have accumulated substantial knowledge with the primary goal of warning populations about the dangers posed by active seismic zones to save lives and protect assets. Earth scientists have continuously informed both the public and policymakers about the likelihood of significant earthquakes along mapped fault zones, which span most continents and, to some extent, oceanic regions (Bilham 2009, 2019; Kahn 2005; Ambraseys and Bilham 2011; Erdik 2013; Holzer and Savage 2013). While science has provided evidence-based strategies to address earthquake hazards and mitigate risk, earthquakes continue to result in catastrophic outcomes. The first two decades of the twenty-first century have witnessed unprecedented loss of life and destruction due to earthquakes, especially in developing and low-income nations (e.g., Fig. 9.1). This underscores the ongoing challenge of transforming scientific knowledge into effective disaster prevention measures.

Previous studies suggest that the transformation of earthquake hazards into disasters is driven by multiple factors that operate at local, regional, and global levels, disproportionately affecting millions of people (Bilham 2009, 2019; Kahn 2005; Ambraseys and Bilham 2011; Erdik 2013; Holzer and Savage 2013; Doocy et al. 2013). This chapter critically examines the key factors contributing to this transition, drawing on examples from around the world where earthquakes have caused significant damage over the past two centuries. The focus is particularly on South and Southeast Asia, regions of heightened earthquake risk. The chapter concludes by outlining strategic and policy measures aimed at reducing the projected earthquake risks and mitigating their impact.

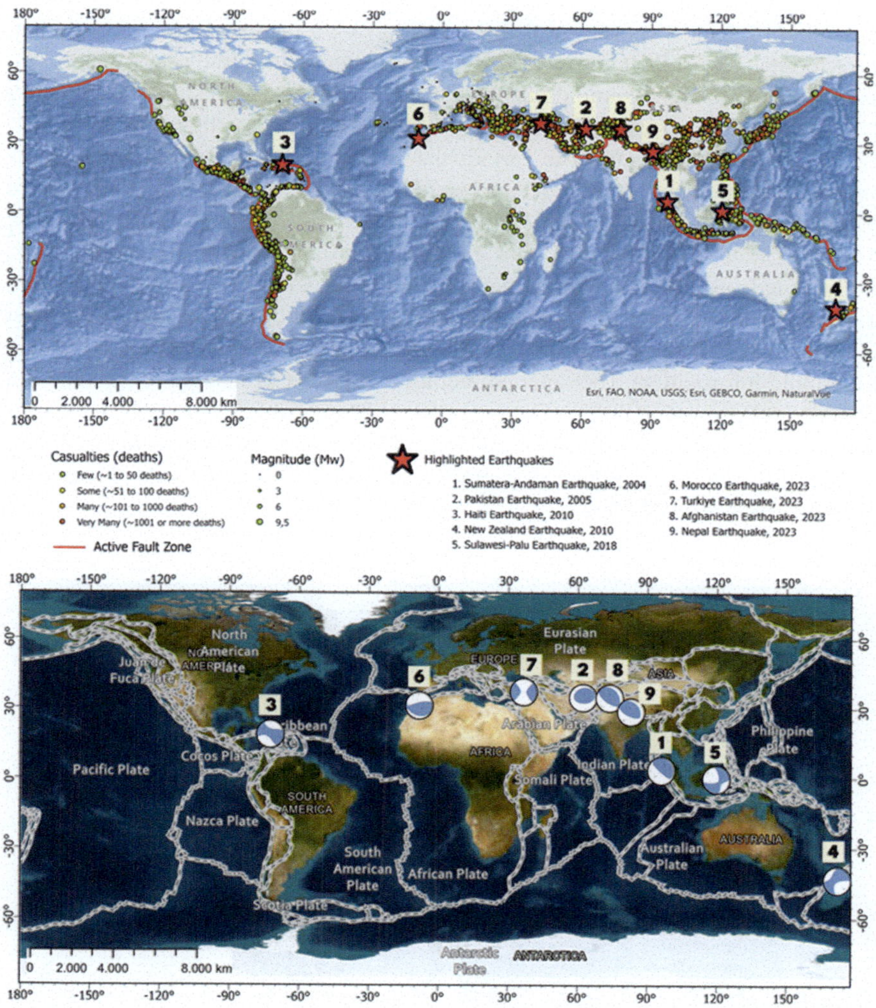

Fig. 9.1 The top portion of the figure shows the distribution of earthquakes classified by color based on the casualties: the highlighted and numbered earthquake events are discussed in detail (see the text). The earthquake centroid moment tensor solutions of the highlighted earthquakes indicate that most of the earthquakes are either reverse or strike-slip (bottom portion of the figure)

9.2 Methodology

We have used published and open-source seismological and satellite data to compile the major sources of earthquakes worldwide (NGDC, WDS 2023), plotted on the topographic maps (Figs. 9.1, 9.2, 9.3, 9.4, 9.5, and 9.6). The Significant Earthquake Database is a comprehensive global record spanning from 2150 BC to the present, containing information on over 5,700 earthquakes that meet specific criteria, such as number of deaths, significant damage and having a magnitude of 7.5 or higher, and Modified Mercalli Intensity (MMI) X or greater, or generating a tsunami. We also used open-source seismological data from the United States Geological Survey (USGS) (2023) to provide detailed information on selected earthquakes (Figs. 9.1, 9.2, 9.3, 9.4, 9.5, and 9.6). The earthquake hazards associated with future large-magnitude earthquake events are interpreted using the updated and open-source population database, the Global Human Settlement Layer (GHSL) (Pesaresi and Politis 2023). The GHSL project has global spatial data, evidence-based analytics, and knowledge to characterize human presence worldwide. It employs spatial data processing technologies for automated analytics and information extraction from diverse geospatial sources, such as satellite images, census data, and crowd-sourced geographic information. The detailed methodology for GHSL data is explained in reference (Commission 2023). The type of fault responsible for the selected earthquake events is also shown (Figs. 9.1, 9.2, 9.3, 9.4, 9.5, and 9.6) compared with the hazards in densely populated regions worldwide. All the datasets above were compiled using ArcGIS Pro.

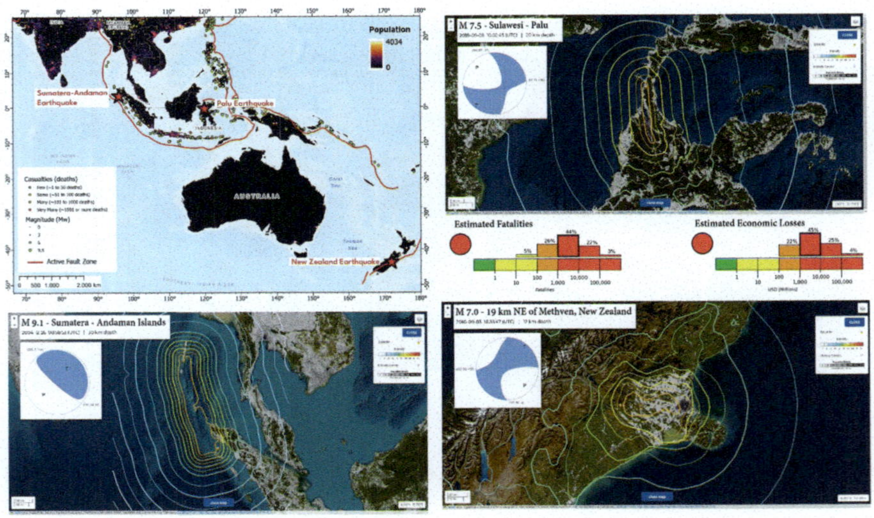

Fig. 9.2 The map illustrates the distribution of earthquakes across Southeast Asia and Australia (top left), highlighting three significant events: the M 7.5 Sulawesi-Palu earthquake and tsunami (top right), the M 9.1 Sumatra-Andaman earthquake and tsunami (bottom left), and the M 7.0 Methven earthquake in New Zealand (bottom right). The differences in magnitude and location led to varying levels of devastation, emphasizing the urgent need for global collaboration to enhance earthquake resilience

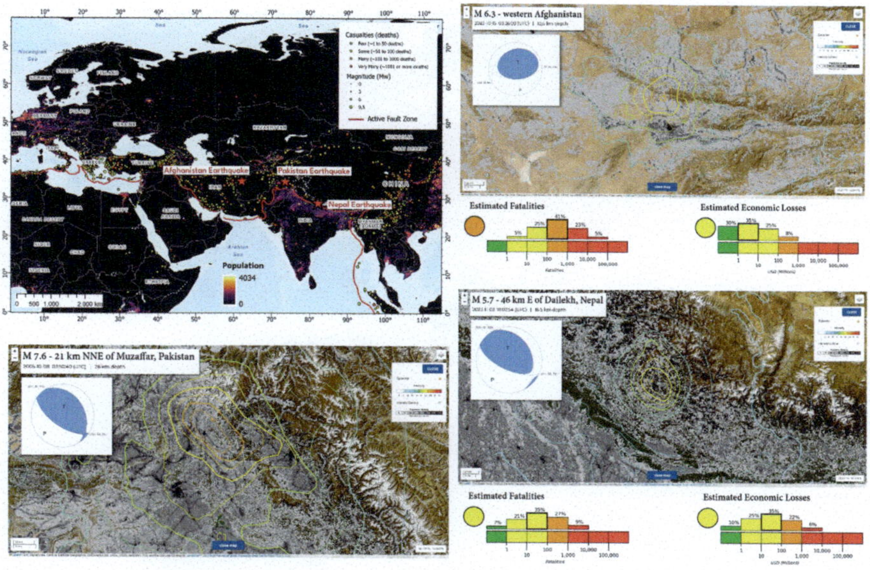

Fig. 9.3 The map displaying the distribution of earthquakes in Asia (top left) highlights three significant events: the M 6.3 Afghanistan earthquake (top right), the M 7.6 earthquake north-northeast of Muzaffarabad, Pakistan (bottom left), and the M 5.7 Dailekh, Nepal earthquake (bottom right). The differences in magnitude and location resulted in varying levels of devastation, underscoring the urgent need for collaboration to strengthen earthquake resilience in South Asia and beyond

9.3 Interpretations and Discussion

9.3.1 The Problem of Earthquake Risk

We have selected several devastating earthquakes (Fig. 9.1), primarily in regions known for significant seismic activity. Our critical analysis of these events raises an important question: Could the death tolls, damage, and asset losses be minimized if decades of established earthquake science had been effectively implemented? Unfortunately, ground preparation remains insufficient, and earthquake hazards often become disasters due to the failure to enforce earthquake safety regulations (see below).

The current tectonic setting of the Earth focuses earthquake hazards primarily along plate boundaries, as reflected in the distribution of losses and damage associated with both plate boundary and plate interior fault systems (Fig. 9.1). Our analysis of selected earthquake disasters suggests that while earthquake hazards are unavoidable due to their tectonic origins, the risks can be mitigated. The structural setting of a region largely dictates the potential for earthquakes, making complete avoidance can be mitigated. The structural setting of a region largely dictates the potential for earthquakes, making complete avoidance impossible (Fig. 9.1).

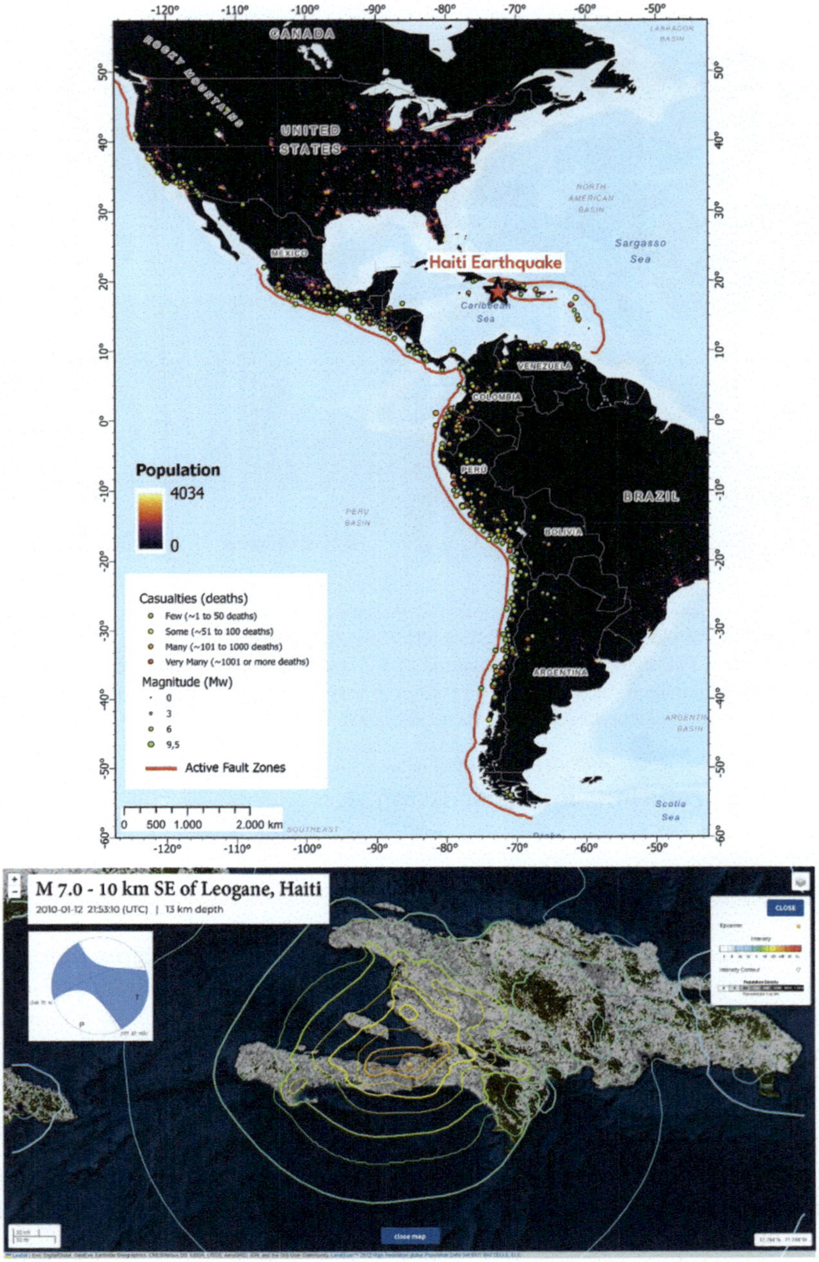

Fig. 9.4 The map displays the distribution of earthquakes in the USA and South America (top). The M 7.0 earthquake southeast of Leogane, Haiti, is highlighted as a shallow event with a depth of only 13 km (bottom), causing unprecedented devastation. This was primarily due to the country's lack of preparedness for such disasters, underscoring the urgent need for international funding and collaboration to enhance earthquake resilience in regions like Haiti

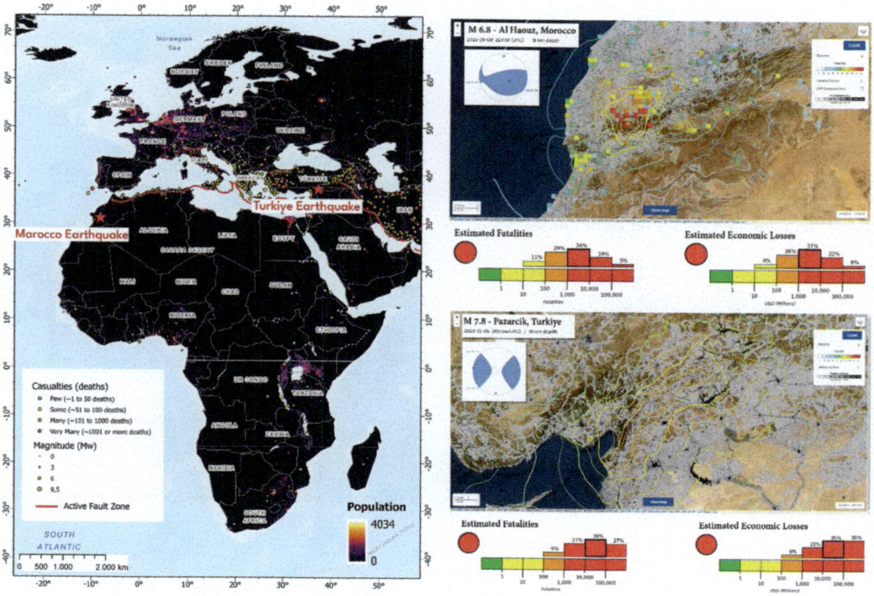

Fig. 9.5 The map on the left illustrates the distribution of earthquakes in Europe and Africa, highlighting two significant events: the M 6.8 earthquake in Al Haouz, Morocco (top right) and the M 7.8 earthquake in Pazarcik, Türkiye (bottom right). The highlighted earthquakes correspond to estimated fatalities and economic losses based on USGS data. The damage and losses from the Türkiye-Syria earthquakes were primarily attributed to poor-quality structures, a recurring issue in many devastating seismic events. This situation underscores the urgent need for international funding and collaborative efforts to enhance earthquake resilience globally. While the Morocco earthquake may have seemed unexpected, it was, in fact, anticipated; proactive measures should be implemented to mitigate future devastation

One striking example from the early twenty-first century is the intraplate earthquake in the Indian subcontinent, far from an active plate boundary. Known as the Republic Day Earthquake of January 26, 2001, in Bhuj, Gujarat, it resulted in the loss of over 9,000 lives. This event surprised scientists, policymakers, and the public, and it sparked a nationwide campaign to advance earthquake research and outreach efforts (Bendick et al. 2001; Jain 2016; Shah et al. 2020a, 2018). This intraplate event underscored that even regions considered tectonically stable can experience significant earthquakes.

The magnitude 9.3 earthquake that struck the Sumatra-Andaman region in 2004 shocked coastal and scientific communities alike. This event, which caused the death of approximately 156,000–178,000 people across more than 11 nations, with 26,500–142,000 suspected dead from the resulting tsunami (Liu et al. 2005), raised global awareness about tsunami hazards. At the time, the region lacked critical knowledge about the potential for tsunamigenic earthquakes, reflecting the inadequacy of tsunami science. The 2004 earthquake ruptured part of the Sunda megathrust fault system, one of Southeast Asia's most active faults (Figs. 9.1 and 9.2). This disaster

Fig. 9.6 The construction patterns of older houses in Kashmir, India (top portion of Fig. 9.6), exhibited superior structural design that minimized earthquake damage. However, these have now been replaced by modern construction methods (bottom portion of Fig. 9.6) that are highly susceptible to seismic activity. Most contemporary designs significantly increase vulnerability to earthquakes, putting over 10 million residents in the Jammu and Kashmir region at risk

highlighted the urgent need for improved tsunami warning systems and preparedness in coastal areas.

Another deadly earthquake occurred in 2005, this time in the interior region of the Himalayan front, once again catching scientists off guard. The magnitude 7.6 (Bilham 2019; Doocy et al. 2013) Kashmir earthquake devastated Muzaffarabad, Pakistan (Fig. 9.3), resulting in over 80,000 deaths (Shah et al. 2018; Bilham

2019). The surprise stemmed from a widespread misconception that earthquakes in the Himalayan region were confined to the Himalayan frontal fault system. A re-examination of seismic data and topographic analysis revealed that earthquakes can also occur in the interior regions, though they are less frequent compared to those along the frontal fault, which exhibits more aggressive tectonic activity (Shah et al. 2021). This event exposed significant gaps in both scientific understanding and administrative preparedness for earthquake hazards. The devastation continued with the 2008 M 7.9 Wenchuan earthquake in China, which claimed over 85,000 lives, further emphasizing the need for better planning and ground preparation for such catastrophic events.

Our analysis of the devastating M 7.0 earthquake that struck Haiti in 2010 (Figs. 9.1 and 9.4), resulting in over 200,000 deaths (Bilham 2010) and causing approximately USD 8 billion in damages, reveals that earthquake hazards were severely mismanaged due to several factors. These include widespread poverty, a lack of resources, insufficient political commitment, and a lack of international support for resilient-building efforts. The shallow depth of the fault (Fig. 9.4), combined with Haiti's densely populated areas and the absence of adequate preparedness, allowed the earthquake to escalate into a major disaster—one that could have been largely prevented with proper hazard preparedness measures in place. What was surprising about the Haiti earthquake was not its occurrence, but the fact that it was linked to a previously unknown oblique-strike-slip fault zone, running parallel to the well-known and hazardous Enriquillo–Plantain Garden fault. The earthquake ruptured the causative oblique fault, which exhibited both left-lateral strike-slip and reverse faulting (Calais et al. 2010). This underscores the reality that our understanding of fault systems can still surprise us, and these uncertainties must be considered when mapping earthquake hazards and risks. Similarly, the 2011 M 9.0 earthquake in Tohoku, Japan, struck one of the most earthquake-resilient nations in the world, yet it triggered a devastating tsunami that claimed over 20,000 lives (Crowley and Elliot 2012). This demonstrates that even in scientifically advanced countries, hazards can still lead to massive destruction. This paints a concerning picture for many developing and densely populated nations, where earthquake hazards remain largely unaddressed (Fig. 9.1). This highlights the urgent need for global collaboration to improve earthquake preparedness and resilience, especially in vulnerable regions.

New Zealand is situated at a critical juncture where a plate boundary subduction-related megathrust system connects to another subduction-related fault via a transform fault. As a result, the country's tectonic and structural framework makes it highly susceptible to earthquakes. Both historical and recent seismic events have led to significant loss of life and damage to infrastructure. For instance, in September 2010, a magnitude 7.1 earthquake struck near the town of Darfield on a previously unidentified fault (Fig. 9.2a and d). This incident underscores the challenge even well-prepared nations face due to unmapped faults (Fig. 9.2a and d)—a cautionary lesson for countries that are less equipped to handle earthquake hazards. While the Darfield earthquake caused no fatalities, it resulted in substantial damage to infrastructure, particularly to unreinforced masonry buildings (Ingham and Griffith 2010). Unfortunately, just a few months later, in February 2011, a magnitude 6.3 earthquake hit 5 km southeast of Christchurch at a shallow depth of 5 km, causing significant destruction. Occurring during lunchtime, this event led to 185 fatalities, most of which were caused by collapsing buildings, including two multi-story office complexes (Potter et al. 2015). Despite being one of the most prepared nations, New Zealand's struggle to mitigate the impact of earthquake hazard highlights the dire situation for countries with inadequate earthquake preparedness.

Earthquake-related hazards frequently occur worldwide, often in complex and cascading sequences. This was demonstrated by the magnitude 7.5 earthquake that struck Minahasa, Sulawesi, Indonesia, on September 28, 2018 (Shah et al. 2020a), causing over 4,000 casualties (Shah et al. 2018) and displacing large number of people (Rubin et al. 2017). This earthquake revealed new insights into the secondary hazards triggered by seismic activity, particularly how such hazards are exacerbated by agricultural practices dependent on continuous water supplies. Studies (Shah et al. 2018; Rubin et al. 2017) have shown that irrigation practices, such as those used in rice cultivation, can significantly worsen secondary effects like liquefaction and landslides.

The Palu-Koro fault, which ruptured during this earthquake, had been previously mapped (Navakesh et al. 2019), but despite the scientific understanding of its potential for destruction, the necessary precautions were not implemented on the ground, leading to significant loss of life and property. Indonesia, one of Southeast Asia's most densely populated regions, remains underprepared for large seismic events, highlighting the urgent need for collaboration on multiple fronts to build earthquake resilience.

More recently, the devastating earthquake in Morocco on September 8, 2023, resulted in extensive damage and over 2,000 deaths (Fig. 9.5a and b). Although the earthquake had a moderate magnitude of Mw 6.8, the damage it caused exposed serious gaps in Morocco's earthquake preparedness. The quake also led to destruction of significant cultural heritage, including the Tinmel Mosque in the High Atlas, a UNESCO World Heritage site (Marshall 2023). This tragedy underscores the persistent challenges many nations face in executing effective earthquake resilience programs.

The unpreparedness was again witnessed during the February 2023 doublet earthquake in Turkey and Syria, causing 52,000 casualties (Fig. 9.5a and c) and billions in economic losses (Zilio and Ampuero 2023). The causative and tectonically active fault system had been well mapped, and seismic hazard has been communicated to the public for decades, yet the unfortunate loss of life and assets is a question of what went wrong and why. The enormous number of collapsed buildings, 10,200 (Hussain et al. 2023), suggests that seismic regulation and monitoring of buildings for hazards was not properly done. It is estimated that about 100,000 buildings were severely damaged (Cetin et al. 2023), including critical infrastructures such as hospitals, schools, etc. Recent studies (e.g., Karasozen et al. 2023) have shown that corruption has played a negative role in building earthquake resilience in Türkiye, which caused substantial damage during the 2023 quake due to substandard building material, poor construction design, and lack of seismic vulnerability mapping and enforcement of building codes. It is linked with what has been suggested earlier for other countries where earthquake-related casualties and damage to assets have remained high due to various forms of corruption (Ambraseys 2013; Shah et al. 2018; Zilio and Ampuero 2023). Decades ago, Professor Nicholas Ambraseys said: "Earthquakes don't kill people; buildings do." "The collapse of buildings in earthquakes are not acts of God. All too often nowadays they are acts of criminal negligence" (Ambraseys 2013), which, even today, remains a vital challenge for us to answer.

Over 100 people lost their lives in the twin quake sequence of magnitude 6.3 that struck the war-torn nation of Afghanistan on October 7, 2023 (Fig. 9.3a and b). Such a small-intensity quake will normally cause no harm to a country prepared to handle earthquake hazards, signifying the dangers turning into disasters because of unpreparedness, which is linked to overall political and economic stability. Yet again, the twin medium-magnitude earthquake caused substantial damage and loss of more than 100 people in Nepal on October 3, 2023 (Fig. 9.3a and c). The loss of life and damage to assets indicate a lack of preparedness to tackle the expected large-magnitude earthquake hazards in the region, which risks millions. The large-magnitude (Mw~7.8) earthquake of 2015 that resulted in the loss of more than 8000 people also caused extensive damage to buildings in the rural areas as compared to the urban areas, which suggests a greater need to relook at the construction standards of the urban earthquake of 2015 that resulted in the loss of more than 8000 people also caused extensive damage to buildings in the rural areas as compared to the urban areas, which suggests a greater need to relook at the construction standards of the urban regions with a motivation to improve it (Sapkota et al. 1934).

9.3.2 *Transformation of Earthquake Hazards into Disasters*

Examining damaging earthquakes plotted on the shuttle radar topography (Fig. 9.1) suggests hazards turning into disasters in countries ill-prepared to handle earthquakes. All these events show that the transformation of earthquake hazards into disasters primarily occurred because the fundamental steps in the earthquake safety procedures of the buildings and building sites are needed to be followed (Erdik 2013; Holzer and Savage 2013; Bilham 2019, 2010; Zilio and Ampuero 2023). The disasters struck across different economic backgrounds, including the poor and rich, and affected urban and rural populations and critical civil infrastructure such as schools, roads, hospitals, bridges, and jails. For example, (Bilham 2010) reports that 80% of all schools collapsed in the Port-au-Prince area during the 2010 quake, which was mimicked by the 2005 Kashmir earthquake, suggesting preparations to address earthquake hazards are inadequate in the developing world. It includes the global South, where millions are directly risking their lives. The more than 80,000 casualties during the Kashmir earthquake of 2005 in Pakistan occurred primarily due to the poor quality of buildings that failed to comply with the codes for earthquake-resistant building designs (Bilham 2010). The lack of adequate earthquake-building safety mechanisms remains the leading cause of earthquake-related losses and damages, such as in Haiti in 2010 (Bilham 2010) and Türkiye in 2023 (Zilio and Ampuero 2023). The enormous scale of losses and damage associated with earthquakes in the twenty-first century makes it again evident that the ground preparations were inadequate, which questions the earthquake safety policy procedures at the national and international levels. It asks questions about the non-implementation of earthquake-building codes, earthquake resilience procedures, and what to do about it when earthquake disasters seem inevitable in the coming years. For example, citing the latest devastation in Turkey and Syria in 2023, which occurred on a well-known fault system with known risk yet caused widespread destruction. It remains the latest example of how earthquakes pose severe threats even though enormous scientific and technological achievements are available to perform better and raises questions on the earthquake safety preparations of countries to risk the lives of millions during the forecasted events.

Contrary to most of the world, where future earthquakes are predicted to be severe and damaging, the earthquake risk profile of many developed countries has dramatically improved, such as Japan, New Zealand, and the USA. These nations have the resources and relatively well-equipped administrations to make that transition possible. It is good news, which means such a transition is possible. The lack of resources and implementation of seismic building codes makes earthquake hazards more lethal in the developing world, which is forecasted to continue (Zilio and Ampuero 2023). The nations in the global South and some in the global North already face an enormous task and risk to save millions from ongoing crises such as poverty, hunger, availability of clean water, and climate change-related hazards, which hinder the preparation to cope with earthquake hazards. Therefore, international community support is needed to address countries' seismic hazards, including organizations such as the United Nations, to collaborate on earthquake safety procedures, policies, funding sources, and human resource development (see the section on the United Nations Framework Convention on Earthquake Hazards).

9.3.3 Political Stability and Winning the War with Earthquake Hazards

The unstable political landscape has added yet another critical dimension to the already ailing conditions of some nations, and Afghanistan in South Asia is the best example to demonstrate that. The high-intensity protracted conflicts have rendered almost all institutions dysfunctional, and disasters have affected virtually all of the country's 34 provinces during the last three decades. The science and action-oriented strategy to address the growing threats of transforming hazards into disaster has failed because of conflict-related issues, economic sanctions, and security lapses, making this country one of the most challenging nations to face the incoming threats of climate change and earthquakes in South Asia. It was demonstrated again in recent times when over 2000 people died in a sequence of magnitude 6.3 and 6.2 events that struck the northern parts of Afghanistan near Herat in October 2023 (Fig. 9.3).

Such a small magnitude earthquake is expected to cause no to little harm in the modern scientific age, and we often witness it in seismically resilient places such as California and Tokyo.

The devastation was not unexpected in a tectonically active region (Shah et al. 2020b) where reverse and strike-slip fault-related earthquakes are expected. Astonishingly, earthquake science wisdom is more than 200 years old, yet it has not reached people, which is worrying. The earthquakes in Afghanistan were not that large to cause considerable damage to life and assets, which raises an important question about the ground preparedness realities to tackle earthquake hazards in a tectonically active area and what to do about it. The important realization that we confront on a daily basis is to map the consequence of the next earthquake and the ways to avert the expected damage to innocent lives and assets, which could potentially cascade into other problems in conflict-driven societies such as poverty, malnutrition, and migrations. The nation is one of the least prepared to handle the forecasted earthquake risks, demanding how international institutions and researchers could contribute to the betterment of the societies that fail to provide adequate safety and security to their people. Framing earthquake safety policy at international institutions such as the United Nations could provide the impetus for securing life in future earthquakes, which requires work on national and international levels, as outlined in Shah (2023).

9.4 Discussion

9.4.1 The Forecasted Risk and What to Expect in South and SE Asia

The forecasted large earthquakes may strike the highly urbanized and populated regions in South and SE Asia, including Afghanistan, Pakistan, India, Nepal, Bangladesh, Myanmar, and Indonesia, which will be a significant challenge. These countries share the active plate boundary fault system, severely threatening the millions occupying the seismically active zones (Fig. 9.1). The fault system runs under the Indo-Gangetic plains and the delta system of Bangladesh, which has an additional risk of liquefaction, making the region extremely vulnerable to earthquake-induced shaking, liquefaction, and secondary hazards such as floods and landslides.

It remains an uphill task for developing nations to prepare for earthquake hazards because an enormous amount of work still needs to be completed, which may take decades. The real reason for the lack of implementation of earthquake-building codes remains fuzzy in some countries, such as India, Pakistan, Bangladesh, and Indonesia, where significant portions of the population can afford to do so but are not doing it. The future risks to millions of people living near the active fault zones make it essential for the world to offer help and open discussions on what makes resilience an uphill task for some, which remains unexplored. The major challenge is that the decades of inaction on the ground may continue in the future, risking millions, as was observed recently in Turkey, which raises important questions about the safety and welfare of people and how to improve it. One of the major steps in this regard is to rank the structural integrity of buildings to withstand earthquakes, which could be done at the national level (see below).

9.4.2 Ranking the Efficiency of Built Areas to Withstand Damaging Earthquakes

Earthquake science is more than two centuries old, which has improved the scientific methodology, understanding, and implications to make living with earthquake hazards possible. Architects and engineers worked on innovations in designing earthquake-proof buildings through innovative methodologies that include, for example, strengthening of the foundations, reinforcements, damping, vibrations deflection technology, cross braces, diaphragms, and so on. The traditional approach of typical buildings involves designs to handle vertical loads from the weight of the structure and the forces of gravity. However, the horizontal forces were not part of the design, which is why most of the buildings often collapse during earthquakes because of the horizontal vibrations.

The built area in South and SE Asia substantially lacks earthquake safety procedures, similar to what we have witnessed in Kashmir, India, in the NW Himalayas. The built area is prone to earthquake hazards, which could morph into cascading hazards, particularly liquefaction and landslides. The built area follows a traditional design mostly by non-specialist local engineers, often not trained in engineering schools. It is a huge risk because the structures may not be feasible even to withstand vertical loads, and horizontal forces will cause severe damage. There has not been any comprehensive campaign to check the structural integrity of buildings and rank them accordingly so that the proper procedures are taken to make them earthquake-resistant, sustainable, and risk-free.

Our fieldwork shows that traditional buildings in Kashmir, now a rare sight, used timber-based vertical and horizontal beams in masonry walls, which were reasonably better equipped to cause less damage than modern designs (Fig. 9.6). The now widespread building designs are like a pack of cards ready to tumble during an earthquake. The lack of structural integrity in design is reflected by the concrete masonry walls with no support to withstand loads during horizontal motion, which paints a grim picture of the region. What is more worrying is the lack of planning in designing earthquake-resistant structures and the impact of liquefaction, particularly in regions that are prone to such effects such as buildings on unconsolidated sediments, such as flood plains and paleo-lake deposits.

9.4.3 Management of Earthquake Hazards

1. **Earthquake Hazard Vulnerability Index (EHVI) Mapping**

At the national level, we propose that countries start a comprehensive campaign to check and rank the structural integrity of built areas to withstand earthquakes with a priority in regions that are prone to earthquake risks, such as the area near the Himalayan frontal fault system in most of the South Asian countries such as India, Bangladesh, Pakistan, and Nepal (Fig. 9.1). It will be an essential step toward minimizing the losses and damages expected from a large-magnitude earthquake anticipated to occur anytime on the Himalayan frontal fault and the interior faults (Shah et al. 2021, 2018; Bilham 2019; Survey 2023). The ranking campaign would need engineering and geological investigations and should ideally involve other departments such as Municipality and Town Planning. The high population densities near the active fault systems are particularly vulnerable to severe seismic shaking, which could impact millions in a future earthquake. It will have additional problems of disruptions in road networks, liquefaction, flooding, landslide hazards, etc., which is why early interference would help to minimize the expected scale and intensity of destruction.

The Earthquake Hazard Vulnerability Index (EHVI) mapping involves mapping and ranking built areas on a scale of 0–10, with 0 as the most vulnerable to earthquake hazards and 10 as the most stable. The mapping will need a thorough methodological approach, which we will provide in a separate article because of space limitations here. As a conceptual ranking system, it will, however, initiate an interest in people to think about managing hazards and minimizing losses and damages in future disasters.

2. **What makes resilience an uphill task for some nations?**

At the international level, there is a desperate need to initiate a process to frame a body that will solely work on earthquake hazards. It is because earthquake science is more than two centuries old, and it was realized from the very start that earthquakes cause shaking of the foundations, which causes the collapse of structures such as buildings because their structural integrity lacks resistance to shaking. Therefore, the earthquake engineers worked on the problem and offered solutions mainly related to making buildings tolerable to earthquake-induced shaking, liquefaction, etc. This science wisdom is widely known and has reached most of the world, yet it has not been practised, which makes earthquake hazards lethal in many countries such as Haiti, Indonesia, India, Bangladesh, Pakistan, Turkey, and Nepal (Fig. 9.1).

It raised important questions about what makes resilience an uphill task for many nations. The answer to that is not a straightforward one, yet we provide a glimpse into what we have understood working in South and SE Asia. People have little sense of dealing with earthquake hazards, and worryingly, they are not even asking or looking for solutions. We have witnessed this in many parts of India, Indonesia, Bangladesh, Malaysia, Pakistan, Nepal, and so on. The first author of this Chapter is from Kashmir, India, one of the most tectonically active regions in the world, yet his family members have not followed the earthquake safety procedures, nor anyone from the close and distant relatives and friends. It paints a grim picture of what could happen if a major earthquake occurs. Most of the people who have built their houses are not poor, particularly in towns such as Anantnag and Srinagar, which means they have the resources to do better, but why they have chosen not to follow the earthquake safety protocol seems completely reckless practice, which is unfortunately, the culture in that region that needs to be reversed. Fieldwork observations in the region suggest that older houses used to be structurally better than the new construction patterns (Fig. 9.6).

3. **United Nations Framework Convention on Earthquake Hazards (UNFCEH)**

Effectively managing earthquake disasters, especially in the developing world, is achievable through the establishment of a well-coordinated international forum. Resource shortages, political conflicts, social inequalities, and pervasive poverty are significant roadblocks on the path to swiftly addressing earthquakes and related hazards (Shah 2023). Globally, these challenges can be effectively addressed by establishing an international platform that unites the Global North and South. This collaborative forum would foster discussions, deliberations, and the exchange of innovative ideas, all aimed at enhancing safety protocols. With a shared mission to transform earthquake hazards into manageable risks, this initiative would facilitate mutual cooperation and empower nations to prevent devastating disasters. Together, we can pave the way for a more resilient future. Moreover, it will directly address Goal 11 of the United Nations Sustainability Development Goals (SDGs). Currently, such a platform only exists for climate change threats, which has mobilized the world to act against the impacts of climate change. It didn't happen overnight. The global warming threat led to the United Nations Intergovernmental Panel on Climate Change (IPCC) foundation in 1988, tasked with looking at greenhouse warming and global climate change. In the next decade, IPCC discourse on climate change reached most of the world and created the United Nations Framework Convention on Climate Change (UNFCCC) in 1992, and a climate change treaty took effect in 1994. Today, the UNFCCC has a worldwide reach with 198 parties; it is the only organization that works globally to improve the world from climate disasters through collaborations, funding, integrations, deliberations, and other helping mechanisms (Shah 2023). The UNFCCC takes decisions where it asks countries to mitigate, adapt, and care for losses and damages from climate change; such agreements are democratic and nonbinding, which gives it power.

Earthquake hazards disproportionately impact the poor and developing nations, which makes it imperative for a global organization such as the United Nations to create a parallel treaty for earthquake hazards (Shah 2023). It will give a unique opportunity and platform for the global North and South to collaborate, discuss, and exchange ideas about funding, human resources, workshops, training, and study opportunities on earthquake hazards. Therefore, the United Nations Framework Convention on Earthquake Hazards (UNFCEH) is an ideal forum to fill the gaps between developing and developed nations (Shah 2023), improving global earthquake hazard vulnerability research, mapping, and preparations. Annual UN conferences such as the Conference of Parties (COP) could become the annual bridge to deal with earthquake hazards and plans.

4. **Public engagement, earthquake science education, and outreach**

Earthquake science education and outreach programs provide one of the best platforms to initiate interest in people about the science of hazards and how to manage them. Currently, public outreach programs are not part of the scientific discourse in most of the countries prone to earthquake hazards, which we argue is one of the problems that must be addressed. The scientific community can, for example, directly engage and communicate with schools, universities, and state administrators about the hazards and how to prepare for any eventuality. The response to earthquake hazards changes from place to place, and therefore, it is important to map the ground realities before adopting any scientific methodology, which requires work on the ground, and engaging with locals is one of the best options. For example, most of the earthquake hazards in India are restricted to the Himalayan frontal faults, however, the interior faults in the hinterlands (Shah et al. 2018) as well as the Indian continent (Kothyari et al. 2016) are not focused, which could prove dangerous, and require public engagement programs.

References

Ambraseys NN (2013) Nicholas Neocles Ambraseys 1929–2012. J Earthquake Eng 17(3):301–303
Ambraseys N, Bilham R (2011) Corruption kills. Nature 469(7329):153–155
Bendick R, Bilham R, Fielding E, Gaur VK, Hough SE, Kier G, Mukul M (2001) The 26 January 2001 "Republic Day" Earthquake, India. Seism Res Lett, 72(3):328–335
Bilham R (2009) The seismic future of cities. Bull Earthq Eng 7:839–887
Bilham R (2010) Lessons from the Haiti earthquake. Nature 463(7283):878–879
Bilham R (2019) Himalayan earthquakes: a review of historical seismicity and early 21st century slip potential. Geol Soc Lond 483(1):423–482
Calais E, Freed A, Mattioli G (2010) Transpressional rupture of an unmapped fault during the 2010 Haiti Earthquake. Nat Geosci 3:794–799
Cetin KO, Ilgac M, Can G, Cakir E (2023) Preliminary Reconnaissance Report on February 6, 2023, Pazarcik Mw = 7.7 and Elbistan Mw = 7.6, Kahramanmaras-Turkiye Earthquakes. Earthquake Engineering Research Center, Turkiye
Commission E (2023) GHSL data package 2023. Publications Office of the European Union, Luxembourg
Crowley K, Elliot JR (2012) Earthquake disasters and resilience in the global North: lessons from New Zealand and Japan. Geogr J 178(3):208–215
Dal Zilio L, Ampuero JP (2023) Earthquake doublet in Turkey and Syria. Commun Earth & Environ 4(1):71
Doocy S, Daniels A, Packer C, Dick A, Kirsch TD (2013) The human impact of earthquakes: a historical review of events 1980–2009 and systematic literature review. PLoS Curr, 5
Erdik M (2013) Earthquake risk in Turkey. Science 341(6147):724–725
Holzer TL, Savage JC (2013) Global earthquake fatalities and population. Earthq Spectra 29(1):155–175

Hussain E, Kalaycioglu S, Milliner CW, Cakir Z (2023) Preconditioning the 2023 Kahramanmaras (Turkiye) earthquake disaster. Nat Rev Earth & Enironment, 1–3
Ingham JM, Griffith MC (2015) The performance of unreinforced masonry buildings in the 2010/2011 Canterbury earthquake. Report to the Royal Commision of Enquiry
Jain SK (2016) Earthquake safety in India: achievements, challenges and opportunities. Bull Earthq Eng 14:1337–1436
Kahn ME (2005) The death toll from natural disasters: the role of income, geography and institutions. Rev Econ Stat 87:271–284
Karasozen E, Buyukakpinar P, Ertuncay D, Havazli E, Oral E (2023) A call from early-career Turkish scientist: seismic resilience is only feasible with "earthquake culture". Seismica, 2(3)
Kothyari GC, Rastogi BK, Morthekai P, Dumka RK, Kandregula RS (2016) Active segmentation assessment of the tectonically active South Wagad fault in Kachchh, Western Peninsular India. Geomorphology 253:491–507
Liu PLF, Lynett P, Fernando H, Jaffe BE, Fritz H, Higman B, Synolakis C (2005) Observations by the international tsunami survey team in Sri Lanka. Science 308(5728):1595–1595
Marshall M (2023) Why was the Morocco earthquake so deadly? Nature, 14. [Online]. Available https://www.nature.com/articles/d41586-023-02880-3. [Accessed 27 Nov 2023]
NGDC/W.D.S. (NGDC/WDS) NCEI/WDS global significant earthquake database. NOAA National Centers for Environmental Information. Accessed 3 Nov 2023
Navakesh B, Shah AA, Prasanna MV (2019) Earthquake education trough the use of documentary movies. Front Earth Sci, 7(42)
Pesaresi M, P. Politis P (2023) GHS built-up surface grid, derived from Sentinel2 composite and Landsat, multitemporal (1975–2030). European Commission, Joint Research Centre (JRC)
Potter SH, Becker JS, Johnston DM, Rossiter KP (2015) An overview of the impacts of the 2010–2011 Canterbury earthquakes. Int Hournal Disaster Risk Reduct 14:6–14
Rubin CM, Horton BP, Sieh K, Pilarczyk JE, Daly P, Ismail N, Parnell AC (2017) Highly variable reccurence of tsunamis in the 7400 years before the 2004 Indian Ocean Tsunami. Nat Commun 8(1):16–19
Sapkota SN, Bollinger L, Perrier F (1934) B Fatality rates of the Mw~8.2, 1934, Bihar-Nepal earthquake and comparison with the April 2015 Gorkha earthquake. Earth Planets Space 68(40)
Shah AA, Khwaja S, Shah BA, Reduan Q, Jawi Z (2018) Living with earthquake and flood hazards in Jammu and Kashmir, NW Himalaya. Frontiers Ind Earth Science 6:179
Shah AA (2023) Opinion: Do we need a United Nations Framework convention on earthquake hazards? Temblor. [Online]. Available https://temblor.net/temblor/united-nations-framework-convention-on-earthquake-hazards-15263/. [Accessed 11 11 2023]
Shah AA, Navakesh B, Sheeba K, Reduan Q, Jawi Z, Shah BA, Malik JN (2020) Making living with earthquake hazards a culture. In: Proceeding of the Geoscience & Society Summit, AGU
Shah AA, Batmanathan NM, Manan N, Ali A (2020) Mapping of active deformation domains in the west of Hazara-Kashmir Syntaxis. In: 3rd conference of the Arabian Journal of geoscience
Shah AA, Sahari S, Navakesh B (2021) Tectonic geomorphology of the Jhelum fault zone and its Contigous regions in Western Himalaya. Bull Geol Soc Malays, 73
U. G. Survey (2023) Earthquake lists, maps, and statistics. U.S. Geological Survey. [Online]. Available: https://www.usgs.gov/natural-hazards/earthquake-hazards/lists-maps-and-statistics. [Accessed 11 Nov 2023]

Chapter 10
Site-Specific Seismic Hazard Assessment of Gorakhpur City, Uttar Pradesh, India: A Holistic Approach

Nazia Khan, Rajesh Chaturvedi, Bishakha Prasad, and Ram Jivan Singh

Abstract A comprehensive strategy incorporating inputs from geological, geotechnical, and geophysical studies was carried out to account for the local site effects on earthquake ground motion at various sites in and around Gorakhpur City, Uttar Pradesh, India, falling in Seismic Zone-IV of Seismic Zoning Map of India (BIS 2002, 2016). Gorakhpur City, one of the biggest and rapidly developing cities is located on the middle Gangetic Plain, which comprises fluvial sediments from the Older Alluvium (Middle to Late Pleistocene age) and the Newer Alluvium (Holocene age). The M 7.8 (Mw), 2015 Gorkha/Nepal Earthquake, located around 200 km northeast of Gorakhpur, has jolted the city at > VI seismic intensity, with vertical and horizontal peak ground acceleration (PGA) of 0.25733 g and 0.385995 g, respectively. The horizontal-to-vertical-spectral-ratio (HVSR) approach of ambient noise survey and microtremor study is used to evaluate the seismic site amplification of Gorakhpur city. The fundamental soil frequency and HVSR peak amplitude maps demonstrate peak soil amplification in the range from 0.7 to 3.5. The scatter plot between predominant frequency and peak amplification shows that the bulk of sites in Gorakhpur City (65.42%) fall into the medium damage potential zone, followed by the low damage potential zone (31.53%), and the high to extremely high damage potential zone (3.05%). In accordance, with the multichannel analysis of surface waves (MASW) and Standard Penetration Test (SPT) studies, the average shear wave velocity at different locations in and around Gorakhpur city varies between 202.5 m/s and 357.4 m/s, which corresponds to soil type D (stiff soil) of the site classes of the NEHRP Code Provisions (1997). In Gorakhpur, the prevalence of high liquefaction domains can be seen at low-lying places where the groundwater is shallow (= < 10 m depth), as evidenced by liquefaction susceptibility safety factor values for different locations. The Analytical Hierarchy Process and GIS-based integration were applied to produce a final seismic hazard microzonation map of Gorakhpur City. The southern part of Gorakhpur city, particularly the low-lying area, where

N. Khan (✉) · R. Chaturvedi
Geological Survey of India, Northern Region, Lucknow, India
e-mail: nazia.khan@gsi.gov.in

B. Prasad · R. J. Singh
Geological Survey of India, Western Region, Jaipur, India

B. K. Rastogi et al. (eds.), *Natural Hazards and Risk Mitigation*, Springer Transactions in Civil and Environmental Engineering, https://doi.org/10.1007/978-981-97-7658-0_10

newer alluvium is exposed, around Ramgarh Tal/lake, has the highest seismic hazard index. A few isolated high seismic microzone patches have been identified in the northern region in Chilua Tal/Lake, Sonbarsa, and Balapar regions. The present study is beneficial for seismic disaster preparedness, urban planning, and the preservation of cultural heritage in the city while designing future major engineering projects, multi-story buildings, etc. to withstand the incoming seismic ground motion caused by the large earthquakes in the region.

Keywords Peak amplification · Shear wave velocity · Liquefaction · Seismic microzonation · Seismic disaster preparedness

10.1 Introduction

The Indo-Gangetic Plains, which span an area of around 300,000 square kilometers in northern India and are 150–500 km wide, developed as a result of the foredeep basin filling up in front of the rising Himalaya, the seismically active region of the globe. The Himalayan mobile belt continues to be seismically active because of the continuous collision of the Indian and Eurasian plates from 50 Ma (Dewey and Bird 1970; Bilham 2019). The significant stress building of the Himalayan collision belt as a result of the Indian plate's ongoing movement against Eurasia in a northerly direction finally led to mild to big earthquakes. Historical earthquakes (such as the 1934 Bihar-Nepal, 2005 Kashmir, and 2015 Nepal catastrophic seismic events) provide evidence for it. Any seismic activity in the Himalaya area has a substantial negative impact on the IGB's (Indo-Gangetic Belt) foreland basin's economy and population. The IGP (Indo-Gangetic Plains) is underlain by thick Quaternary alluvial sediments deposited by the Ganges River and its tributaries. These sediments unconformably overlie Precambrian basement rocks and are characterized by significant variations in thickness and composition. Studies by Rao (1973) and Valdiya (1976) highlight the presence of subsurface structures like faults and folds that may influence seismic wave propagation and amplification. Raiverman et al. (1983) and Dasgupta et al. (1987) further suggest the presence of active transverse features in the region, potentially increasing seismic vulnerability. These numerous weak zones, faults, lineaments, and ridges (e.g., subsurface faults/lineaments, West Patna Fault, East Patna Fault, Moradabad Fault, Faizabad Ridge, etc.) in the basement rocks of the IGB periodically reactivate to readjust or accommodate the regional tectonic stresses of the Indian plate's movement to the north, which are also causative factors of seismic events in the Indo-Gangetic region Parkash and Kumar (1991); Singh et al. (1996); GSI (2000); Thakur et al. (2009); Jayangondaperumal et al. (2010); Goswami (2012).

The Study area, Gorakhpur City, U.P. (Fig. 10.1) forms a part of the Gangetic Alluvial Plains (i.e., plain *terai* area). The outer foothills of the Himalaya are located in the north, approximately 87 km from Gorakhpur City. The city is located on the Indo-Gangetic plain, terai belt, which is widely farmed and has a few patches of forestland surrounding it. The favorable subsistence environment coupled with rapid urbanization has contributed to making this area one of the most inhabited/populated cities in eastern Uttar Pradesh. Gorakhpur city is a perfect fit case for seismic microzonation investigation due to its considerable population growth rates, proximity to the Himalayan Frontal Thrust, and its location within Seismic Zone III-IV (BIS 2016). Keeping in mind the paramount significance of seismic risk mitigation and disaster preparedness in the study area, an area of 300 sq km was taken up for seismic microzonation studies of Gorakhpur city as per the Master Plan Map of Gorakhpur City, under the Field Season Programme of Geological Survey of India, Northern Region (FS: 2018–19 & 2019–2020) for an integrated seismic evaluation study by Earthquake Geology.

Seismic microzonation investigations conducted in areas covered by alluvium deposits have the potential to contribute significantly to the reduction of seismic risk. Goswami (2012) highlights geomorphic evidence of active faulting in the north-western IGP, emphasizing the importance of microzonation. Recent studies by Nitish and Puri (2019) and Sankar et al. (2022) demonstrate the successful application of microzonation in other parts of the IGP, providing valuable insights for Gorakhpur. These investigations furnish crucial information regarding the velocity structure of the near-surface, which is indispensable for the estimation of ground motion expected at the scale of a human settlement. The shear wave velocity (Vs) profile and its

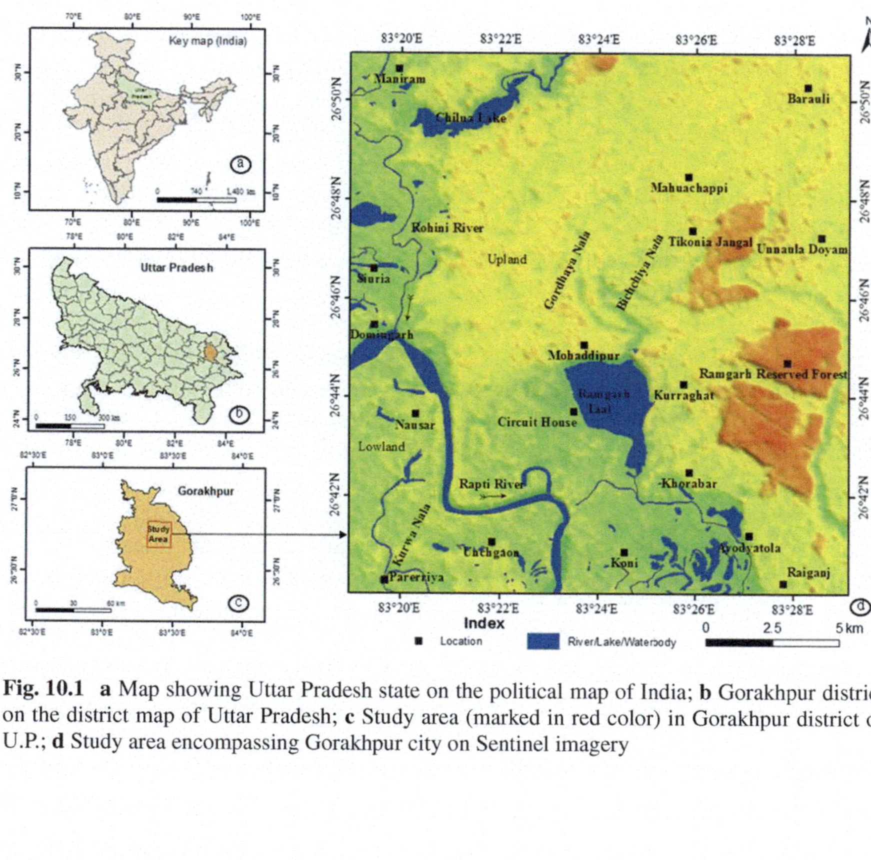

Fig. 10.1 **a** Map showing Uttar Pradesh state on the political map of India; **b** Gorakhpur district on the district map of Uttar Pradesh; **c** Study area (marked in red color) in Gorakhpur district of U.P.; **d** Study area encompassing Gorakhpur city on Sentinel imagery

correlation with the soil characteristic enable the preliminary evaluation of the effects of seismic amplification at a site (Gino et al. 2022). Seismic microzonation study is a must task for every growing city to facilitate the classification of site conditions by identifying seismic hazard vulnerable areas, thereby empowering local governing bodies to incorporate seismic hazard mitigation into their spatial planning efforts (Mase et al. 2021). The integration of conventional H/V techniques and other methods can further enhance the precision of microzonation studies (Keil et al. 2021). The best use of resources for seismic risk reduction may be made possible by sharing experiences and creating common techniques in seismic microzonation for nations vulnerable to destructive earthquakes and this can be done by creating a seismic hazard map by integrating all the geofactors contributing to it of every vulnerable site.

Seismic microzonation hazard maps have been developed for various cities in India by Puri and Jain (2019), Hamid (2018), Kumar et al. (2022), and Mishra (2020). These maps offer details regarding the possible seismic risks in particular locations, taking into factors like peak ground acceleration, site characteristics, potential for liquefaction, and site amplification ratio. The maps are produced by combining probabilistic and geotechnical hazard evaluations with information from fault networks, seismicity patterns, and ground motion prediction equations. The results of these studies show that certain cities in India, such as Srinagar and the Bengal Basin, are susceptible to high or severe seismic hazards. It is recommended that comprehensive seismic microzonation studies be conducted for strategic cities in South Asia, including those in India, to develop risk-resilient building design codes and ensure the safety and sustainability of urban infrastructure. Few attempts have been made to analyze the liquefaction potential around Ramgarh Taal (Chaudhary et al. 2015) and Taramandal area (Kumar 2020) of Gorakhpur city following the simplified procedure of Idriss and Boulanger (2005) using SPT data. Tripathi and Zafar (2016) carried out a seismic risk assessment for Gorakhpur city, in terms of peak horizontal acceleration which was estimated to be 0.312 g using the attenuation model by Sharma (2000) and 0.032 g using the attenuation model by Iyenger and Raghukanth (2004). The calculated PGA values are consistent with observed values of Nepal earthquakes and similar to standards reported in other studies. However, these studies could not delineate the high seismic risk areas and they recommend further studies to be conducted in order to assess the specific vulnerabilities of infrastructure and buildings in the Gorakhpur region to seismic hazards, and to develop appropriate mitigation strategies.

The current study has attempted to highlight areas that are most likely to experience serious damage during an earthquake, allowing for targeted measures to be taken to reduce vulnerability and value at risk associated with earthquakes, and are an important tool for disaster mitigation planning, by focusing on the integration of multi-criteria decision analysis (MCDA) techniques, specifically the Analytical Hierarchical Process (AHP) with GIS for earthquake hazard analysis to produce seismic microzonation hazard maps for Gorakhpur city, to aid in earthquake disaster monitoring, analysis, and risk management. This study also highlights the importance of considering various criteria, such as Geology/Lithology,

Geomorphology/Topography, Land Use-Land Cover, Liquefaction, Peak ground acceleration, Predominant frequency, Peak amplification, and Shear wave velocity.

10.2 Study Area

10.2.1 Geology and Geomorphology

The Gorakhpur City, situated in seismic zone IV of the Seismic zonation map of India, lies on the Quaternary sediments of the middle Gangetic Plain (i.e., a geographical category of the Gangetic Plain) along the left bank of Rapti River within interfluve of Ghaghra River in the west and Gandak River in the east. These fluvial Quaternary sediments are represented by the Older Alluvium and Newer Alluvium (Fig. 10.2). The Older Alluvium comprises a polycyclic sequence of oxidized, yellow to brown colored silt–clay and micaceous sand forming an upland surface. The Newer Alluvium consists of Terrace cum Channel Alluvium and Paleochannel Alluvium. Terrace cum Channel Alluvium comprises unoxidized sand, silt, and clay occurring all along Rapti and Rohini Rivers forming the lowland surface. Paleochannel fill deposits are distinctly less oxidized brownish to unoxidized dark grayish to blackish fine sands with sorted-rounded grains lying on the Older Alluvium upland surface as low-lying tracts in a linear fashion within the main city.

Geomorphologically, the northern and south-western parts of the area comprise the upland surface of the Older Alluvium along with the lowland surface of the Newer Alluvium surrounding the Rapti and Rohini rivers, Ramagarh Tal, and other tals/lakes. The upland surface comprises the distal part of the mega fan (i.e., Gandak fan) or older floodplains of the east migrating Gandak River (Pati et al. 2019). These are located mainly in the northern, north-central, southwestern, and isolated patches in the western part of the area. This surface is generally higher in elevation (73–82 m above msl) and flood-free. A few small streams/nalas (i.e., Gordhiya nala, Bichhia nala, Tura nala, etc.) drain the upland surface connecting to present streams/lakes of the lowland surface. Main Gorakhpur city is settled on this flood-free upland surface. Many small lakes/ponds/tals (i.e., Chilua Tal and others) are located within the upland surface mainly formed along paleochannels/abandoned channels and others. At places, small mounds are also observed at the upland surfaces which are generally occupied by the habitations. The lowland surface comprises low-lying areas of mainly stream courses (i.e., Rapti, Rohini Rivers, and their tributaries). Active channels and their flood plains or T_1 terraces have many tals/lakes/ponds (i.e., Ramgarh Tal, Chilua Tal, Semri Tal, etc.) formed as abandoned channels, oxbow lakes, or meander cut-offs. The Older Flood Plain (erosional T_1) is marked by a distinct scarp of 1–2 m height with respect to the upland surface of the Varanasi Plain. The Older Flood Plain (depositional) of Rapti and Rohini rivers (paired T_1 terrace surface) is characterized by a distinct scarp of 1.5–3 m with respect to the upland surface of the Varanasi Plain. The Active Flood Plain of Rapti and Rohini rivers are restricted within

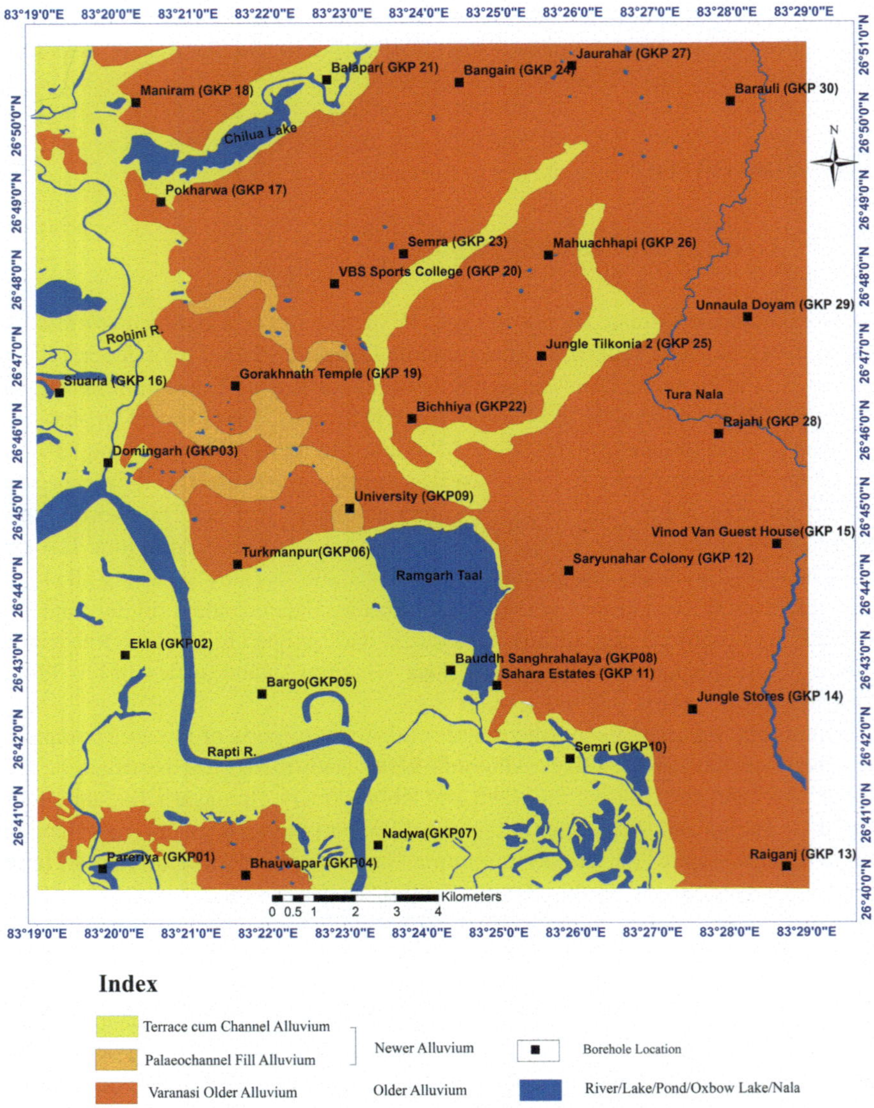

Fig. 10.2 Geological Map of Gorakhpur City and adjoining areas

the present-day bank limits of the river. The presence of point bars and occasional channel bars is common in Rapti River. On the whole, the lowland and highland surfaces of the area have been delineated into four geomorphic units on the basis of the geomorphic characteristics observed on digital DEM, imageries, and field observations which are as highland surface, lowland surface, active channel, and lake/tal. The elevations and slope of the area is much affected by the course of Rapti River and its abandoned channels which is bordered by a chain of depressions and lakes in the city. The upland surface may be relatively safer having low ground motion potentiality due to firm ground compared to moist and softer ground/sediments of the lowland surface having plenty of water bodies.

10.3 Seismo-Tectonic Setting and Seismicity

Gorakhpur city lies in the seismic zone IV of the Seismic Zoning Map of India (2016) having a zone factor of 0.24. According to the Vulnerability Atlas of India (BMPTC 2006; 3rd edition, 2019), the civil constructions in Gorakhpur fall in medium vulnerability to earthquake forces and may suffer effects of intensity VII to VIII in the event of a destructive earthquake, which in turn implies that besides the non-structural damage, some structural damage to the constructions leading to injury or even death may take place. The poorer constructions are likely to get damaged even more severely.

According to the seismo-tectonic map of India (GSI 2000), Gorakhpur lies in part of Indo-Gangetic plain, bounded by the Himalaya in the north (at ~ 100 km) and the terai region of Middle Gangetic Plains in the south. The Indo-Gangetic plains are traversed by a few subsurface faults (Fig. 3a and b). These subsurface faults exhibit two prominent trends (i) NNE-SSW to NE-SW and (ii) East–West. The Gorakhpur lies in between NNE-SSW and East–West trending subsurface faults in the north of subsurface Mirganj Graben (Fig. 3a). According to the Fault map of India (Gupta et al. 2016), Gorakhpur Fault is 120 km in length and pass through Mirganj Graben (Lat: N26.4° to N27.1°; Long: E82.8° to E83.8°). An ENE-WSW trending subsurface Gorakhpur-Padrauna Fault lies toward the north of Gorakhpur city (Pati et al. 2011, 2019) (Fig. 3b). The HFT is located about 85 km north around the exit of Gandak River.

The historical seismic events around a 300 km radius, keeping Gorakhpur in the center were cataloged and processed after homogenization to calculate the highest yielding peak ground acceleration at the study area. The maximum magnitude (M_{max}) is an important variable which affects the site as it reflects the maximum potential of energy that an earthquake can release. Maximum Credible Earthquake (MCE), i.e., the magnitude of the largest possible earthquake is considered for a defined area based on the highest magnitude, PGA values, the intensity at the site, and epicentral distance from the site. In the current study, using Deterministic Seismic Hazard seismic event yielding the highest PGA value (Sharma 2000) within the area bounded by 06° × 06° coordinates around the study area is the 2015 Gorkha/Nepal earthquake of magnitude 7.8, a M_{max} of 8.3 (7.8 + 0.5) is taken as the scenario earthquake (Nuttli et al. 1979) which is further used for evaluation of other geofactors—Liquefaction and Shear Wave Velocity.

The Peak ground acceleration (PGA) values estimated at the center of the study area are calculated through NGA-West 2 GMPE in Gorakhpur city which vary from 0.0457 g to 0.03394 g. This GMPE is utilized for the preparation of a PGA map (Fig. 10.4) as it also takes into consideration local site conditions, i.e., the value of

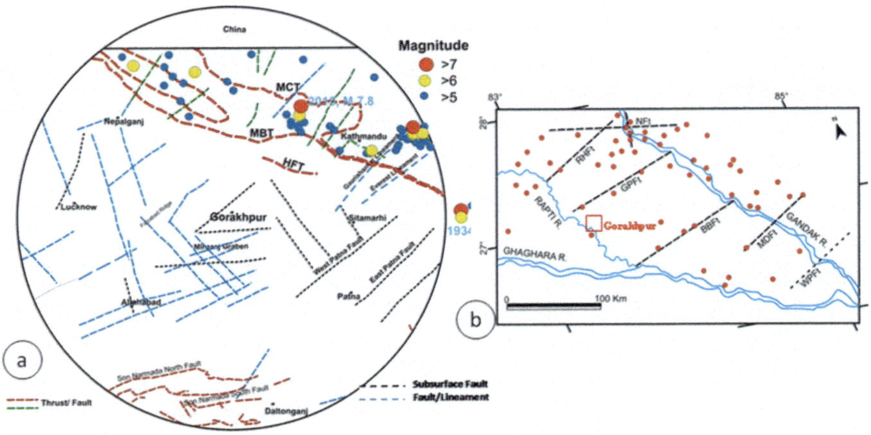

Fig. 10.3 A Fault and lineament map within a 300 km radius around Gorakhpur with seismic events > 5 M (USGS) (based on Seismo-tectonic Atlas and its Environs, GSI 2000) **b** Map showing Gorakhpur-Padrauna Fault (GPFt) (After Pati et al. 2019)

shear wave velocity at the site of interest along with distance (R) and magnitude of the earthquake (MCE) of the source zone (i.e., 2015, Gorkha earthquake).

10.4 Methodology

The site-specific seismic hazard assessment includes data collection, seismic hazard modeling, vulnerability assessment, and a hazard map based on the hazard index. The present study follows the broad methodology laid down as per the Standard Operating Procedure (SOP) of GSI for Level 'B' seismic microzonation, 2017. A total of 300 sq km was taken up for assessing the seismicity, estimation of M max, and Intensity of the maximum credible earthquake using deterministic seismic hazard analysis. Seismo-tectonic maps (GSI 2000) have been used for information on the basement depth and its structural features. The first phase involves creating a number of thematic maps, such as geological, geomorphological, land use land cover, and seismo-tectonic ones, based on earlier studies and literature on the region as well as additional field data. The area was subjected to the drilling of 30 vertical boreholes, each 30 m deep, utilizing an SPT-based geotechnical approach, as well as core logging for the investigation of geotechnical aspects such as grain size analysis and others. Finally, the geotechnical characteristics of the subsurface profile at the drill sites, including the shear wave velocity, were determined using the N values (i.e., the number of blows/strokes by the SPT assembly hammer to penetrate a 15 cm vertical soil column) and corrected N value (N1)60 of the subsurface soil layers obtained by Standard Penetration Test (SPT) in the drill holes at 1.5 m intervals. Shear Wave Velocity Maps were created using geotechnical SPT-derived Vs 30 values up to

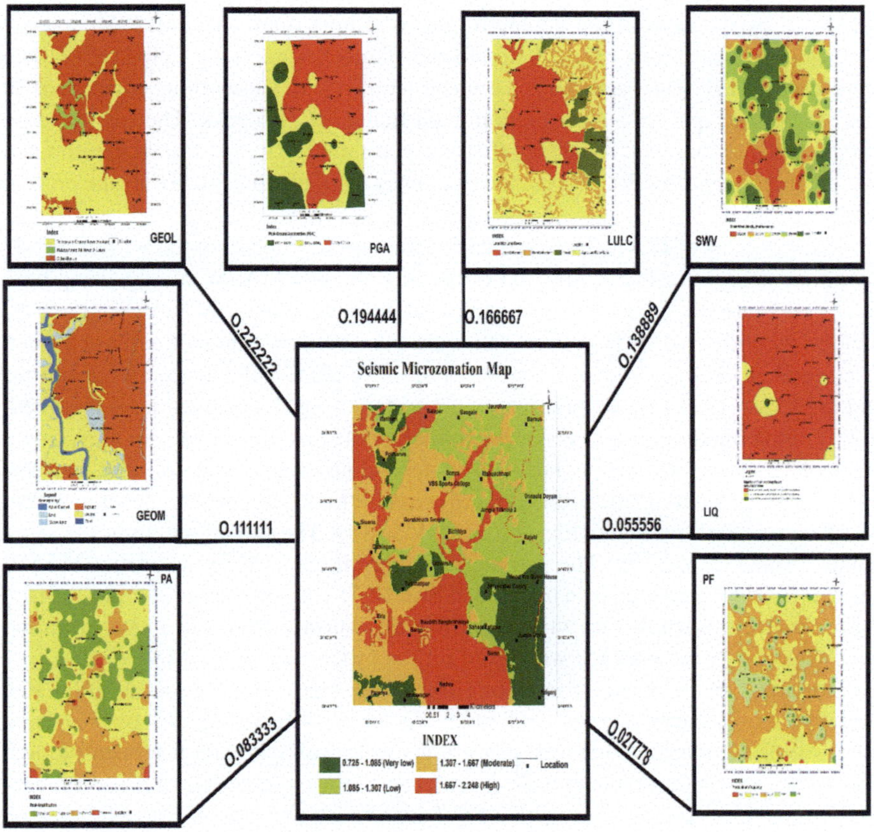

Fig. 10.4 Seismic-microzonation map of Gorakhpur city along with other thematic maps (i.e., GEOL-geological map, GEOM-geomorphological map, LU-land use map, SWV-shear wave velocity map, PA-peak amplification map, LQ-liquefaction susceptibility map, and PF-predominant frequency map)

30 m depth, as well as geophysical instrument-derived Vs 30 values using MASW. Geophysical investigations comprising an Ambient Noise/microtremor survey at 300 sites, and a Multichannel Analysis of Surface Waves (MASW) survey at 150 sites have been carried out with the objectives of determining the predominant frequency and site amplification from noise survey site response study, average shear wave velocity up to 30 m depth from MASW survey. Noise recordings have been done at a sampling rate of 100 Hz through moving stations at different places for an average period of 40–60 min by deploying short-period digital seismographs (ALTUS, K-2, Kinemetrics Inc., USA) coupled to three numbers of short-period seismometers (RANGER, SS-1, Kinemetrics Inc., USA, and Ambient noise data was processed using the Geopsy (Geophysical signal database for noise array) software suite. The recorded ambient noise data have been split into 30 s time windows. The H/V is

computed by the ratio of the geometric mean of horizontal (NS and EW) components with vertical components. A Konno and Ohmachi smoothing technique (Konno and Ohmachi 1998) was applied to the spectral ratios with a smoothing constant of 30 and 5% cosine taper for a clear predominant frequency (fo) peak. The site response parameters, predominant frequency (fo), and peak amplification (Ao) were estimated from ambient noise for 300 sites in the frequency band of 0.5 to 10 Hz for the present work.

The Stratavisor NZXP signal enhancement seismograph coupled with a 4.5 Hz geophone is used for the acquisition of the seismic data/MASW survey, using both active and passive micro-tremor sources. Data acquisitions in MASW (Passive) include recording of surface waves that are generated by ambient activities, such as cultural noise, e.g., highway traffic, wind movement, and construction activities and in MASW (Active), seismic energy generated using sledge hammer is recorded. In this active method, a 10 kg sledgehammer is used as the active source. A linear array of 12–24 4.5 Hz geophones is used depending on the availability of space in the area. Based on the evaluation of initial measurements, a geophone spacing of 5 m or 3 m is selected for conducting the MASW survey. The spread length of the array ranges from 115 to 33 m. The parameters and settings used for acquiring MASW (Active/Passive) data are given in Tables 4.6 and 4.7. The software programs, Surface Wave Analysis Wizard (Pickwin) and Wave Eq (Surface Wave Analysis) from the SeisImager software package were used for the processing of the MASW data.

Estimation of liquefaction susceptibility of the study area, based on grain size analysis, its derivative curves, and safety factor at different depths by Seed and Idriss (1982) empirical equations using a geotechnical approach has been plotted. The multi-criteria decision analysis specifically the Analytical Hierarchical Process (AHP) was utilized to determine the weights and rank of all geofactors/parameters. Finally, integration of all the normalized weight and ranks data of thematic maps' (viz., geological, geomorphic, land use- land cover, average shear velocity, peak amplification, liquefaction susceptibility, predominant frequency, and PGA maps) has been carried out for preparation of integrated seismic microzonation hazard map of the area on Arc GIS platform along with its statistical validation using ROC curves.

10.5 Data Integration: Unveiling Hazard Zones

Earthquake hazard is principally considered as a matter of spatial and temporal constraint where the area under threat is classified into different hazard levels and the objective tends to predict the qualitative/quantitative spatial–temporal hazard of effect of a future earthquake. Seismic microzonation, aided by AHP, provides a robust framework for mapping earthquake hazard zones and will serve as a testament to Gorakhpur's commitment to earthquake preparedness, and paves the way for a safer future for its citizens.

Seismic Hazard Map of Gorakhpur City is generated by using multi-criteria assessment, i.e., Analytical Hierarchical Process (AHP), and integrating eight thematic maps depicting various aspects of seismic vulnerability: geology, peak ground acceleration, land use-land cover, average shear wave velocity, geomorphology, peak amplification, liquefaction susceptibility, and predominant frequency.

The instrumental derived geofactors/thematic maps, i.e., Peak amplification and Predominant Frequency by ambient noise survey and the average shear wave velocity down to 30 m depth by MASW survey have been determined using Seisimager software. The overall variations of average shear wave velocity (obtained by MASW survey) down to 30 m depth of sites within Gorakhpur city range from 202.5 m/s (Champa Devi Park, Taramandal site) to 357.4 m/s (Chawri, SE of Ayodhya Tola). According to the NEHRP standard, all the studied sites fall in site class D belonging to the stiff soil type. The predominant frequency contour map of Gorakhpur city is shown in Fig. 10.4. The peak amplifications obtained from the geophysical/MEQ site response curves of the surveyed area vary in the range of 0.7–3.5. The overall variation in predominant frequencies of sites within Gorakhpur city ranges from 1.0 to 9.3 Hz.

Geology, Geomorphology is validated through field checks, and Land use of the study area is demarcated on the basis of visual interpretation of Survey of India's toposheets (2015 updated), Google Imagery, LISS III satellite imageries and Gorakhpur Master Plan 2021. The area is divided into settlements (including residential, industrial, commercial, public, road network, etc.), further divided into urban and rural settlements, forest cover, and agricultural land including water bodies. About 60% area of Gorakhpur is under habitation (including both urban and rural settlement), ~ 35% is under agricultural and fallow land and ~ 5% is under forest cover. Over time, the city is growing in the north, northeast, and east directions (Fig. 10.4).

Each thematic map, further divided into units, was assigned a rank based on its influence on seismic hazard, following the Geological Survey of India's (GSI) standard operating procedure. The AHP method then determined the relative importance of each map, assigning appropriate weights. Pair-wise comparisons were used to determine the relative importance of each alternative (i.e., themes) with respect to each other in terms of each criterion following a scale from 1 to 9 and its reciprocals. This gives a normalized weight for each criterion as shown in Table 10.1.

GEOL: Geology; PGA: Peak Ground Acceleration; LULC: Land Use-Land Cover; GEOM: Geomorphology; SWV: Average Shear Wave Velocity; PA: Peak Amplification; LIQ: Liquefaction susceptibility in terms of safety factor; PF: Predominant frequency.

Geographical Information System (GIS) provides a suitable platform for the logical integration of inputs, their analysis, and retrieval or development of a hazard index map based on weighted seismic hazard parameters coupled with ground information using the Analytic Hierarchy Process (AHP). AHP approach is used to construct seismic hazard maps by establishing the weights of several characteristics that affect how much an earthquake affects a particular site (Erden and Karaman 2021). According to their significance in the hazard assessment, several parameters such as, soil classification/ lithological/ geological maps, peak ground acceleration, shear wave velocity, predominant frequency, topography/geomorphology, liquefaction, and various other factors are given weights. These weights are computed using

Table 10.1 Pair-wise comparison matrix of themes and their normalized weights

	GEOL	PGA	LULC	SWV	GEOM	PA	LIQ	PF	Normalized weight
GEOL	1	1.142857	1.333333	1.6	2	2.666667	4	8	0.222222
PGA	0.875	1	1.166667	1.4	1.75	2.333333	3.5	7	0.194444
LULC	0.75	0.857143	1	1.2	1.5	2	3	6	0.166667
SWV	0.625	0.714286	0.833333	1	1.25	1.666667	2.5	5	0.138889
GEOM	0.5	0.571429	0.666667	0.8	1	1.333333	2	4	0.111111
PA	0.375	0.428571	0.5	0.6	0.75	1	1.5	3	0.083333
LIQ	0.25	0.285714	0.333333	0.4	0.5	0.666667	1	2	0.055556
PF	0.125	0.142857	0.166667	0.2	0.25	0.333333	0.5	1	0.027778

the Analytical Hierarchical Process (AHP) and then used as input to create seismic hazard maps (Mokarram et al. 2021). The generated hazard maps include details on the seismic danger's geographical distribution in the area. The hazard maps may also be used for risk assessment and management by assessing the elements at risk, such as population and structures, by combining AHP with Geographic Information Systems (GIS) (Wang and Rathje 2015).

In this analytical approach to mitigate earthquake damage, three interacting factors, viz., geological, peak ground acceleration (PGA) values as well as land use-land cover pattern have a significant role. The role of peak ground acceleration based on the source and site characteristics cannot be overlooked. Significant importance has been given to assigning weightage to all those variables that are major influencing factors. The shear wave velocity, peak amplification, and liquefaction susceptibility of the area have direct control in deciding the intensity of the input seismic ground motion. The average shear wave velocity (Vs30) map is prepared based on site-specific geotechnical (SPT- $(N_1)_{60}$ values) and geophysical (MASW) studies. All these thematic layers have been integrated into the ArcGIS platform to prepare the final seismic hazard microzonation map as per the standard operating procedure (SOP) of GSI (2017) for the same purpose. AHP rating has been assigned to each sub-unit/ polygon of the layers/thematic maps. Values of each source layer have been extracted in the centroid of the mapping unit/grid. The addition of all source layer values has given the final value (AHP_{total}) of the microzonation map over each mapping unit. The extraction of the area has been carried out considering the assigned importance/ weightage and rank of the themes.

The integrated map gives the Hazard Index (HI), which is the integrated factor depending on the weights and ranks of the geoscientific, and geotechnical attributes, ground motion parameters, and liquefaction susceptibility. Finally, the integration of all thematic layers is achieved through the following formula on the GIS platform for the preparation of a seismic hazard microzonation map of Gorakhpur city:

$$HI = GEO_W.GEO_R + PGA_W.PGA_R + LULC_W.LULC_R$$

$$+ SWV_W.SWV_R + GEOM_W.GEOM_R + PA_W.PA_R$$
$$+ LIQ_W.LIQ_R + PF_W.PF_R/\sum W$$

where, HI represents the Hazard Index, and $_W$ and $_R$ represent the assigned weight and rank, respectively.

The final seismic susceptibility microzonation map is prepared based on the addition of normalized weightage and rank values of different thematic maps. The thematic maps (themes/geofactors) used for integration (Fig. 10.4) are as geology peak ground acceleration land use- land cover, shear wave velocity, geomorphology, peak amplification, liquefaction susceptibility, and predominant frequency. The seismic hazard microzonation map prepared by weighted overlay/ integration of the above-mentioned thematic maps (themes/geofactors) is shown in Fig. 10.5. The final product, the seismic microzonation hazard map, paints a clear picture of Gorakhpur's varying seismic hazard zones, categorized as low, moderate, high, and very high.

10.6 Results and Discussion

The Hazard Index thus obtained (Fig. 10.5) varies from 0.725 to 2.248, which have been divided as per geometrical interval statistical parameter into four groups as follows—(i) Susceptibility Index 0.725–1.085, (ii) Susceptibility Index 1.085–1.307, (iii) Susceptibility Index 1.307–1.667, and (iv) Susceptibility Index 1.667–2.248.

On the basis of the seismic Susceptibility Index, the studied part of Gorakhpur city and adjoining areas has been broadly classified into four zones of seismic susceptibility. The Susceptibility Index 0.725–1.085 is attributed to very low seismic susceptibility; 1.085–1.307 is attributed to low hazard, 1.307–1.667 is attributed to moderate hazard, and 1.667–2.248 is attributed to high seismic susceptibility. Keeping in view of the low range of variability of ranks/ values of input thematic maps, the difference between high-hazard and moderate-hazard zones is relative and is not so high. However, damages may be caused due to variations in magnitude, distance, depth, and time of future earthquakes along with the type of structures/ buildings. The nearby, high-magnitude earthquake at night may cause more damage and loss to the poorly constructed structures/buildings (i.e., traditional *Kuchcha* houses and poorly constructed masonry buildings, high-rise buildings, etc.) compared to well-engineered RCC buildings.

As per the seismic hazard microzonation map, high seismic hazard zones are mainly present in the low-lying areas of the Newer Alluvium of Rapti River and Rohini River tracts including Ramgarh Tal/lake in the southern and western portions of Gorakhpur city. High seismic zones are also present in the southern part of the city around Ramgarh Tal/lake area, Bargo, Choti Semri, Koni, Khairatola, Gehuasagar, Pipra, Jhorwan, Naukapar, Nadwa, Bishunpur Buzurg, Permanpur, Bhaghagra, Nausar and Dangipar areas. In the northern extension of the city, high seismic susceptibility zones are present along Chilua Tal/lake around Sonbarsa and

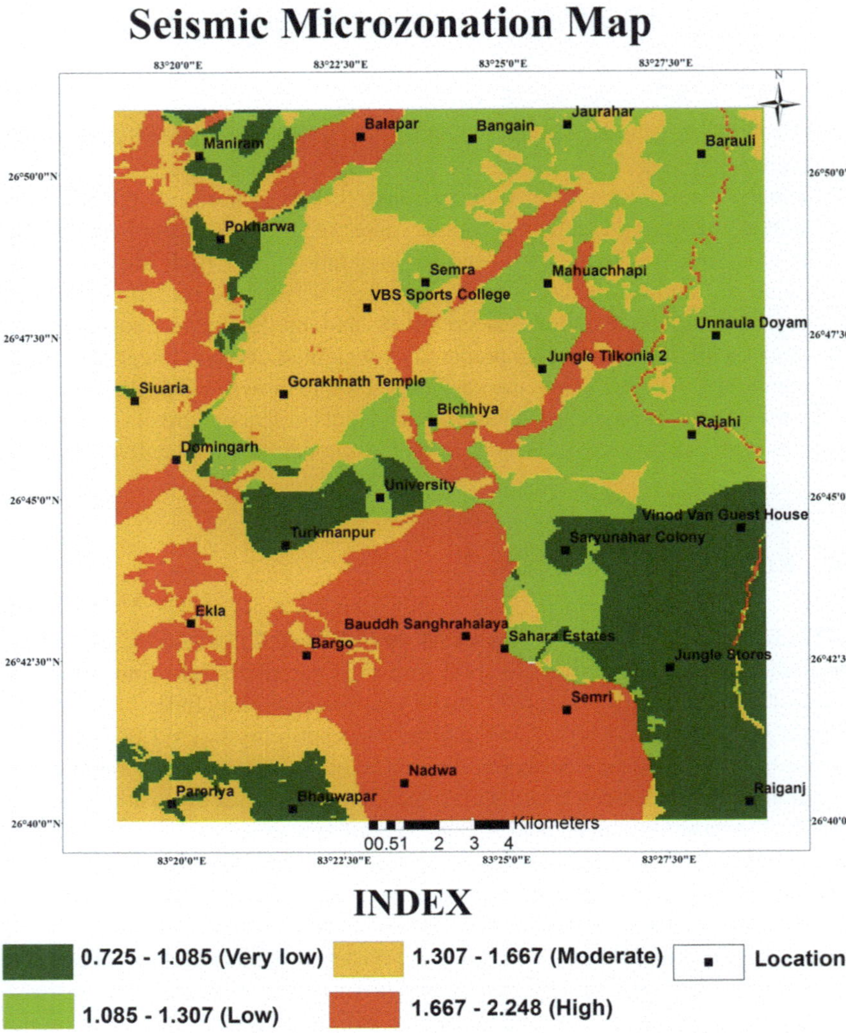

Fig. 10.5 Seismic-microzonation map of Gorakhpur city, Uttar Pradesh

Balapar. A few isolated seismic high zones are present around Semra, Bicchiya, Shahpur, Lalganj, Jangaltola, Dhusar, Sanjhai, and near Jangal Tikonia (Fig. 10.5).

The "high" seismic susceptibility zones are observed mainly due to geological components like the presence of Newer Alluvium (i.e., Channel/Terrace/Paleochannel Alluvium) where the shear wave velocity is low and peak amplification is high. Out of the 300 sq km studied area, 44% area falls under the low and very low seismic susceptibility zone, 30% area falls under the moderate susceptibility zone, and 26% of the total area falls under the high susceptibility zone (Fig. 10.6).

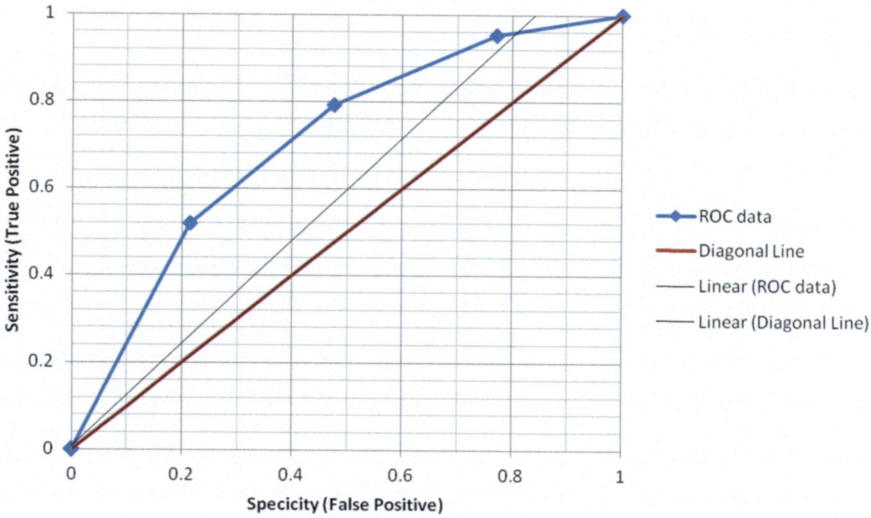

Fig. 10.6 ROC curve of the Seismic Hazard Zonation Index map of Gorakhpur city

Validating the accuracy of seismic hazard zonation maps is crucial for ensuring their effectiveness in disaster mitigation planning. This process involves comparing model predictions with observed data, leveraging the Receiver Operating Characteristic (ROC) methodology (Peterson et al. 1954). ROC analysis provides a robust framework for evaluating model performance by plotting sensitivity (true positive fraction) against specificity (true negative fraction) across various thresholds Deleo (1993).

In this study, the ROC analysis confirmed the validity of our integrated seismic hazard zonation map for Gorakhpur city. The ROC curve showed no spatial overlaps between successive hazard indexes (Fig. 10.6), demonstrating the model's ability to distinguish between stable and hazardous areas Bonham-Carter (1994a, b); Soeters and van Westen (1996). Quantitative validation using the Area Under Curve (AUC) measure further confirmed the model's accuracy. The AUC value of 0.714, calculated as the sum of individual trapezoidal areas as described by Pradhan and Lee (2010), indicates a fairly good prediction accuracy. Brenning (2005) suggests that the integrated hazard zonation map can effectively identify areas at risk from earthquakes. This validation emphasizes the importance of rigorous accuracy assessment, especially considering the inherent probabilistic nature of predictions (Holliday et al. 2006). The ROC methodology offers a valuable tool for researchers and practitioners to assess the effectiveness of seismic hazard zonation models, ultimately contributing to more robust and effective earthquake preparedness strategies.

10.7 Conclusion

The salient points of conclusive remarks/inferences drawn from this Level-B, integrated seismic microzonation study of Gorakhpur city are given below:

i. M 7.8, 2015, Gorkha/Nepal earthquake located at a distance of about 200 km from Gorakhpur city within 06 deg by 06 deg coordinate bounded area produced the highest peak ground acceleration (PGA) in the city. Accordingly, computation based on Shebalin (1958) and Karnik (1969) relationships for maximum credible earthquake (MCE) of magnitude 8.3 (M7.8 + 0.5) at a depth of 8.2 km occurring at the location of 2015 Gorkha/ Nepal earthquake lying at about 200 km distance gives maximum intensity > VI at Gorakhpur city (which has also been recorded during ground macroseismic survey just after 2015 Gorkha Earthquake (7.8 Mw) by GSI for preparation of isoseismal map; Joshi et al. 2015). Further, on the basis of the attenuation relationship for vertical peak ground acceleration for the Himalayan Region (Sharma 2000), we get vertical PGA as 0.25733 g and corresponding horizontal PGA (i.e., 2/3 of vertical PGA) as 0.385995 g (by deterministic seismic hazard assessment/ DSHA methodology). However, the strongest seismic event within 06 deg by 06 deg coordinate bounded area is 1934 (M 8), Bihar-Nepal Earthquake (at a distance of 319 km NNE of Gorakhpur) having a maximum credible earthquake of magnitude 8.5 at a depth of 15 km which has yielded vertical and horizontal PGA values as 0.236194 g and 0.354291 g, respectively, (GMPE by Sharma 2000). PGA has also been calculated at 30 sites of the area as 0.0457 g to 0.03394 g by NGA-West2 ground motion prediction equation/ GMPE (Boore et al. 2014).

ii. The study area comprises fluvial deposits of the Gangetic system and is subdivided into the Older Alluvium and Newer Alluvium. The Older Alluvium/Varanasi Alluvium of the Middle to Late Pleistocene age comprises an oxidized, polycyclic sequence of reddish brown, oxidized sand, silt, and clay with ferruginous and calcareous concretions, occasionally with distinct buff-colored stains due to leaching. It forms an upland surface in the northern and southwest parts. The Newer Alluvium of the Holocene age comprises white to light gray riverine, calcareous, moist, unoxidized sand, silt, and clay horizons. The Newer Alluvium occupies the lowland surface, mainly along Rapti River and its tributaries, and comprises the Terrace-cum-Channel Alluvium with light gray, fine to medium grained, unoxidized sand, silt, and clay horizons.

iii. The geotechnical, grain size data curves suggest the presence of shallow subsurface (<10 m depth) liquefiable sand horizons in the area. Based on the methodology of Seed and Idriss (1982) considering scenario earthquake of M8.3, h 8.2 km and corrected SPT $(N_1)_{60}$ values, liquefaction safety factor has been calculated estimating seismic demand (CSR) and capacity (CRR) ratios. A relatively high liquefaction domain (<1 safety factor) has been identified in almost all over Gorakhpur city except around Bangain, Rajahi, Turkmanpur, Siuaria, and Vinod Van areas.

iv. By geotechnical studies (i.e., SPT- $(N_1)_{60}$ values) and MASW survey, the shear wave velocity/Vs30 values range from202.5 m/s to 357.4 m/s corresponding to Soil Type - D (i.e., stiff soil) of the Site classes of NEHRP Code Provisions (1997). The average shear wave velocity is relatively lower over the sites occurring on the Newer Alluvium Surface.
v. Based on ambient noise survey and average shear wave velocity empirical data, the peak amplification is up to 3.5 times (ranging from 0.7 to 3.5) and is generally toward the lower side where Vs30 has comparatively higher values, which suggests the higher potential for damage during earthquake shaking.
vi. Based on the ambient noise survey, the predominant frequency varies from 2 Hz to 9.5 Hz, predominantly between 2 Hz and 4.5 Hz having mainly medium damage potential. The localities of the Older Alluvium upland surface show relatively higher predominant frequency like around Maniram, Auraiya, Lakshmipur, Jharkhandi, Siktauli, near north of Ramgarh Tal, and Pareriya (i.e., with > 7 Hz).
vii. The integrated seismic microzonation map of a 300 sq km area of Gorakhpur city prepared by integrating all the thematic layers on the Arc GIS platform suggests 44% area under low and very low seismic susceptibility zone, 29% area under moderate susceptibility zone and 27% under high susceptibility zone, out of the total area. The integrated seismic microzonation map depicting "very low" to "high" susceptibility zones basically suggests relative seismic susceptibility levels based on low variability of input parameters.
viii. The validation of the seismic microzonation model using the ROC (receiver operating procedure) curve gives a calculated value of 0.714 with an estimated standard error of 0.0036 which shows that the integrated seismic hazard microzonation map of Gorakhpur city holds fairly good prediction accuracy.
ix. The high seismic zones are also present in the southern part of the city around Ramgarh Tal/ lake area, Bargo, Choti Semri, Koni, Khairatola, Gehuasagar, Pipra, Jhorwan, Naukapar, Nadwa, Bishunpur Buzurg, Permanpur, Bhaghagra, Nausar and Dangipar areas. In the northern extension of the city, high seismic susceptibility zones are present along Chilua Tal/lake around Sonbarsa and Balapar. A few isolated seismic high zones are present around Semra, Bicchiya, Shahpur, Lalganj, Jangaltola, Dhusar, Sanjhai, and near Jangal Tikonia.

10.7.1 Recommendations

It is suggested that while designing the construction of major engineering projects/ multi-storied buildings in high and moderate seismic micro-zones, adequate seismic inputs should be incorporated to withstand the incoming seismic ground motion from a big earthquake in the surrounding region, as a part of seismic hazard preparedness adopting seismic safe construction practices.

References

Bonham-Carter GF (1994) Geographic information systems for geoscientists: Modelling with GIS. Pergamon. eBook ISBN: 9781483144948

Bonham-Carter GF (1994) Geographic information systems for geoscientists: Modelling with GIS. Pergamon

Boore DM, Thompson EM, Cadet H (2014) NGA-West2 equations for predicting PGA, PGV, and 5% damped PSA for shallow crustal earthquakes in western North America. Earthq Spectra 30(3):1057–1085

Brenning A (2005) Landslide susceptibility mapping using optimal classification trees. Doctoral dissertation, Technische Universität München

Chaudhary R, Kumar S, Kumar P (2015) Liquefaction susceptibility analysis of Ramgarh Taal area, Gorakhpur district, Uttar Pradesh, India: a GIS based approach. Intern J Remote Sens GIS 4(1):33–42

Dasgupta S, Mukhopadhyay M, Ghosh DB (1987) Seismicity of the Himalayan region and its relation to the structure. Tectonophysics 138(1–2):177–193

Deleo GA (1993) Receiver operating characteristic (ROC) analysis: sensitivity and specificity revisited. Acad Radiol 1(2):147–152

Dewey JF, Bird JM (1970) Mountain belts and new global tectonics. J Geophys Res 75:2625–2685

Erden T, Karaman H (2021) Earthquake hazard assessment of the city of Trabzon, Turkey, by using analytical hierarchy process (AHP). Bull Earthq Eng 19(10):4439–4469

Frattini P et al (2010) Statistical methods for the validation of quantitative landslide susceptibility maps. Nat Hazard 10(5):1055–1064

Gino D, Del Monaco M, Sbarra P (2022) Seismic microzonation: a brief review of current practice and perspectives. Bull Earthq Eng 20(2):699–720

Goswami SB (2012) Seismic microzonation of Guwahati city, India, using microtremor measurements. Nat Hazards 63(2):1235–1252

GSI (2000) Seismicity and related studies in the Himalayas. Geol Surv India, Special Publication 56:1–215

SOP GSI (2017) Standard Operating Procedure for Seismic Microzonation Studies in India, Geological Survey of India, Government of India, Kolkata

Hamid A (2018) Seismic microzonation of Srinagar city, Kashmir Valley, India: a GIS-based approach. Nat Hazards 90(3):1009–1028

Holliday JN, Abedin MZ, Godwin RJ (2006) The use of ROC curves in the validation of liquefaction susceptibility maps. Nat Hazards 39(1):215–228

Idriss IM, Boulanger RW (2005) Standard penetration test (SPT) and liquefaction evaluation. Earthquake Engineering Research Institute

Iyengar RN, Raghukanth ST (2004) Attenuation of strong ground motion in peninsular India. Bull Seismol Soc Am 94(1):281–292

Jayangondaperumal R, Elangovan K, Srinivasamurthy B (2010) Site characterization for seismic microzonation of Chennai city, India. J Asian Earth Sci 37(3):256–268

Keil C, Wylegalla K, Lehujeur M (2021) Integrating ambient vibration and full-waveform inversion for improved microzonation of the city of Strasbourg, France. Geophy J Intern 226(2):1255–1274

Kono T, Ohmachi T (1998) Evaluation of site amplification effects based on S-wave velocity and spectral ratio of microtremors. Soil Dyn Earthq Eng 17(1):39–48

Kumar A (2020) Liquefaction susceptibility mapping of Taramandal area, Gorakhpur city, India using SPT data and GIS. Intern J Recent Technol Eng 8(3):2277–2281

Kumar S, Singh R, Rana SN, Bhandari RK (2022) Seismic microzonation of Shimla city, Himachal Pradesh, India. J Seismol 26(1):1–18

Mase S, Otake K, Satoh T (2021) Seismic microzonation using the spectral ratio method: a review. Geosciences 11(11):459

Mishra A (2020) Seismic microzonation of Delhi city, India. Nat Hazards 102(2):665–685

Mokarram M, Asadi SH, Karimzadeh S (2021) Seismic microzonation of Ahvaz city using AHP and GIS techniques. Nat Hazards 109(1):425–445
NEHRP Code Provisions (1997) National earthquake hazards reduction program recommended provisions for seismic regulations for new buildings and other structures, Part 1: Provisions, FEMA 302/303. Federal Emergency Management Agency, Washington, D.C.
Parkash B, Kumar S (1991) Earthquakes in northern India. Curr Sci 60(3):193–204
Pati S, Kumar D, Sah MP (2019) Morpho-tectonic characteristics of the Gandak megafan, north Bihar plains, India: implications for neotectonics and paleoseismology. Geomorphology 327:146–160
Peterson WW, Birdsall TG, Fox WC (1954) The theory of signal detectability. Trans IRE Prof Group Inform Theory 4(4):171–212
Pradhan AM, Lee CT (2010) Regional landslide susceptibility mapping using an ensemble of AdaBoost, support vector machines and decision trees. Landslides 7(1):111–125
Puri VK, Jain SK (2019) Seismic microzonation of Amritsar city, India. Nat Hazards 98(1):215–233
Raiverman V, Jain AK, Rastogi BK (1983) Seismicity and tectonics of the northeastern Himalayas. Tectonophysics 96(1–2):49–66
Rao R (1973) A review of the seismicity of the Himalayan region and adjoining areas. Bull Seismol Soc Am 63(5):1785–1817
Seismotectonic Maps, GSI 2000. Seismotectonic Atlas of India, Geological Survey of India, Government of India, Kolkata
Sharma ML (2000) Attenuation of strong ground motion in India. Curr Sci 79(5):619–627
Singh DP, Kumar A, Rastogi BK (1996) Source parameters of earthquakes in the Garhwal-Kumaon Himalaya and their tectonic implications. J Seismolog 1(1):23–33
Soeters R, Van Westen CJ (1996) Slope instability recognition, analysis and zonation. Landslides: Investigation and mitigation
Thakur VC, Suresh G, Kumar R, Srivastava PK (2009) Geophysical investigations for microzonation of Lucknow city, India. J Appl Geophy 68(3):402–419
Tripathi B, Zafar A (2016) Seismic risk assessment of Gorakhpur city, India. Intern J Eng Res Techn 5(1):261–267
Valdiya KS (1976) Himalayan tectonics and seismicity. In Himalayan Geology (pp. 87–114). Springer, Berlin, Heidelberg
Wang Y, Rathje EM (2015) Seismic microzonation for building earthquake risk assessment: a case study of Portland. Oregon. Earthquake Spectra 31(4):1787–1807

Chapter 11
Assessment of Liquefaction Potential of Bridge Site Located at Koyalajan Kamla River in Rasalpur and Koyalajan Road, Bihar

Anurag Goyal, Anjali Gupta, and Brijesh Kumar

Abstract In seismically active zones like bridge project sites located at Koyalajan Kamla River in Rasalpur and Koyalajan Road at Block Biraul, Bihar, India structural safety necessitates assessing liquefaction potential. An analysis for a bridge project in these areas, located in seismic zone V, revealed vulnerability due to a previous magnitude 7.5 earthquake with peak ground acceleration (PGA) of 0.36. With groundwater near the surface, two 50-m boreholes were drilled to evaluate site-specific conditions and assess the likelihood of liquefaction. The soil's liquefaction susceptibility was assessed using the Standard Penetration Test (SPT), measuring N-value and fines percentage in line with IS 1893-2016 Annexure-F guidelines for liquefaction potential analysis. The soil investigation report indicated that the site mainly consists of poorly graded silty sand with medium to very dense characteristics (SM). Given the significant foundation load, a well foundation was deemed necessary, and the gross allowable load-bearing pressure at the well tip for bridge abutments and piers was recommended in a range from 85 to 90 T/m^2. The study's methodology highlighted the importance of being cautious and attentive as the analysis indicated that liquefaction is likely to occur up to 11 m depth below the ground level.

Keywords Earthquake · Bridge project site · Liquefaction · SPT · Site investigation

---

A. Goyal
Geotechnical Consultant, Liniva Consultants, M.Tech, New Delhi 110029, India

A. Gupta (✉)
Department of Civil Engineering, Manav Rachna International Institute of Research and Studies, Faridabad 121003, India
e-mail: anjaligupta.set@mriu.edu.in

B. Kumar
Department of Computer Science and Engineering, Manav Rachna International Institute of Research and Studies, Faridabad 121003, India

B. K. Rastogi et al. (eds.), *Natural Hazards and Risk Mitigation*, Springer Transactions in Civil and Environmental Engineering, https://doi.org/10.1007/978-981-97-7658-0_11

11.1 Introduction

Numerous devastating earthquakes have demonstrated the widespread occurrence of liquefaction and associated ground failures. During seismic events, rapid loading leads to insufficient time for natural drainage, resulting in excess pore-water pressures. This further reduces the shear strength and stiffness of soil deposits (Silahtar et al. 2023; Gurung and Chatterjee 2023; Kumar et al. 2023). Liquefaction typically occurs in loose saturated sands and is dependent on various factors, such as earthquake magnitude, peak ground acceleration, ground motion intensity and duration, distance from the earthquake source, soil type and thickness, fines content, degree of saturation, etc. (Youd and Perkins 1978; Kramer 1996).

The assessment of the possibility of liquefaction at a particular site is usually indicated by factors of safety against liquefaction (FS). In 1971, Seed and Idriss introduced a simplified method for assessing the liquefaction resistance of soils (the approach is adopted in IS 1893: 2016), which entails comparing the cyclic resistance ratio (CRR) of a soil element to the cyclic stress ratios (CSR) induced by seismic activity to determine its ability to withstand liquefaction (Seed and Idriss 1971; IS 1893 2016). The use of SPT data, which is widely employed to assess the liquefaction resistance of soils is integral to this procedure. FS values throughout the depth of a soil profile are usually determined by taking into account several factors such as PGA, earthquake magnitude (Mw), N-value, and overburden pressure (σv). A soil layer is deemed susceptible to liquefaction if $FS<1$ (Liao and Whitman 1986).

The Koyalajan Kamla River and Koyalajan Road region in Darbhanga, Bihar, India, is known to be seismically active and has a history of earthquakes. In this context, the liquefaction potential of soil strata below the proposed bridge between Rasalpur and Koyalajan Road needs to be assessed to ensure its structural safety. The paper discusses in detail a case study of the bridge, the procedure for liquefaction analysis, various parameters to be considered for its calculation, impact of liquefaction on SBC (Safe Bearing Capacity) for the well foundation at the proposed bridge site.

The findings of this study can provide valuable insights into the liquefaction susceptibility of soils in the region and can help in developing effective mitigation measures to ensure the safety of the proposed bridge.

11.2 Study Area

Darbhanga is situated in the northern part of the state of Bihar, India, and is situated on the Indo-Gangetic plain. The geology of Darbhanga is primarily characterized by Quaternary alluvial deposits, which are mainly composed of sand, silt, and clay. These deposits were formed by the sedimentation of river and stream channels, and are generally unconsolidated and poorly compacted. The thickness of these deposits varies from a few meters to several tens of meters, and the grain size distribution is variable, depending on the depositional environment and source of sediments.

The region around Darbhanga is seismically active and has a history of earthquakes (Thakur and Roy 2022; Rajendran et al. 2016; Umar et al. 2018). The seismic activity is mainly attributed to the collision between the Indian and Eurasian plates. The seismic hazard in the region is mainly associated with the occurrence of shallow earthquakes, which can cause significant ground shaking and liquefaction of unconsolidated soils. The geology and the associated hazards in the region have important implications for infrastructure development, particularly for designing and constructing structures like bridges, dams, and buildings, which need to be resilient and resistant to seismic and hydrological hazards (Capacci et al. 2022). The proposed bridge site spread from Koyalajan Kamla River in Rasalpur to Koyalajan Road, Darbhanga, Bihar, India. The project site is situated in an area designated as "Zone V" on the latest seismic hazard map, indicating that it is in a seismically active region and is susceptible to earthquakes (Kamble and Bahadure 2023; Thakur 2004). The site has experienced a magnitude of 7.5 earthquake in the past. Additionally, groundwater is present near the surface. The zone factor, as defined by the seismic code 1893 (Part 1): 2016, is 0.36, reflecting the perceived maximum seismic risk. Two boreholes, each 50 meters deep, were drilled at the proposed bridge site across the Koyalajan Kamla River. Figure 11.1 shows the project site and borehole locations.

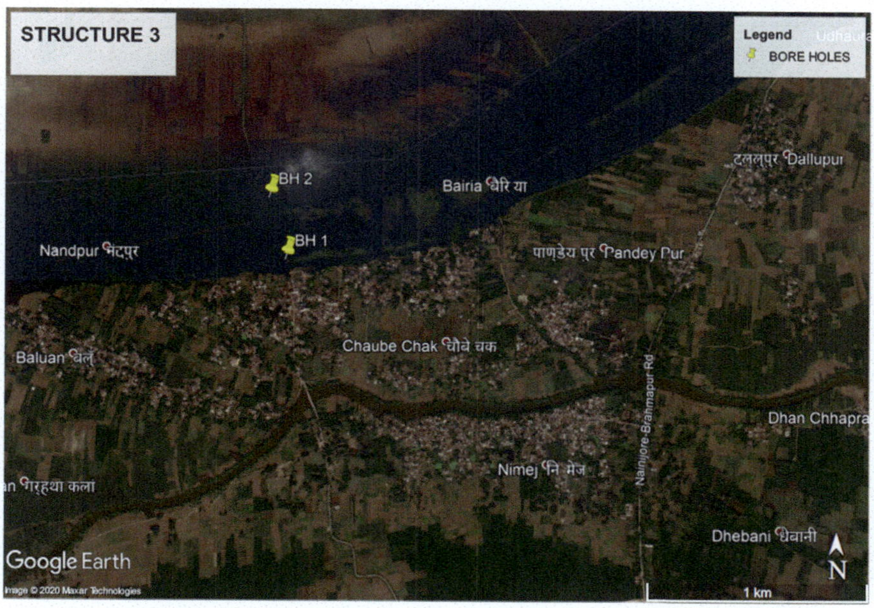

Fig. 11.1 Project site and location of boreholes

Table 11.1 Summary of anticipated sub-soil conditions

Borehole no	Sub-surface conditions		
	Approximate depth of soil strata (m)		
	Medium Stiff to Stiff Silty Clay / Clayey Silt (CL/CI/MI) - Stratum I	Silty Sand/Fine Sand medium Loose to Medium dense (SM/SP-SM) - Stratum II	Silty Sand/Fine Sand medium dense to very dense (SP/SP-SM)-Stratum III
BH-1(A)	0.00–13.40	13.40–21.50	21.50–50.00
BH-2(P)	0.00–8.90	8.90–15.50	15.50–50.00

11.3 Field and Laboratory Test Results

SPTs were performed in compliance with IS 2131, to gather samples from various depths. The samples acquired were subjected to laboratory testing for index and engineering properties in accordance with IS 2720.

The sub-surface conditions at the project site have been identified based on field and laboratory testing data collected during the exploration. The subsoil can be broadly divided into the following three main strata:

- Stratum I: Medium Stiff to Stiff Silty Clay / Clayey Silt (CL/CI/MI).
- Stratum II: Silty Sand/Fine Sand medium Loose to Medium dense (SM/SP-SM).
- Stratum III: Poorly graded Silty Sand medium dense to very dense (SM/SP-SM).

The location predominantly consists of Poorly graded Silty Sand medium dense to very dense (SM). Table 11.1 summarizes the foundation soil.

A comprehensive soil profile and pictorial summary of borehole data for the proposed bridge are presented in Fig. 11.2.

For cohesion-less soil, shear parameters used are based on laboratory test results and in-situ SPT. Table 11.2 shows the ranges of properties of each soil identified at various locations based on laboratory test results.

11.4 Groundwater

Groundwater was encountered at a depth of approximately 3.0 m for BH-1 and 1.0 m for BH-2 (March 2020). Given the site's proximity to the river, the groundwater table design has been assumed to be at the ground surface.

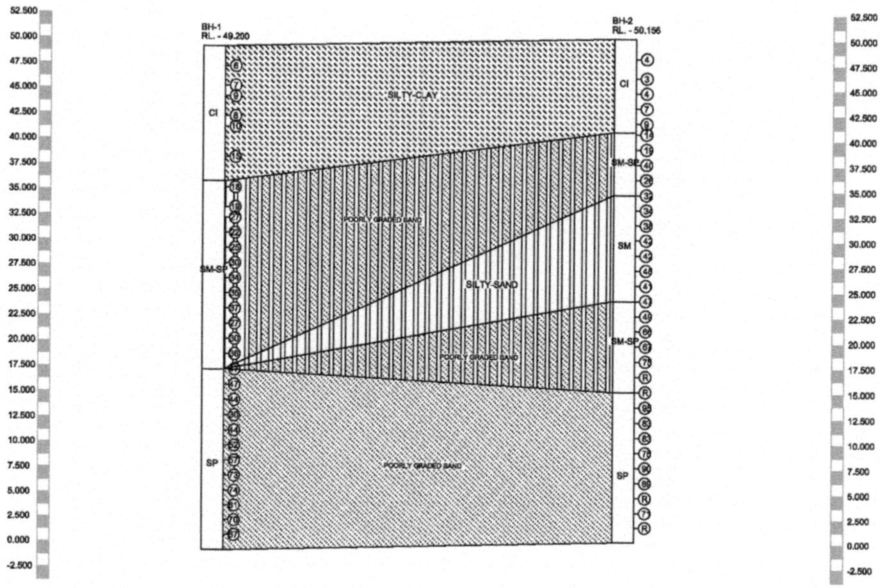

Fig. 11.2 A comprehensive soil profile and pictorial summary of boreholes data

11.5 Methodology

The methodology of analysis includes the following steps (Kramer 1996):

Step 1: Collect sub-surface data to evaluate the potential for liquefaction. This data should include information on the water table location, as well as measurements of soil properties such as SPT blow count N, and fines content (%age by wt. finer than 75μ).

Step 2: Data to assess the total vertical overburden stress (σ_{vo}) and effective vertical overburden stress (σ'_{vo}) for all layers present in the deposit that may be susceptible to liquefaction.

Step 3: Evaluate stress reduction factor reducing:

$r_d = 1–0.00765*Z$ for $0 < Z < 9.15$ m.

$= 1.174–0.0267*Z$ for $9.15 \text{ m} < Z < 23.0$ m.

where Z = depth in meters.

Step 4 - Determine CSR generated by the earthquake:

$CSR = 0.65 \text{ x } (a_{max}/g) * (\sigma_{vo}/ \sigma'_{vo}) \text{ x } r_d$.

Where,

a_{max} = peak ground acceleration preferably in terms of g, g = acceleration due to gravity,

and r_d = stress reduction factor.

In case, PGA is not available a_{max}/ g equal to seismic zone factor, Z could be used instead.

Table 11.2 Range of engineering properties of foundation soil

Description of soil/rock	Boreholes No	For soil							
		SPT (N)	LL/PL (%)	% (gm/cc)	NMC (%)	C (t/m^2)	(degree)	e	c
Medium stiff to stiff silty clay/clayey silt (CL/CI/MI) -Stratum I	BH-1 (A)	6–15	37.00–46.00/ 21.00–23.00	1.98–2.01	19.60–21.50	3.50–5.00	7–8	0.587–0.607	0.120–0.123
	BH-2 (P)	3–9	39.00–44.00/ 22.00–23.00	1.91–1.99	20.50–21.80	3.50–4.00	4–5	0.598–0.643	0.123–0.127
Silty sand/fine sand medium loose to medium dense (SM/SP-SM)-Stratum II	BH-1 (A)	18–30	Non Plastic	1.88–1.91	19.80–21.90	0.00	31–33	-	-
	BH-2 (P)	14–26	Non Plastic	1.89–1.90	19.60–22.20	0.00	33–34	-	-
Silty sand/fine sand medium dense to very dense (SM/SP-SM)-Stratum III	BH-1 (A)	27–81	Non Plastic	1.89–1.94	17.50–20.90	0.00	34–36	-	-
	BH-2 (P)	32–94	Non Plastic	1.92–2.00	17.60–19.60	0.00	35–36	-	-

Zone	II	III	IV	V
Z	0.10	0.16	0.24	36

Step 5: Obtain CRR by correcting $CRR_{7.5}$ for earthquake magnitude, high overburden stress level, and high initial static shear strength using:

$CRR = CRR_{7.5} * (MSF) * K_{\sigma} * K_{\alpha}$.

Where, $CRR_{7.5}$ = standard cyclic resistance ratio for 7.5 magnitude earthquake obtained using values of SPT, MSF = magnitude scaling factor given by.

$MSF = 10^{2.24}/M^{2.56}{}_{w}$.

This correction factor is necessary when the earthquake magnitude differs from 7.5. The adjustment for high overburden pressure is significant and can be calculated using the following equation:

$K_{\alpha} = (\sigma'_{vo} / P_a)^{(f-1)}$

where P_a is atmospheric pressure, f is an exponent which is dependent upon relative density D_r. For Dr = 40 ~ 60%, f = 0.8 ~ 0.7 and for Dr = 60 ~ 80%, f = 0.7 ~ 0.6. The correction factor for static shear stresses, Kα, is only necessary when dealing with sloping ground and is not typically used in standard engineering procedures. Therefore, for the purposes of this standard, the value of Kα will be assumed to be 1.

Step 6: Obtain cyclic resistance ratio $CRR_{7.5}$

Evaluate the SPT blow count N_{60}, (hammer efficiency 60%).

$N_{60} = C_N N_{60}$,

where.

$C_N = \sqrt{(P_a / \sigma'_{vo})} < 1.7$,

To calculate $CRR_{7.5}$ $(N_1)_{60}$ is used and obtained by using the equation:

$(N_1)_{60cs} = \alpha + \beta (N_1)_{60}$,

where.

$\alpha = 0$ $\beta = 1$ For $FC \leq 5\%$

$\alpha = e^{1.76-(190/F.C^{*2})}$ $\beta = 0.99 + FC^{1.5}/1000$ for $5\% < FC < 35\%$ and.

$\alpha = 0.5$ $\beta = 1.2$ for $FC \geq 35\%$

For $(N_1)_{60}$ values less than 30, the equation proposed by Rauch in 1998 is used to estimate the clean-sand base curve for determining CRR through the corrected SPT N-values.

$$CRR_{7.5} = \frac{1}{34 - (N_1)_{60}} + \frac{(N_1)_{60}}{135} + \frac{50}{[10.(N_1)_{60} + 45]^2} - \frac{1}{200}$$

Step 7: Calculate FS against liquefaction:

FS = CRR/CSR.

When the design ground motion is considered to be conservative, any permanent ground deformation caused by the earthquake is usually minimal if the FS value is equal to or greater than 1.2.

Step 8: If FS < 1, it indicates that the soil is susceptible to liquefaction.

11.6 Result and Discussions

As per IS:1893 (Part-1):2016 (IS 1893 2016), Liquefaction occurs when saturated soil loses its shear strength due to vibration induced by the earthquake and flow-like liquid mass. It is often observed in saturated loose (SPT N < 15) sandy soil mainly "poorly graded" SP and silty sand SM. Since the project area is in seismic zone V which is the most vulnerable seismic zone. The magnitude of the earthquake is 7.5, PGA is 0.36, and groundwater is near the surface. So, a detailed liquefaction analysis is carried out and calculations are presented in Tables 11.3 and 11.4.

The analysis suggests that liquefaction is probable at a depth of 11 m from ground level. Therefore, due to the potential risk of liquefaction and the heavy load requirements, utilizing open foundations on natural soils as a foundation system is impractical. Hence, it is suggested to have well foundation of diameter ranging from 7.0 m to 10.0 m. These well foundations should be installed to extend well below the scour level and liquefiable zone to transfer the loads to more stable soil strata. The gross allowable load-bearing pressure at the well tip for bridge abutments and piers was recommended in the range from 85 to 90 T/m^2 at the founding level of 25–35 m.

11.7 Conclusion

As a geotechnical engineer, it is essential to recognize that liquefaction can significantly impact the design of foundations in many parts of India. The ability to predict, evaluate, and address liquefaction susceptibility properly is critical to ensuring the long-term success of a project while balancing project costs and schedules.

From the results, it is clear that liquefaction is a major concern for the project area due to its location in seismic zone V and the characteristics of the soil. The calculated factors of safety against liquefaction show that most of the topsoil layers are unsafe, with only a few bottom layers having a factor of safety greater than 1.0. Therefore, utilizing open foundations is impractical, and it is suggested to use well foundations with diameters between 7.0 m and 10.0 m. These foundations should be installed well below the liquefiable zone and scour level to transfer loads to more stable soil strata. This recommendation should be taken into consideration for the design of the foundation system to ensure its long-term success and safety.

Table 11.3 Calculations of FS at BH-I location

BH-1 (A)											
Layer No	Soil description	Depth (m)	Total overburden pressure (T/m^2)	Correction for SPT hammer, weight of fall, CHW	Correction for SPT borehole diameter, CBD	N_{60}	N_{160} CS	CRR	CSR	Computed factor of safety against liquefaction	Comments
1	Silty Clay	0.0–5.0	5.0	1	1.05	4	12	0.24	0.30	0.80	Unsafe
2	Silty Clay	5.0–11.0	15.9	1	1.05	7	13	0.26	0.29	0.91	Unsafe
3	Silty Clay	11.0–13.4	24.3	1	1.05	14	19	0.37	0.26	1.42	Safe
4	Fine sand	13.4–15.0	28.3	1	1.05	18	15	0.28	0.24	1.17	Safe
5	Fine sand	15.0–20.0	34.7	1	1.05	24	18	0.36	0.22	1.64	Safe
6	Fine sand	20.0–30.0	48.9	1	1.05	28	19	0.36	0.16	2.28	Safe
7	Fine sand	30.0–50.0	66.0	1	1.05	38	20	0.40	0.08	4.72	Safe

Table 11.4 Calculations of FS at BH-2 location

BH-2 (P)											
Layer No	Soil description	Depth (m)	Total overburden pressure (T/m2)	Correction for SPT hammer, weight of fall, CHW	Correction for SPT borehole diameter, CBD	N60	N_{160} CS	CRR	CSR	Computed factor of safety against liquefaction	Comments
1	Silty clay	0.0–6.5	6.2	1	1.05	2	9	0.19	0.31	0.62	Unsafe
2	Silty clay	6.5–9.3	15.2	1	1.05	7	15	0.28	0.30	0.95	Unsafe
3	Fine sand	9.3–11.0	19.6	1	1.05	13	14	0.27	0.29	0.95	Unsafe
4	Fine sand	11.0–12.5	22.6	1	1.05	18	19	0.36	0.27	1.32	Safe
5	Fine sand	12.5–18.5	29.9	1	1.05	33	28	0.68	0.24	2.83	Safe
6	Fine sand	18.5–24.0	41.0	1	1.05	38	33	2.26	0.19	11.73	Safe
7	Fine sand	24.0–50.0	51.2	1	1.05	43	28	0.67	0.15	4.50	Safe

References

Capacci L, Biondini F, Frangopol DM (2022) Resilience of aging structures and infrastructure systems with emphasis on seismic resilience of bridges and road networks. Rev Resil Cities Struct 1(2):23–41. https://doi.org/10.1016/j.rcns.2022.05.001

Gurung L, Chatterjee K (2023) Evaluation of liquefaction potential of kolkata city, india: a deterministic approach. Pure Appl Geophys 180:439–474. https://doi.org/10.1007/s00024-022-03193-8

IS 1893 (2016) Criteria for earthquake resistant design of structures, part-1 general provisions and building sixth revision. New Delhi, India: Bureau of Indian Standards

Kamble T, Bahadure S (2023) Assessing building vulnerability of Indian cities in Earthquake Zone V. Fifth World Congress on Disaster Management: IV, Proceedings of the International Conference on Disaster Management, November 24–27, 2021, New Delhi, India. Taylor & Francis

Kramer SL (1996) Geotechnical earthquake engineering. Pearson Education Pvt. Ltd., Singapore

Kumar S, Muley P, Syed NM (2023) Soil liquefaction potential of Kalyani Region, India. Indian Geotech J 53:139–153. https://doi.org/10.1007/s40098-022-00658-4

Liao SSC, Whitman RV (1986) Catalogue of liquefaction and non-liquefaction occurrences during earthquakes. Res Rep Dept Civil Engrg. MIT, Mass

Rajendran CP, John B, Rajendran K, Sanwal J (2016) Liquefaction record of the great 1934 earthquake predecessors from the north Bihar alluvial plains of India. J Seismolog 20(3):733–745. https://doi.org/10.1007/s10950-016-9554-z

Seed HB, Idriss IM (1971) Simplified procedure for evaluating soil liquefaction potential. J Soil Mech Found Divis 97:1249–1273

Silahtar A, Karaaslan H, Kocaman K (2023) Site characterization and liquefaction hazard assessment for the Erenler settlement area (Sakarya Province, Turkey) based on integrated SPT-Vs data. Sustainability 15:1534. https://doi.org/10.3390/su15021534

Thakur VC (2004) Active tectonics of himalayan frontal thrust and seismic hazard to Ganga plain. Curr Sci 86(11):1554–1560

Thakur IC, Roy LB (2022) Soil liquefaction potential in different seismic zones of Bihar, India. Eng Technol Appl Sci Res 12(6):9471–9476

Umar SK, Samui P, Kumari S (2018) Reliability analysis of liquefaction for some regions of Bihar. Intern J Geotech Earthq Eng 9(2):23–37. https://doi.org/10.4018/IJGEE.2018070102

Youd TL, Perkins DM (1978) Mapping liquefaction-induced ground failure potential. J Geotech Eng Divis 104:443–446

Chapter 12
Time–Frequency Analysis of Strong Ground Motions from the 1994 Northridge Earthquake

Faisal Mehraj Wani, Chaturya Ganne, Jayaprakash Vemuri, and Chenna Rajaram

Abstract Accelerograms recorded in an earthquake are time-domain signals and do not convey information on the frequency content of the wave. The frequency domain representation of the accelerogram shows the distribution of the frequency content but hides information on the time characteristics. Since both the amplitude and frequency of the seismic wave contribute to structural damage, a comprehensive understanding of the possible structural damage can be understood using the three-dimensional approach of continuous wavelet transform (CWTs). Due to damage in an earthquake, the stiffness of a structure degrades, and the natural period elongates. The vulnerability of the structure thereby also depends on the sequence of arrival of damaging frequencies. In this paper, the strong ground motions, recorded within a fault distance of 20 km, in the devastating 1994 Northridge earthquake (Mw 6.7) are characterized in both the time domain and the frequency domain. The CWTs for these ground motions are generated to understand the time sequence of the arrival of high amplitude frequencies. The information obtained from the CWTs is analyzed and correlated with the observed structural damage, in a variety of buildings having a varied range of natural periods, due to the earthquake.

Keywords Accelerograms · Frequency content · Time–frequency analysis · Continuous wavelet transforms · 1994 Northridge earthquake

F. M. Wani (✉) · C. Ganne · J. Vemuri
Ecole Centrale College of Engineering, Mahindra University, Hyderabad, Telangana 500043, India
e-mail: faisal20pcie006@mahindrauniversity.edu.in

C. Rajaram
Department of Civil Engineering, Rajeev Gandhi Memorial College of Engineering and Technology, Nandyal, Andhra Pradesh 518501, India

B. K. Rastogi et al. (eds.), *Natural Hazards and Risk Mitigation*, Springer Transactions in Civil and Environmental Engineering, https://doi.org/10.1007/978-981-97-7658-0_12

12.1 Introduction

The past two decades have witnessed several earthquakes, damaging life and property tremendously (Bird and Bommer 2004). The Northridge Earthquake (*hereafter read as* NRE) with a magnitude of 6.7 shook the San Fernando Valley of Los Angeles at 4.31 a.m. on January 17, 1994, due to a previously unknown thrust fault located at least 9 miles beneath the valley floor. The intensity of the earthquake was felt From Las Vegas to San Diego (Hauksson et al. 1992). The earthquake epicenter was reported 30 km northwest of Los Angeles (Proctor et al. 1972), where the motion of the Pacific Plate moving north-westward along the San Andreas Fault and interacting with the North American Plate generated the earthquake. This motion led tension to build up along the Northridge blind thrust fault, resulting in an earthquake (Yeats et al. 1988).

The overall estimated damage due to NRE surpassed $40 billion (Mitchell et al. 1995), and even after several years, buildings remain unrepaired and insurance claims are disputed. The 1994 NRE was one of the costliest natural disasters in U.S. history, both in terms of monetary losses and the human toll. Following the earthquake, preliminary investigations revealed that many of the buildings that sustained the most damage or collapsed altogether were older structures built before applying contemporary seismic building rules. During an earthquake, these buildings frequently lack ductile detailing, which helps structures to absorb energy and distort in a controlled manner (Bruneau 1995; Iverson and Hawkins 1994; Tremblay et al. 1995).

The analysis of the time–frequency characteristics of seismogenic energy signals is commonly performed using wavelet transform. The concept of wavelet analysis was first introduced in the 1980s by Morlet et al. (1982) and later by Morlet (1983) for the analysis of seismic data. Since then, wavelet analysis has become a popular tool for studying non-stationary signals in many different applications. Wavelet transform is a powerful tool for analyzing non-stationary signals at different frequencies, making it suitable for studying earthquake ground motion signals (Morlet 1983).

In this paper, the 1994 NRE ground motion within 20 km from the fault is compiled from the Centre for Engineering Strong Motion Data (COSMOS) database (Archuleta et al. 2006) and later characterized to examine different Intensity Measures (IM). Moreover, Fast Fourier Transform (FFT) and Continuous Wavelet Transform (CWT) have been implemented to understand the frequency content of ground motions during an earthquake and how this relates to observed structural damage. The overall methodology, followed in this paper is depicted by the flowchart in Fig. 12.1.

12.2 Ground Motion Data

In this study, a set of 50 Near-Fault Ground Motions (NFGM) collected from the COSMOS database were compiled from various stations as shown in Fig. 12.2 (Archuleta et al. 2006). NFGM, which occurs close to the earthquake source, differs

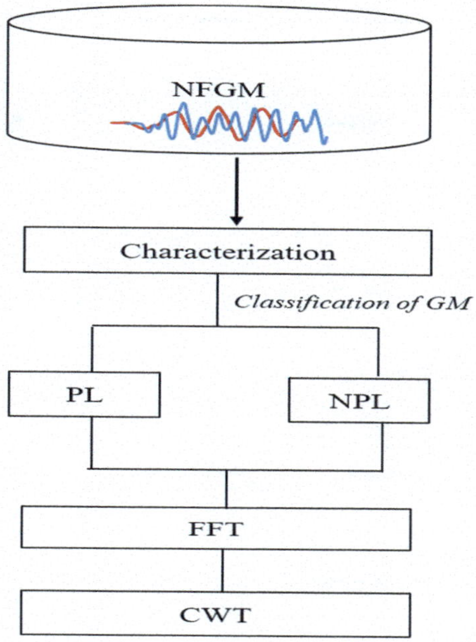

Fig. 12.1 General layout of the current study. *Note* NFGM: Near Fault Ground Motion, PL; Pulse Like, NPL: Non-Pulse like, FFT; Fast Fourier Transform, CWT; Continuous Wavelet Transform

from far-field ground motion, which occurs further away from the source (Morlet et al. 1982). The presence of these highly impulsive pulses or high-frequency energy in the waveform is one of the most noticeable indicators of near-fault ground motion (Somerville 2005; Wei et al. 2006).

The data used in the present study is restricted to a few stations located within 20 km, as intense pulse velocities beyond this distance are rare. Moreover, the ground motions were characterized to investigate the different intensity Measures (IM) such as amplitude parameters (Peak Ground Acceleration (PGA) and Peak Ground Velocity (PGV) (Kramer 1996), energy parameters (Arias intensity (Arias 1970) and Cumulative Absolute Velocity (CAV)) (Campbell and Bozorgnia 2023) to understand the nature and severity of the ground motion. Additionally, the ground motion were classified into Pulse-Like (PL) and Non-Pulse-Like (NPL) using the Baker approach (Baker 2007). Baker uses wavelet-based signal processing to identify pulse-shape signals in the area with the largest velocity (PGV). The two predictable variables used by Baker were the PGV ratio and energy ratio for the classification of ground motion. The record with PI values above 0.85 qualifies as pulse-like, whereas PI values below 0.15 are classified as non-pulse. Other records with PI in the range of 0.15 and 0.85 may not be classified with confidence. The Predictive equation for the classification of records by Baker is given as

$$\text{Pulse indicator} = \frac{1}{1 + e^{-23+14.6(PGVratio)+20.5(energyratio)}} \tag{12.1}$$

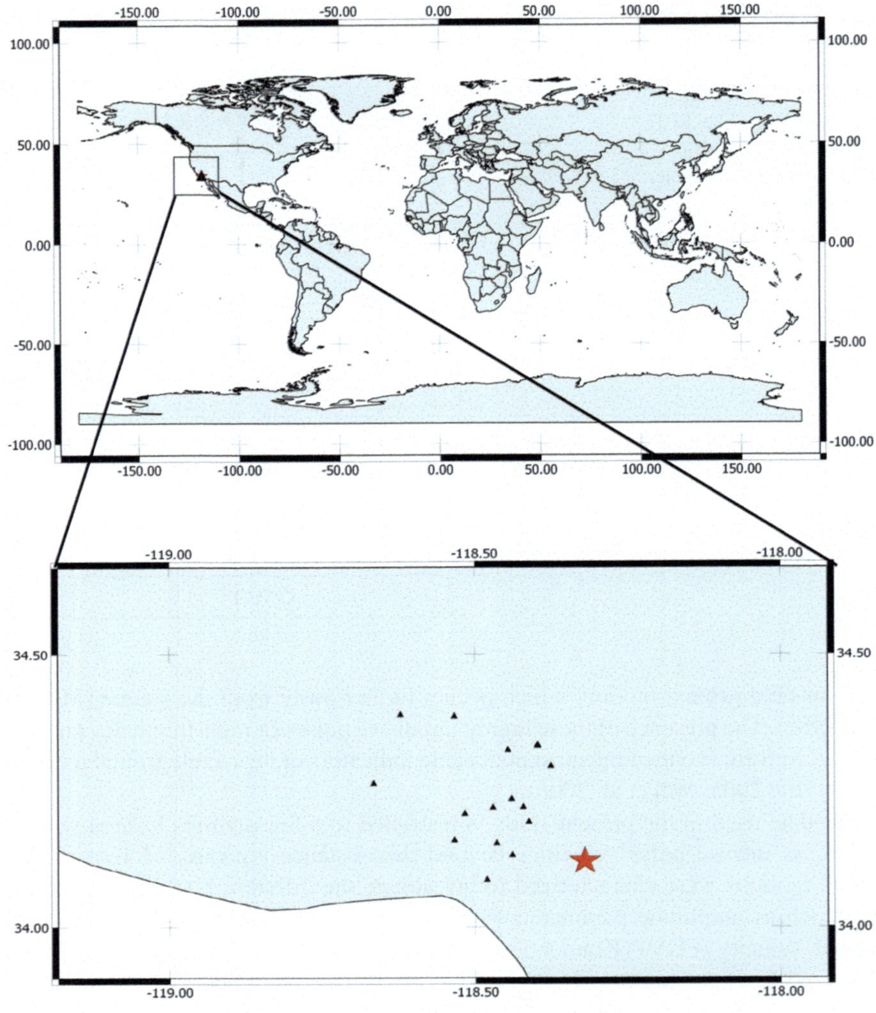

Fig. 12.2 Distribution of ground motion stations used in the present study. *Note* Red star depicts the epicenter whereas small triangles denote the recording stations

Figure 12.3 shows the PL and NPL ground motions distribution for 1994 NRE. It is observed that IM is slightly higher for PL ground motion as compared to NPL ground motion. The acronyms AI, CI, HI, VSI refer to Arias Intensity, Characteristic Intensity, Housner Intensity, and Velocity Spectrum Intensity, respectively. SMA, MIV, V/A refer to sustained maximum acceleration, maximum incremental velocity, and the ratio of PGV/PGA, respectively. The single parameter estimate, V/A, is used by researchers to estimate the frequency content in strong ground motions.

Moreover, the Fast Fourier Transform (FFT) has been utilized to understand the frequency content of a ground motion for PL and NPL ground motion and how they

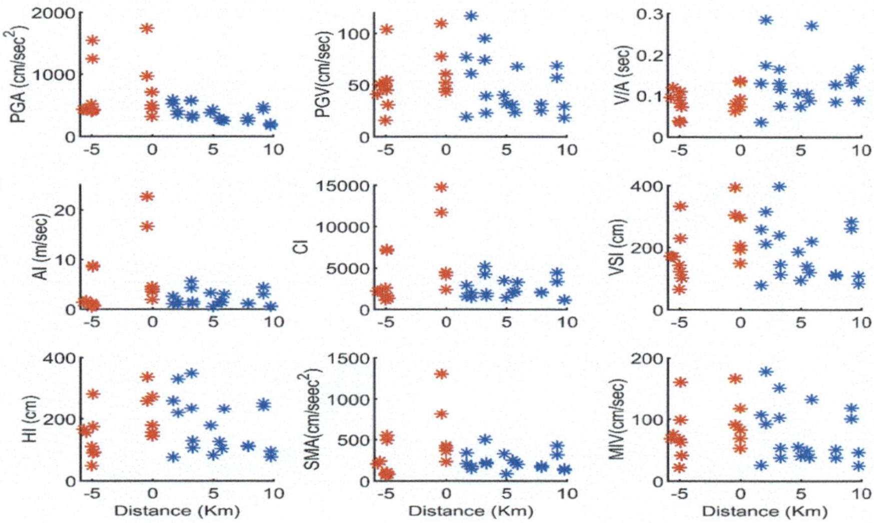

Fig. 12.3 Distribution of PL and NPL ground motion characteristics. *Note* Red color highlights the PL characteristics while the blue scatter depicts NPL

may affect structures. Figures 12.4 and 12.5 show the FFT for PL and NPL ground motions. The frequency content is observed to vary from (0–10 HZ), however, certain information regarding the variation of intensity with time is hidden while converting the signal to the frequency domain. The ground motion signals are typically non-stationary, meaning that they exhibit changes in frequency content and amplitude over time, but FFT assumes that the signal is stationary, which is not always the case with ground motion signals. Therefore, to fully understand the time–frequency characteristics of a ground motion signal, more advanced analysis techniques such as wavelet analysis may be necessary, as they can decompose the signal into a series of wavelets at different scales and provide information on the changes in frequency content and amplitude over time.

12.3 Continuous Wavelet Transform (CWT)

The Continuous Wavelet Transform (CWT) has been widely utilized in earthquake engineering to comprehend the complicated behavior of these time-varying signals that carry information about an earthquake's ground motion. The CWT evaluates the seismic signals in both the time and frequency domains, revealing information about the seismic waves' characteristics (Wani et al. 2022; Ifan et al. 2021; Mathew et al. 2022).

In this study, CWT is used to identify the fundamental frequency of structures, which is an important parameter for the design of earthquake-resistant structures.

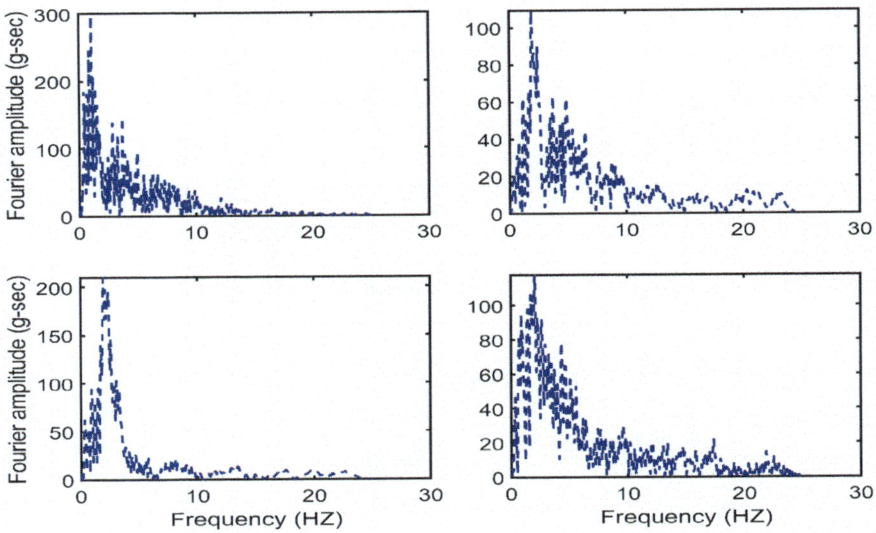

Fig. 12.4 FFT for PL ground motions with station code **a** 13,248, **b** 24207NE, **c** 24,307 SE, **d** 24,907

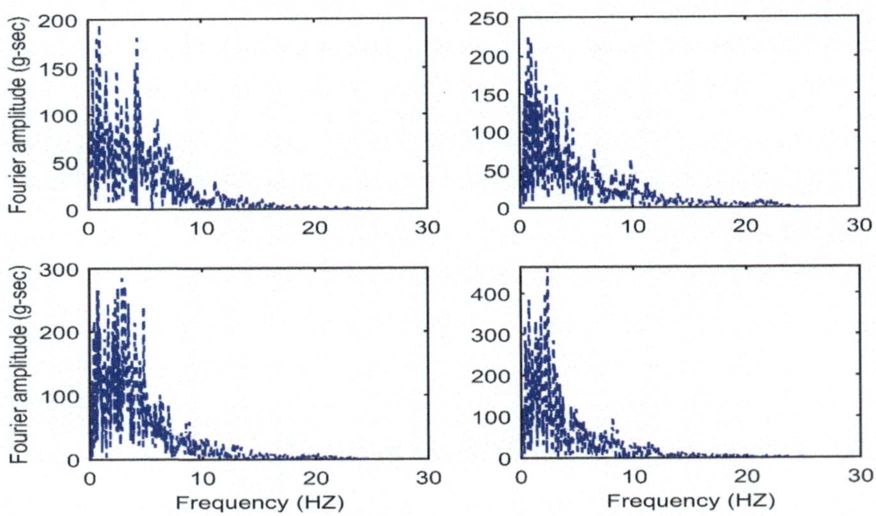

Fig. 12.5 FFTs for NPL ground motions with station code **a** 00,009, **b** 24,087, **c** 24,386, **d** 00,003

The CWT can be used to analyze the response of a structure to an earthquake and determine its fundamental frequency, which is the frequency at which the structure vibrates most strongly. The CWTs are calculated using the analytic Morse wavelet with a symmetry parameter of 3 and a time-bandwidth product of 60. CWT employs ten voices per octave with L1 normalization (Lilly and Olhede 2012). Convolving the signal with the complex conjugate pair of the Morse wavelet and then taking the Hilbert transform of the result yields the analytic signal. Because of its high time–frequency localization capabilities, the Morse wavelet is well-suited for analyzing non-stationary signals. Its parameterization flexibility allows it to be changed to optimize the time–frequency resolution for a specific application (Lilly and Olhede 2008). The lowest and maximum scales are calculated automatically based on the wavelet's energy distribution in frequency and time. In wavelet analysis, a signal is decomposed into different frequency bands using a set of wavelet functions with varying scales. This allows for the analysis of signals at different resolutions in both time and frequency domains. Wavelet analysis is particularly useful for studying non-stationary signals, as it allows for the identification of changes in frequency content over time.

The scalogram represents the absolute value of the CWT plotted against time and frequency. The frequency is represented on a logarithmic scale. The influence cone, which shows where edge effects become prominent, is also plotted. The area outside the dashed white line designates areas with substantial edge effects. The yellow spots present inside the cone highlight the damage potential of the ground motions (Lilly 2017).

Figures 12.6 and 12.7 depict the CWTS for different PL and NPL ground motions. It is observed that the large amplitude of waves plunge (Yellow spots) over a short duration of time for PL ground motions whereas for NPL ground motions the damage spots are scattered in the CWTS, indicating that PL ground motions have a greater potential to inflict structural damage over a shorter period, whereas NPL ground motions have a more gradual and distributed impact on structures. Moreover, the fundamental cause of this distinctiveness of the PL ground motions is seismic source directivity. Seismic waves radiate away from an earthquake source in a pattern determined by the fault's orientation and geometry. In some situations, this can cause energy to be focused in a specific direction or at a specific distance from the source, resulting in the creation of a pulse in the ground motion.

12.4 Conclusion

Earthquakes are unexpected and deadly natural phenomena that pose enormous dangers to life and property. The Northridge earthquake of 1994, which struck Los Angeles' San Fernando Valley, is an illustration of the considerable devastation that earthquakes may wreak. To better understand the nature of ground motion during earthquakes, 50 near-fault ground motions were examined in the time and frequency domain to understand the frequency content of ground motions during an earthquake. CWT analysis was performed to gain a more complete understanding of the nature of the arrival of various frequencies in strong ground motions and how it may affect the response of different types of structures having varied natural periods. During an earthquake, the structure's natural period also elongates due to the damage, and

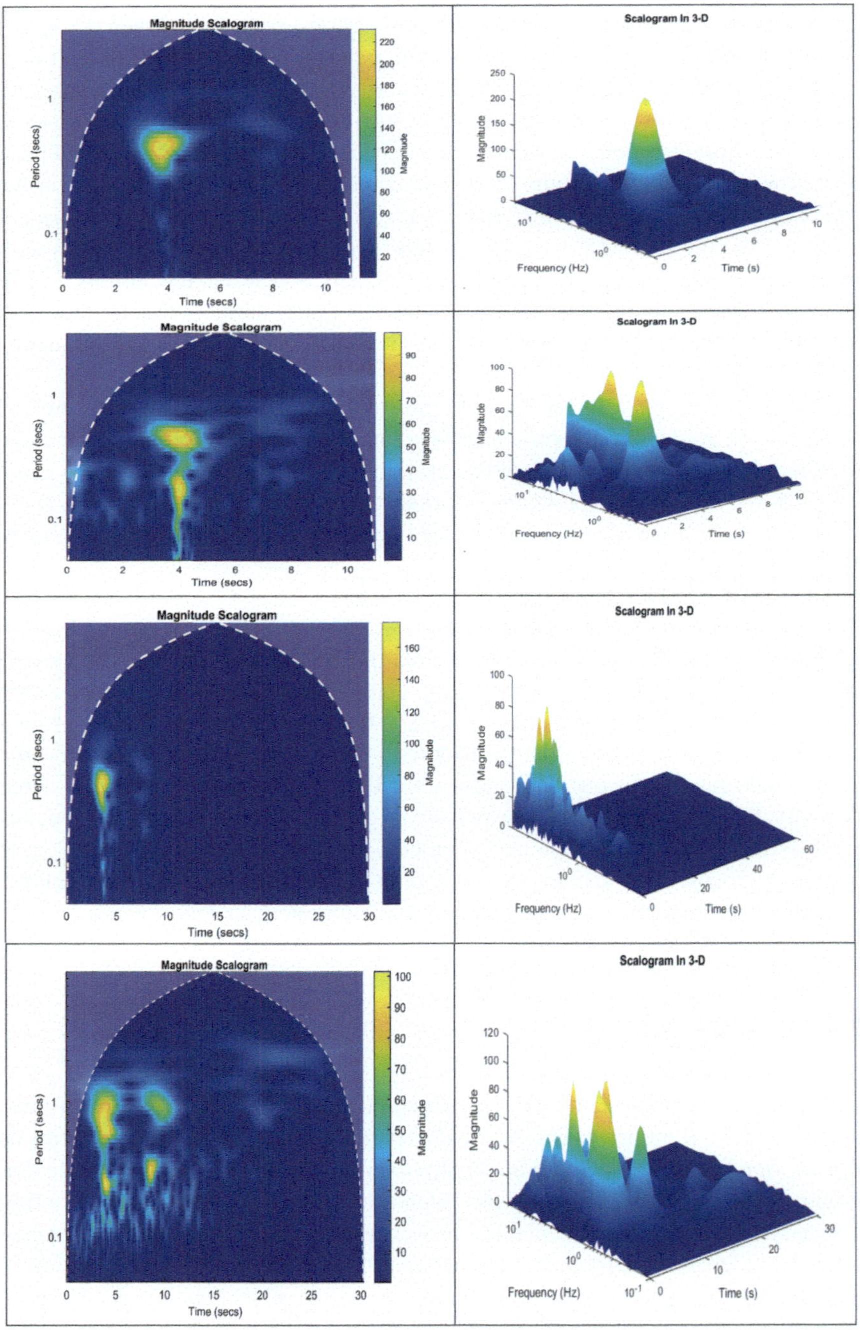

Fig. 12.6. 2D and 3D CWT at several stations from the NRE for PL ground motions with station code **a** 24,207 NE **b** 24,207 SE **c** 24,907 **d** 13,248. *Note* The yellow spots in CWT represent the seismic activity at a certain time period

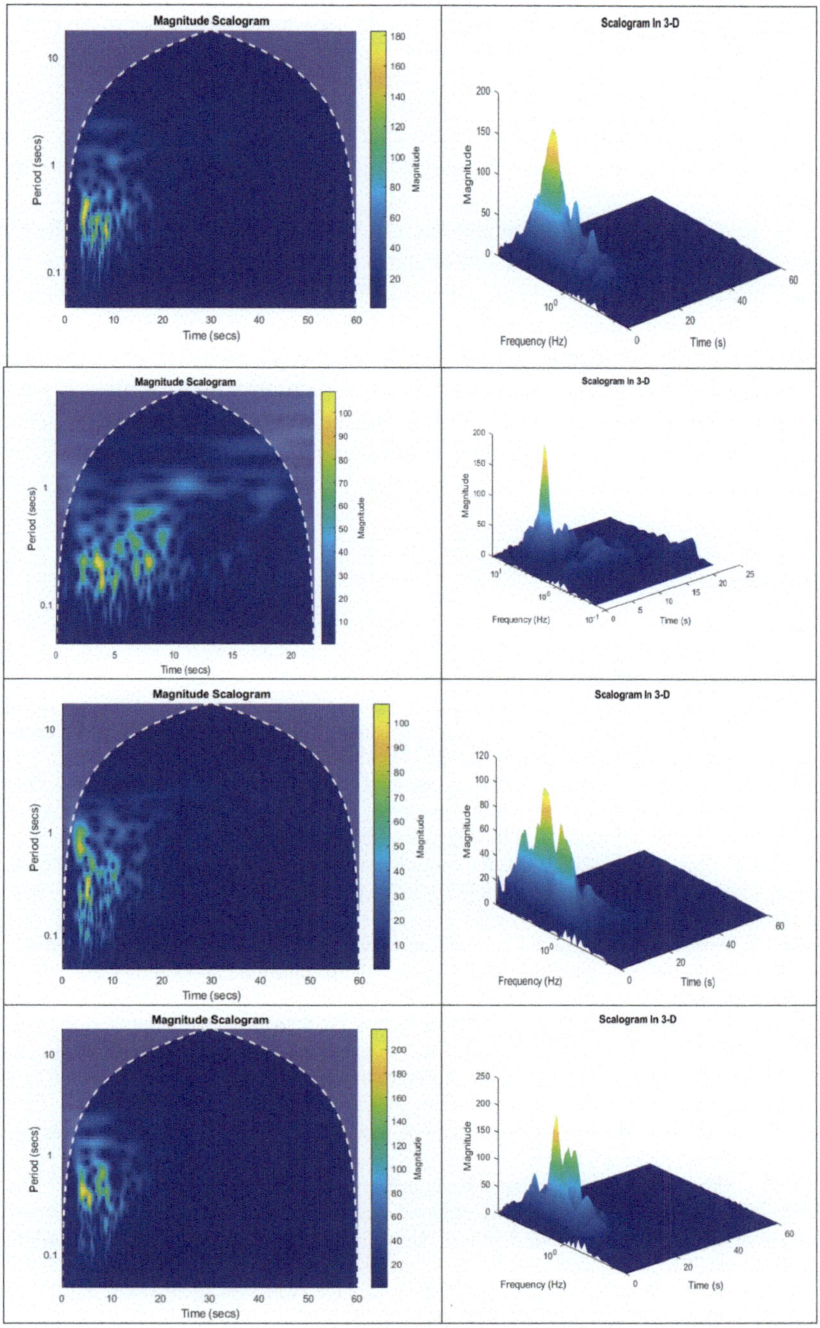

Fig. 12.7. 2D and 3D CWT at several stations from the NRE for NPL ground motions with station code **a** 24,386 **b** 0009 **c** 24,087 **d** 00,003

the time sequencing of damaging frequencies in the seismic wave can often result in further damage at later instants.

- It was observed that the characterization of ground motion and FFT are useful in providing a basic understanding of the characteristics of ground motion signals where as they fail to capture the full complexity and dynamic nature of these signals.
- More advanced analysis techniques such as wavelet analysis may allow the identification of transient features and changes in the frequency content over time, particularly useful in identifying pulse-like features in the ground motion, which may not be apparent in simple frequency analysis.
- PL ground motions tend to have a sudden onset of damage activity over a short duration, which can lead to higher damage to structures with a low natural period. However, NPL motions indicate gradual onset and longer duration, which can lead to higher damage to structures with a high natural period.

References

Archuleta RJ, Steidl J, Squibb M (2006) The COSMOS virtual data center: a web portal for strong motion data dissemination. Seismol Res Lett 77(6):651–658
Arias A (1970) A measure of earthquake intensity. Seismic design for nuclear plants, pp 438–483
Baker JW (2007) Quantitative classification of near-fault ground motions using wavelet analysis. Bull Seismol Soc Am 97(5):1486–1501. https://doi.org/10.1785/0120060255
Bird JF, Bommer JJ (2004) Earthquake losses due to ground failure. Eng Geol 75(2):147–179
Bruneau M (1995) Performance of masonry structures during the 1994 Northridge (Los Angeles) earthquake. Can J Civ Eng 22(2):378–402. https://doi.org/10.1139/l95-048
Campbell KW, Bozorgnia Y (2023) Ground-motion model for the standardized version of cumulative absolute velocity. Earthq Spectra 87552930221144060. https://doi.org/10.1177/875529302 21144063
Hauksson E, Pipkin BW, Proctor RJ (1992) Seismicity, faults, and earthquake potential in Los Angeles, southern California. Eng Geol Pract Southern California Assoc Eng Geol Spec Publ 4:167–179
Ifan MA, Mathew S, Vemuri J (2021) Wavelet analysis of far-field ground motions from the Mw 7.6 2005 Kashmir Earthquake. In Recent Advances in Earthquake Engineering: Select Proceedings of VCDRR 2021 (pp 11–23). Singapore: Springer Singapore. https://doi.org/10.1007/978-981-16-4617-1_2
Iverson JK, Hawkins NM (1994) Performance of precast/prestressed building structures during Northridge earthquake. PCI J 39(2)
Kramer SL (1996) Geotechnical earthquake engineering. Pearson Education India
Lilly JM (2017) Element analysis: a wavelet-based method for analyzing time-localized events in noisy time series. Proc Royal Soc Mathem Phys Eng Sci 473(2200):20160776. https://doi.org/ 10.1098/rspa.2016.0776
Lilly JM, Olhede SC (2008) Higher-order properties of analytic wavelets. IEEE Trans Signal Process 57(1):146–160. https://doi.org/10.1109/TSP.2008.2007607
Lilly JM, Olhede SC (2012) Generalized Morse wavelets as a superfamily of analytic wavelets. IEEE Trans Signal Process 60(11):6036–6041. https://doi.org/10.1109/TSP.2012.2210890
Mathew S, Ifan MA, Vemuri J, Subramaniam KV (2022) Time–Frequency Analysis of Strong Ground Motions from the 2011 Sikkim Earthquake. In Recent Advances in Materials, Mechanics

and Structures: Select Proceedings of ICMMS 2022 (pp 37–47). Singapore: Springer Nature Singapore. https://doi.org/10.1007/978-981-19-3371-4_4

Mitchell D, DeVall RH, Saatcioglu M, Simpson R, Tinawi R, Tremblay R (1995) Damage to concrete structures due to the 1994 Northridge earthquake. Can J Civ Eng 22(2):361–377. https://doi.org/10.1139/l95-047

Morlet J, Arens G, Fourgeau E, Glard D (1982) Wave propagation and sampling theory—part I: Complex signal and scattering in multilayered media. Geophysics 47(2):203–221. https://doi.org/10.1190/1.1441328

Morlet J (1983) Sampling theory and wave propagation. In Issues in acoustic Signal—image processing and recognition (pp. 233–261). Springer Berlin Heidelberg. https://doi.org/10.1007/978-3-642-82002-1_12

Proctor RJ, Crook R Jr, McKeown MH, Moresco RL (1972) Relation of known faults to surface ruptures, 1971 San Fernando earthquake, southern California. Geol Soc Am Bull 83(6):1601–1618. https://doi.org/10.1130/0016-7606(1972)83[1601:ROKFTS]2.0.CO;2

Somerville PG (2005) Engineering characterization of near fault ground motions. In Proc., NZSEE 2005 Conf

Tremblay R, Filiatrault A, Timler P, Bruneau M (1995) Performance of steel structures during the 1994 Northridge earthquake. Can J Civ Eng 22(2):338–360. https://doi.org/10.1139/l95-046

Wani FM, Mujtaba Moid M, Chandrakanth Reddy K, Vemuri J, Chenna R (2022) Wavelet Analysis of Near-Field Ground Motions from the Mw 7.6 1999 Chi-Chi Earthquake in Taiwan. In Recent Advances in Materials, Mechanics and Structures: Select Proceedings of ICMMS 2022 (pp. 311–319). Singapore: Springer Nature Singapore. https://doi.org/10.1007/978-981-19-3371-4_28

Wei T, Zhao FX, Zhang YS (2006) Characteristics of near-fault ground motion containing velocity pulses. Acta Seismol Sin 19:677–686. https://doi.org/10.1007/s11589-006-0677-0

Yeats RS, Huftile GJ, Grigsby FB (1988) Oak Ridge fault, Ventura fold belt, and the Sisar decollement, Ventura basin. Calif Geol 16(12):1112–1116. https://doi.org/10.1130/0091-7613(1988)016%3C1112:ORFVFB%3E2.3.CO;2

Chapter 13
Effect of Soil Structure Interaction on the Response of a Lumped Mass Model

Faisal Mehraj Wani, Sahithi Koduru, Ruthviz Kodali, and Jayaprakash Vemuri

Abstract The dynamic response of the structures, subjected to strong ground motions, depends on the behavior of the soil in and around the structure. The effect of soil-structure interaction is neglected by analysis and designers due to a lack of code provisions. Damage to structures in recent earthquakes has indicated non-uniform distribution of damage, which indicates the effect of soil structure interaction (SSI). In literature, there is no clear consensus on the beneficial or detrimental effect of SSI effect and thus its incorporation in modelling is not prescribed in some seismic codes. In this study, a lumped mass system has been modeled to study the effect of soil-structure interaction. Non-linear time history analysis was performed on this model using ground motions from two major earthquakes in Uttarakhand region. The displacement of the lumped mass was monitored for both cases, i.e., fixed and flexible base conditions. The peak values of displacement time histories are tabulated and plotted versus the corresponding peak ground acceleration. The results indicate that incorporating the effect of soil-structure interaction leads to an increase in the displacement of the structure at certain sites.

Keywords Lumped mass · Non-linear time history · Soil-structure interaction · Peak ground acceleration · Displacements

13.1 Introduction

Civil engineering structures are in direct contact with the ground. The structural and ground displacements are dependent on one another when lateral forces, such as earthquakes, act on these systems. Soil-structure Interaction (SSI) is the process wherein the motion of the structure is affected by the soil's response, and the soil response is affected by the motion of the structure (Lou et al. 2011; Kausel 2010).

F. M. Wani (✉) · S. Koduru · R. Kodali · J. Vemuri
Mahindra University, Bahadurpally, Hyderabad, India
e-mail: faisal20pcie006@mahindrauniversity.edu.in

B. K. Rastogi et al. (eds.), *Natural Hazards and Risk Mitigation*, Springer Transactions in Civil and Environmental Engineering, https://doi.org/10.1007/978-981-97-7658-0_13

The foundation of a building is the primary interface between the structure and the soil. The foundation transfers the weight of the building to the soil and also provides stability to resist lateral loads, such as wind or earthquake forces. The soil properties, including the type of soil, its strength, stiffness, and deformation characteristics, influence foundation and building performance (Mylonakis and Gazetas 2000). SSI is a multidisciplinary field incorporating structural and geotechnical engineering. Even though these methods take a long time and require a lot of work to complete, they have been extensively used to study the impacts of soil-structure interaction (SSI).

The conventional approach presumes that the building's base is fixed, while in reality, it is in contact with the soil (Flexible). The soil in and around the building reduces its flexibility, which alters its response. Previous studies have observed that SSI reduces the base shear and increases the natural period of the structure as the flexibility of the structure increases (Wani et al. 2022). The consideration of SSI is assumed to be beneficial for the response of the structure as it leads to an increase in the natural period of the structure. The necessity for ductility increases with an increase in the natural period. This provides an improved structure when compared to a corresponding rigid structure. The incorporation of SSI effects on the structural design helps in increasing the damping ratio of the structure. Neglecting the SSI in particular for soft soil is considered to be detrimental to the structure making it unsafe. The inclusion of SSI could predict the structural response outcomes close to the real behavior of the structure observed during earthquakes. Several analytical methods have been proposed to predict the behavior of the springs (Oz et al. 2020; Behnamfar and Banizadeh 2016).

In this study, the lumped mass element has been examined to compare the effect of SSI with the fixed base condition. Figure 13.1a, b, c shows the description of the lumped mass model. Non-linear time history analysis from the ground motions for October 1991 Uttarkashi and March 1999 Chamoli earthquakes have been utilized to examine the effect of SSI on the lumped mass model.

13.2 Ground Motion Database

In this study, ground motions from the October 1991 Uttarkashi and March 1999 Chamoli earthquakes have been utilized to examine the effect of SSI on the lumped mass model. The October 1991 Uttarkashi earthquake and the March 1999 Chamoli earthquake were two destructive earthquakes that occurred in Uttarakhand, India. The Uttarkashi earthquake took place on October 20, 1991, while the Chamoli earthquake occurred on March 29, 1999. Both earthquakes had their epicenters located in the Garhwal Himalayas, and both caused widespread damage in the region. The study examines two conditions: fixed base and flexible base, where the ground motions are applied at the base of the structure. The study investigates the impact of soil-structure interaction on the response of the lumped mass model under both conditions. The location of the epicenter and various stations are shown in Figs. 13.2 and 13.3,

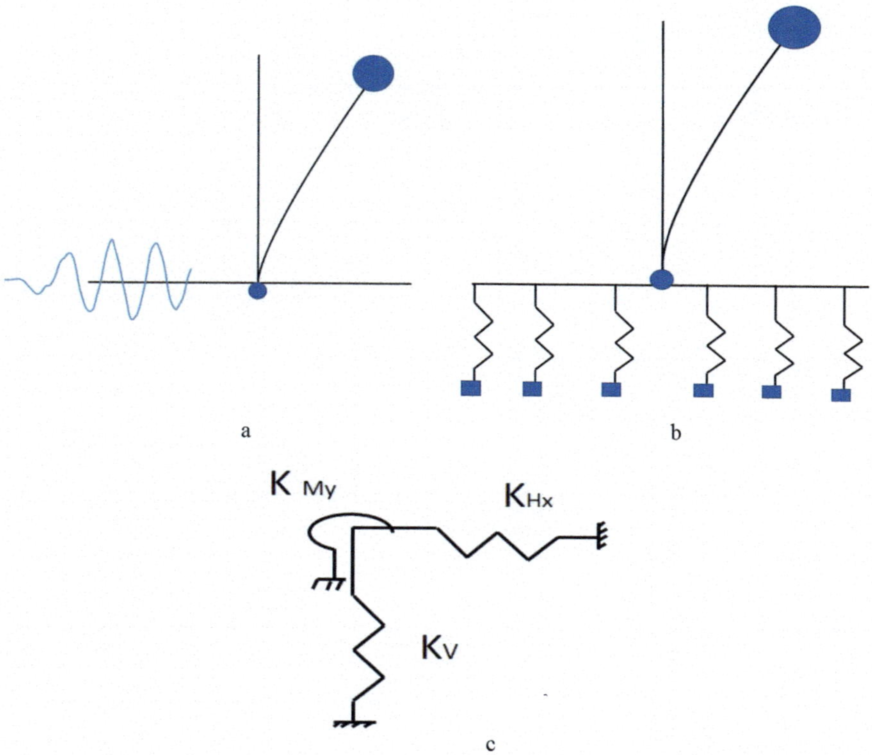

Fig. 13.1. 2-D representation of the lumped mass system with **a** fixed base **b** flexible base **c** Representation of springs at the base of the foundation

respectively. Table 13.1 summarizes some key properties of the ground motions used in the study.

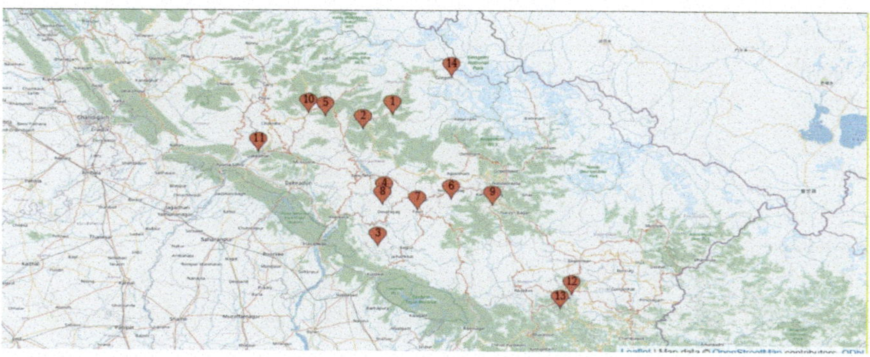

Fig. 13.2 Location of stations at varying epicentral distances for the 1991 Uttarkashi earthquake

Fig. 13.3 Location of stations at varying epicentral distances for the 1999 Chamoli Earthquake

Table 13.1 Key characteristics of ground motions. *Data Source* COSMOS

Station	Hypocentral distance (km)	PGA (cm/s^2)	PGV (cm/s)
1991 Uttarkashi Earthquake			
Bhatwari	21.7	242.00	29.8
Uttarkashi	34.0	237	17
Ghansali	39.3	115	7.82
Tehri	50.6	71.40	4.22
Barkot	55.8	80.5	-4.48
Rudraprayag	56.2	52.30	2.07
Srinagar	58.8	65.4	1.94
Koteshwar	61.3	98.80	5.16
Karnaprayag	69.6	61.00	3.69
Purola	70	73.90	4.81
Koti	98.7	40.9	2.86
Kausani	148.2	28.30	1.88
Almora	153.5	17.4	1.33
1999 Chamoli Earthquake			
Gopeshwar	17.3	195.07	22.5
Joshimath	26	62.28	8.94
Ukhimath	35.6	94.71	5.45
Ghansali	75.3	81.84	4.08
Tehri	89.7	53.43	4.6
Lansdowne	95.8	6.32	0.21
Chinyalisaur	106.9	43.92	3.31
Barkot	123.7	22.27	1.24
Roorkee	160.89	45.89	3.29

13.3 Model Description

In this study, the SSI was examined using a single-degree-of-freedom (SDOF) lumped mass model with a rugby ball bounding surface with a footing size of 5.00 by 2.50 as shown in Fig. 13.4. A lumped mass model is a simplified mathematical representation of a structure that is used to analyze its dynamic behavior, particularly in response to seismic loads. In the case of soil-structure interaction (SSI), a lumped mass model can be used to simulate the interaction between the soil and the structure. Non-linear dynamic time history analysis with a direct approach was implemented on the lumped mass model defined by Behnamfar and Banizadeh (2016) to examine the effect of seismic response of the structure due to SSI. In an SSI lumped mass model, the soil and the structure are modeled as a single system, and the properties of the soil and foundation are represented by a series of springs and dampers. The lumped mass model is typically expressed as a single-degree-of-freedom (SDOF) system, which means that the structure is represented by a single mass that moves in one direction. The SSI model considers soil, structure, and foundation properties as shown in Table 13.2

Fig. 13.4 Finite element model of a lumped mass parameter considered for soil-structure interaction

Table 13.2 Different parameters for incorporating the SSI

Parameters	Values
Footing dimension	5.0×2.50
Footing stiffness (K_N1, K_H2, K_M3, K_M2, K_T)	2.8800E + 006, 2.3045E + 006, 1.3431E + 007, 2.4406E + 006 4.4955E + 006
Footing bearing capacity (Nmax, Hmax2, Mmax3, Hmax3, Mmax2, Tmax)	6147, 923, 3688, 923, 1844, 5000
Uplift initiation parameter	3
Soil footing contact	0.10
Plastic potential parameter	2
Plastic modulus	0.20
Exponent for loading history for loading and unloading	0.1

13.4 Results and Discussion

Non-linear time history analysis of lumped mass systems has been implemented to examine the behavior of fixed and flexible base conditions. The results are discussed in terms of the displacement of the top node of the system. The displacement of node 3 considering the SSI and without SSI for the Uttarkashi and Chamoli earthquakes are depicted in Figs. 13.5, 13.6, 13.7, 13.8 and 13.9. It is observed that the SSI results in an increase in the displacement for certain stations as the flexibility of soil increases due to the incorporation of SSI.

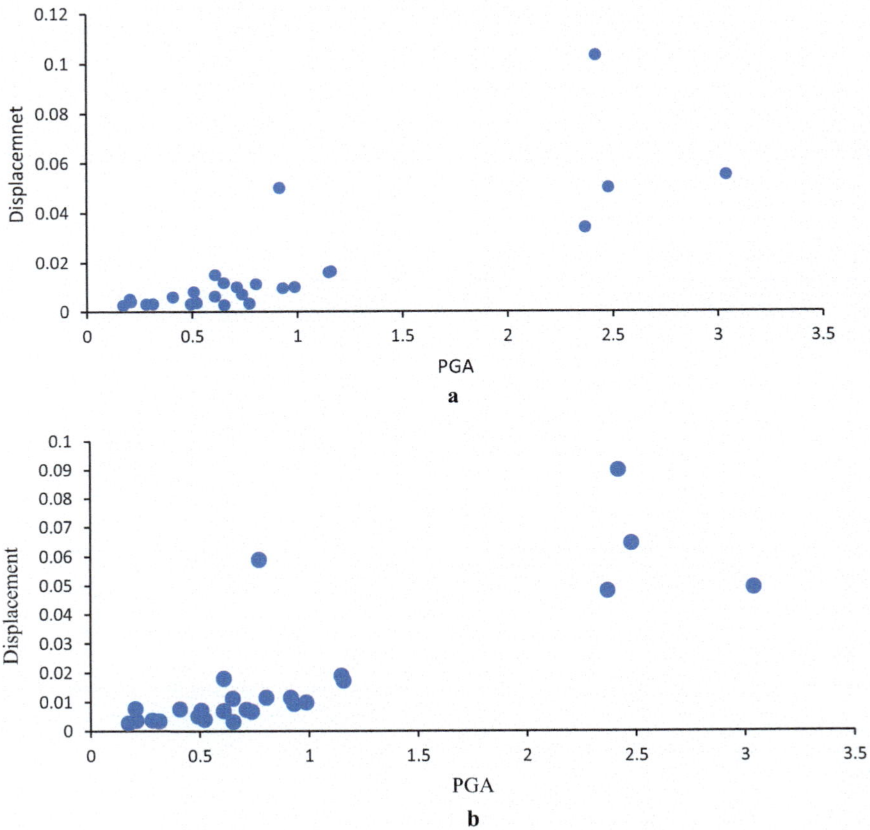

Fig. 13.5 Displacement of top node **a** fixed base **b** flexible base at different stations for the 1991 Uttarkashi earthquake

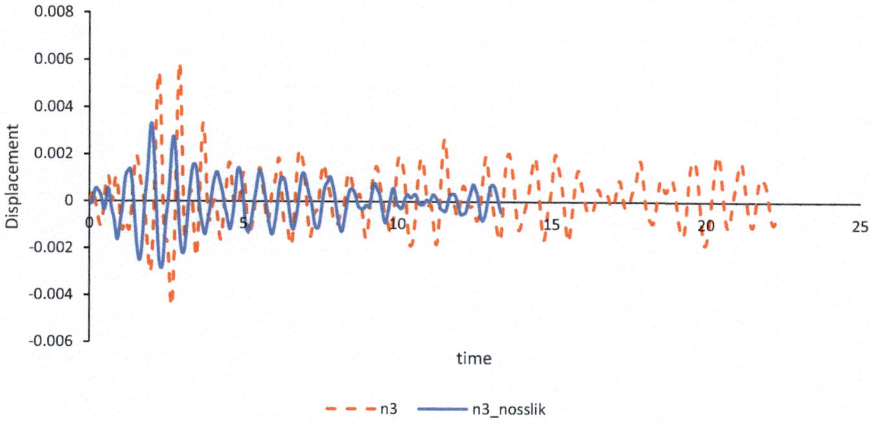

Fig. 13.6 Time history plot for Displacement at node 3 for Karnaprayag-T 1991 Uttarkashi Earthquake

13.5 Conclusions

The response of structures during an earthquake depends on the source characteristics, the travel path of the seismic wave, and also the local site effects. However, despite the non-uniform levels of structural damage observed at several sites in recent earthquakes, the effect of soil structure interaction has been neglected. In the present study, using a lumped mass model and by performing non-linear dynamic time history analysis with a direct approach, a detailed understanding of the SSI effects on the structure is developed. The SSI behaviour is modelled by using macro elements for shallow footings and the rugby-ball bounding surfaces were considered. The study's findings can be used to inform the design and construction of buildings and structures in seismic-prone areas, aiming to improve their seismic performance and reduce the risk of damage or collapse during earthquakes. This chapter examines the impact of SSI using a lumped mass system for two significant earthquakes which struck the state of Uttarakhand in India, i.e. the 1991 Uttarkashi and 1999 Chamoli earthquakes. The observed results are presented as time-history plots of displacement. In particular, the plots of top displacement (node 3) versus the PGA indicate a slight increase in displacement considering SSI. The influence of SSI is also noted to be more pronounced in Barkot and Karnaprayag stations compared to Bhatwari and Uttarkashi.

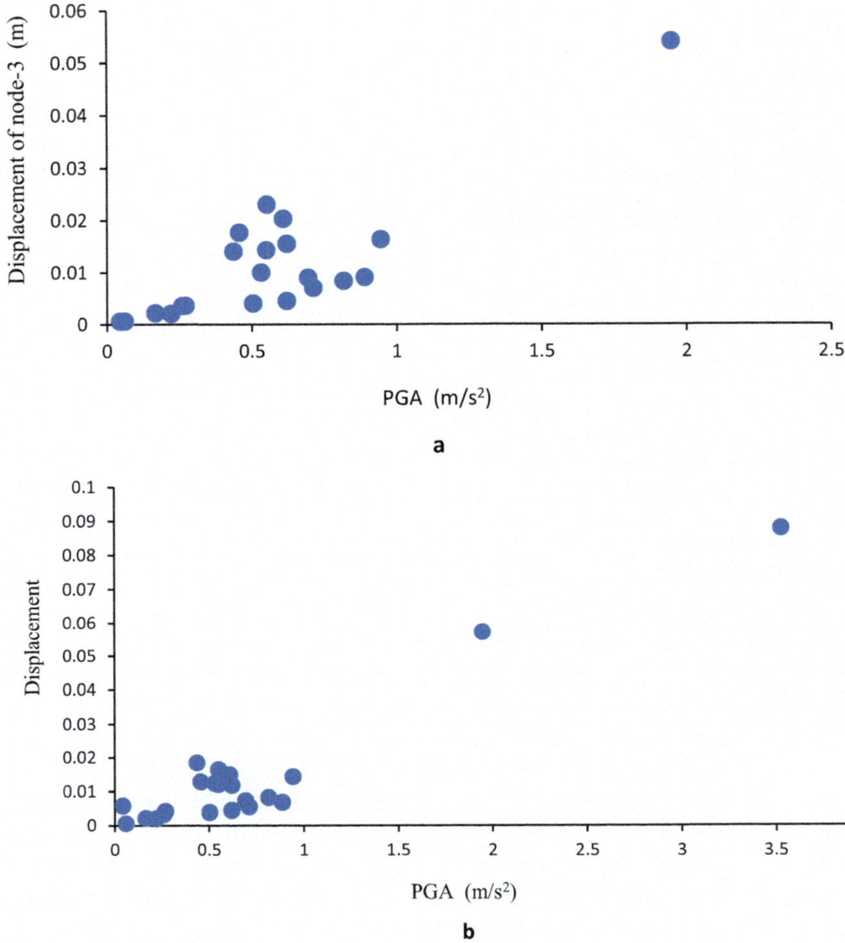

Fig. 13.7 Displacement of top node **a** fixed base **b** flexible base at different stations for the 1999 Chamoli Earthquake

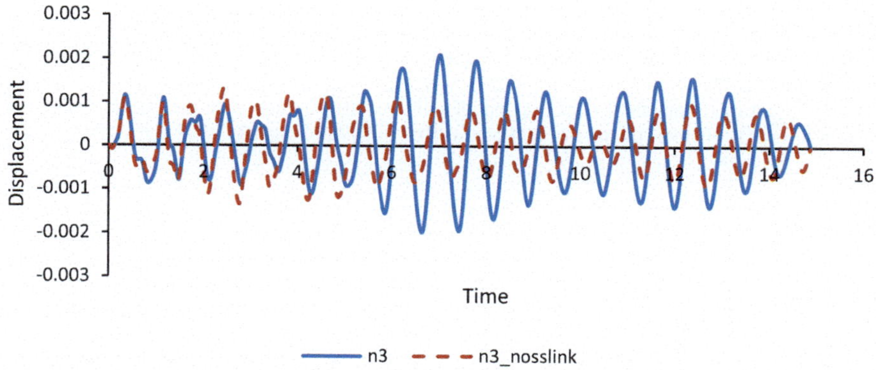

Fig. 13.8 Time history plot for Displacement at node 3 for Barkot T for the 1999 Chamoli Earthquake

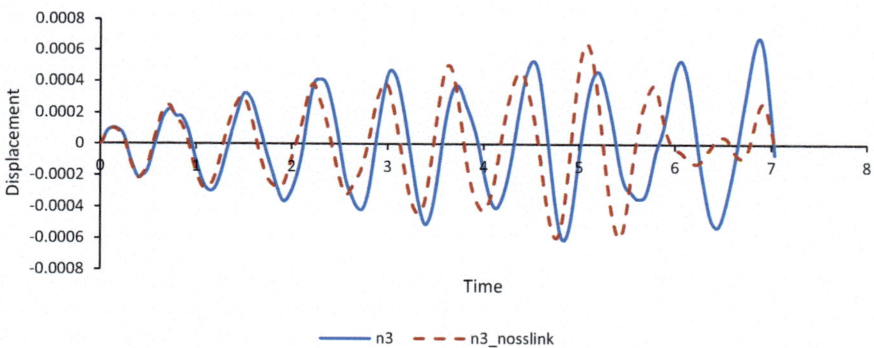

Fig. 13.9 Time history plot for Displacement at node 3 for Lansdowne-T for the 1999 Chamoli Earthquake

References

Allotey Nii, Hesham El Naggar M (2008) Generalized dynamic Winkler model for nonlinear soil–structure interaction analysis. Canad Geotechn J **45.4**:560–573. https://doi.org/10.1139/T07-106

Behnamfar F, Banizadeh M (2016) Effects of soil–structure interaction on distribution of seismic vulnerability in RC structures. Soil Dyn Earthq Eng 80:73–86. https://doi.org/10.1016/j.soildyn.2015.10.007

Kausel E (2010) Early history of soil–structure interaction. Soil Dyn Earthq Eng 30(9):822–832. https://doi.org/10.1016/j.soildyn.2009.11.001

Lou M, Wang H, Chen X, Zhai Y (2011) Structure–soil–structure interaction: literature review. Soil Dyn Earthq Eng 31(12):1724–1731. https://doi.org/10.1016/j.soildyn.2011.07.008

Mylonakis G, Gazetas G (2000) Seismic soil-structure interaction: beneficial or detrimental? J Earthquake Eng 4(3):277–301. https://doi.org/10.1080/13632460009350372

Oz I, Senel SM, Palanci M, Kalkan A (2020) Effect of soil-structure interaction on the seismic response of existing low and mid-rise RC buildings. Appl Sci 10(23):8357. https://doi.org/10.3390/app10238357

Wani FM, Vemuri J, Rajaram C (2022) Effect of soil structure interaction on the dynamic response of reinforced concrete structures. Nat Hazards Res. https://doi.org/10.1016/j.nhres.2022.11.002

Chapter 14
Great Earthquakes of the Central Seismic Gap Through Paleoseismological Perspective

Arjun Pandey and R. Jayangondaperumal

14.1 Introduction

The central Himalayan region encompasses the area of Punjab, Uttarakhand and western Nepal, spanning 500–800 km. Positioned in between the rupture areas of the 1905 CE Kangra and the 1934 CE Bihar-Nepal earthquakes, this section is commonly known as the "central seismic gap" (CSG) (Fig. 14.1) (Khattri and Tyagi 1983; Khattri 1987). There have been five earthquakes along the Himalayan arc during the last two hundred years which have devastated the Himalayan states (1905 CE Kangra, 1934 CE Bihar-Nepal, 1950 CE Tibet-Assam, 2005 CE Muzaffarabad, 2015 Gorkha earthquake). Hence, it is imperative to study the geological signatures of these events to estimate their respective co-seismic slips which will eventually help to calibrate the probabilistic seismic hazard in the region. The active fault studies in India have gained momentum during the last ~ 20 years with numerous paleoseismic trenches dug across the Himalayan front. However, these studies involve uncertainties related to the timing, surface rupture length, and rupture terminations of the historical events identified through paleoseismological studies.

The CSG has generally experienced minimal seismic activity in recent history, numerous significant earthquakes with magnitudes of > 8 have occurred along its western and eastern periphery, while the event in 1803 CE took place within the CSG. On the contrary, the CSG has vast historical archives mentioning many earthquake in the medieval to modern times which have devastated the region (Iyengar et al. 1999; Ambraseys and Douglous 2003).

In this chapter, we examine an expanded historical and paleoseismic dataset covering the central Himalayan region to sum up the earthquake catalog in the most densely populated region in the Himalayan foothills.

A. Pandey · R. Jayangondaperumal (✉)
Wadia Institute of Himalayan Geology, Dehradun, India
e-mail: ramperu@wihg.res.in; ramperu.jayan@gmail.com

B. K. Rastogi et al. (eds.), *Natural Hazards and Risk Mitigation*, Springer Transactions in Civil and Environmental Engineering, https://doi.org/10.1007/978-981-97-7658-0_14

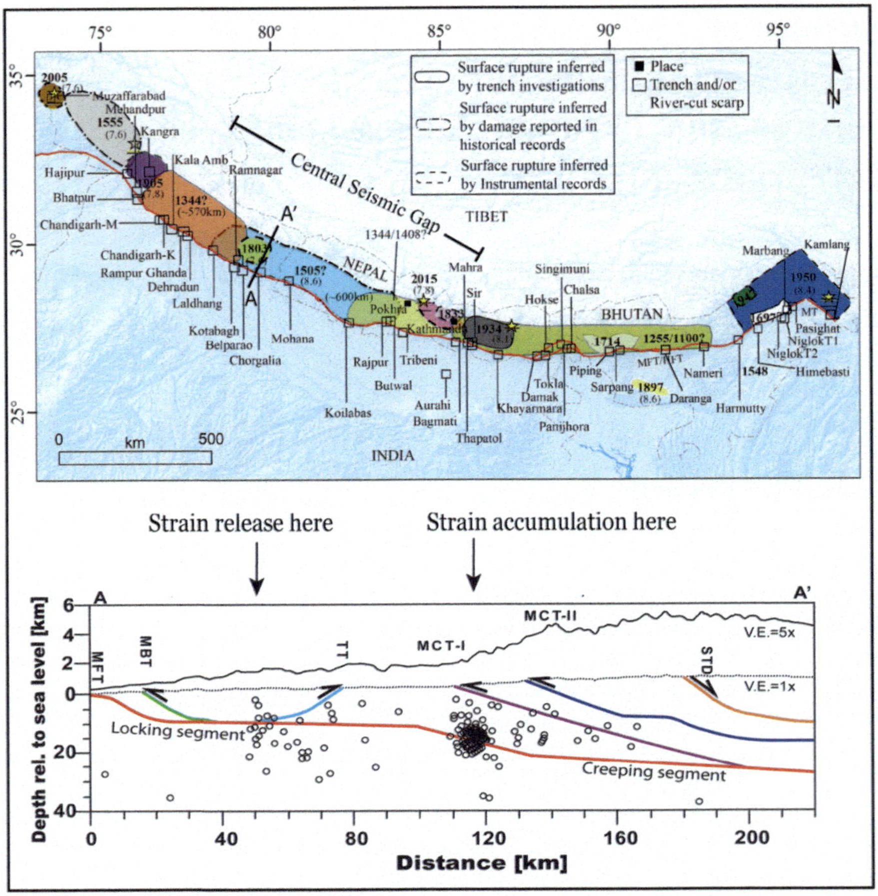

Fig. 14.1 **(Top)** Map showing the rupture zones of great earthquakes in the Himalaya and the locations of paleoseismological trenches excavated along the Himalayan front so far. The studies from west to east are—(Hajipur-Malik et al. 2010; Bhatpur-Kumhara and Jayangondaperumal 2013; Kothyari et al. 2023; Mehandpur-Jayangondaperumal et al. 2017; Pinjor Dun-Arora et al. 2019; Panchkula-Kumar et al. 2006; Kala Amb-Kumar et al. 2001; Rampur Ganda-Kumar et al. 2006; Laldhang-Kumar et al. 2006; Daniels et al. 2023; Ramnagar-Kumar et al. 2006; Padampur-Malik et al. 2016; Belparao-Rajendran et al. 2015; Mohana Khola-Yule et al. 2006; Botechaur-Hossler et al. 2016; Sukhe Taal-Riesner et al. 2021; Koilabas-Mugnier et al. 2005; Bhutwal-Okumara et al. 2018; Tribeni-Wesnousky et al. 2017; Bagmati-Wesnousky et al. 2017; Thapatol-Bollinger et al. 2014; Mahra-Bollinger et al. 2014; Sir-Sapkota et al. 2013; Khayramara-Wesnousky et al. 2019; Damak-Wesnousky et al. 2016; Tokla-Upreti et al. 2000; Hokse-Nakata et al. 1998; Singimuni-Upreti et al. 2007; Panijhora-Mishra et al. 2016; Chalsa-Kumar et al. 2006; Piping-Le Roux-Mallouf et al. 2020;Sarpang-Le Roux-Mallouf et al. 2016; Daranga-Zhao et al. 2021;Nameri-Kumar et al. 2006; Harmutty-Kumar et al. 2006; Himebasti-Pandey et al. 2021; Niglok-Priyanka 2018; Marbong-Jayangondaperumal et al. 2011; Pasighat-Priyanka 2017; Kamlang-Singh et al. 2021). (Bottom) S–N cross section across Himalaya demonstrating principal tectonic zones and faults. This seismotectonic model depicts the locked and ductile creeping segments of the MHT and zones of strain accumulation and release. The location of the cross section is shown as S–N line AA′ in the map above. (From Jayangondaperumal et al. 2018)

14.2 Historical and Pre-Historic Earthquakes in the Central Seismic Gap (CSG)

The central Himalayan region has quite extensive historical archives compiled by different contemporaneous historians and archeologists. Figure 14.2 shows the location of different historic and pre-historic earthquakes adapted from Ambraseys and Jackson (2003) and Iyengar et al. (1999). This section especially focuses on the earthquakes which shook the region in the past.

The 1223 Earthquake in Nepal

The earthquake of 24 December 1223 has been mentioned in Sanskrit and Newari chronicles of the fourteenth century. According to the description, the event occurred in the month of Pusha (December) in Nakshatra 344, corresponding to the date of 24 December 1224 CE. The damages done by the earthquake in Nepal were not mentioned but it was the first earthquake that was taken from authentic historical sources.

The 1255 Nepal Earthquake

The 7th June 1255 CE event is one of the most debated earthquake of the Kathmandu Valley. The earthquake as mentioned in the chronicles perished one-third of the population of the Kathmandu Valley. King Abhaya Malla himself died eight

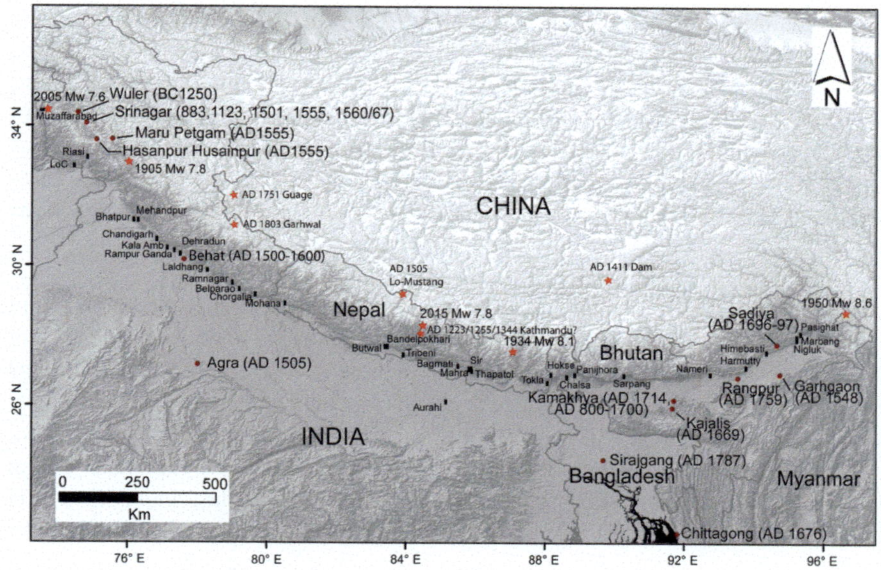

Fig. 14.2 Regional map of India showing Himalayan arc draped over G-Topo (SRTM-Shuttle Radar Topographic Mission) satellite imagery. The Map shows the location of historical earthquakes (red star) and the location of Paleoseismic trenches (Black dot) along with the country boundary of India, Nepal, Bangladesh, and Bhutan. The map has been modified after Iyengar et al. (1999) and locations of the earthquakes were taken from Ambraseys and Jackson (2003) and Iyengar et al. (1999)

days after the main shock. Strong shocks had intermittently shaken the whole Kathmandu valley for 15 days. The aftershocks continued for four months. According to the Newari and Sanskrit documents "the 3rd of the waxing moon of the month of Asadha, when the asterism was Punarvasu and the *yogo* was Dhruva, the Vikramasamvat (V.S.) year was 1310 and the Nepalasamvat (N.S.) year 374" calibrates to the year 1255 CE.

The 1344 Earthquake (Nepal)

The historical Newari chronicle mentions a fatal earthquake in the month of September 1344 CE. The more detailed description is not published or documented anywhere but Newari archives depict the death of the King next day after the calamity.

The 2 October 1411 CE (Damxung/Dam Earthquake Tibet)

The specific date of this earthquake is somewhat ambiguous, with some suggesting it occurred between the 28th and 29th of September in the year 1411 CE in the dBus province of Tibet. Its impact was felt in the region southeast of Lake Namtso and the Nyenchen Tanghla range, approximately 100 km northwest of Lhasa. Much of the information about this historical event is derived from Chinese and Tibetan archives.

The seismic activity commenced with a foreshock on the 11th day of the 9th lunar month around midnight (approximately 28th September 1411 CE), followed by the main shock at dawn the next day, 29th September 1411 CE. Several aftershocks occurred in the region five days later, but they did not cause any significant additional damage.

The majority of the earthquake's impact was concentrated in Tibetan sites within the dBus Province, particularly in Rin-spungs in the eastern gTsang region. In the Dam-gzhung region, the northernmost locality of Tibet, the biography of a prominent Lama recounts that this earthquake was substantial, causing the "running" of mountains, an expression likely associated with widespread landslide movement.

The tremors also inflicted damage in the sTod-lung valley, where both bDe-chen and the mTshur-phu monastery are situated. These locations underwent renovations in the year 1412 CE as a response to the earthquake's aftermath.

The 6 June 1505 Lo-Mustang Earthquake

The earthquake was documented by most of the Indian historians of that time and caused massive destruction up to the Indian Town of Agra in north India. According to medieval scholars, the earthquake caused mountains to crumble down, buildings were ruined and trees were uprooted and moved from their places. In the descriptions, it mentions that specifically Agra was badly affected (Bada uni, 'Abd al Qadir, 1596).

The fatality of the damage at Agra was further confirmed by Firishtah (Qasim Hindu-Shah 1610), who adds that the earthquake hit on a Sunday, which in 911 aH fell on 3rd of Safar or on 6th July, 1505 CE. He mentions that the violent earthquake leveled every lofty building in Agra to the ground while hundreds were buried in the ruins.

However, Babur's memoirs don't mention any earthquake in Agra but rather they spoke about an earthquake that severely affected the Kabul region of Afghanistan. Later on, the debate was settled as there were two separate events- one in Kabul as mentioned in Babur's memoirs (Jackson, D, Tibet 2002). This hypothesis was further been subjugated by several contemporary Tibetan documents. They mention that the earthquake which caused destruction in Agra was located in the Lo-Mustang area of Tibet which was ~ 1500 km from Kabul and ~ 500 km from Agra (Jackson, D, Tibet 2002). In the district of Lo-Mustang as dictated by the eyewitnesses, damages were enormous, almost every monastery and temple was destroyed, and tens of thousands of lives were lost. Damages were extended south of Lo up to the Kali Gandaki river valley in Nepal. The region of Guge, Purang, and the Hilly terrains of western Nepal were heavily devastated.

Several historians like Bada' uni (Bada' uni, 'Abd al Qadir 1596), Firishtah (Qasim Hindu-Shah 1610), Nizam al-Din and others reported violent damages at Agra but during the 1505 earthquake. The event further confirms the reports that Agra was later rebuilt in the same year by Sikandar Lodi (Hasan 1960).

The September 1555 (Srinagar) Earthquake

A violent earthquake occurred in the Srinagar town of present Jammu & Kashmir in the year 1555 CE in the month of September. A detailed report on this was compiled by Iyengar et al. (1999). An eyewitness named Suka mentioned the damage of this earthquake. He mentions that the earthquake occurred in Kashmir in the month of Ashvina of the 30th Laukik year (year 1555) close to midnight. The various travel routes were affected by lots of causalities in the area. The houses were severely damaged and many wooden houses just got scooped off the ground at the bank of Vitasta (Vesha River).

Another historian Nizam (Nizam al-Din Ahmed 1594) mentions the same earthquake and adds that many towns and villages in Kashmir were ruined during this event. The Villages named 'Jalu' and 'Dampur' swapped sites in during this river as they slid down the banks of the Bihat (Jhelam) river.

The earthquake caused severe damage in the Kashmir valley during the year 1555 CE and the damage. The damage extended toward the southeast side of the Kashmir Valley and a village named Mawar was completely destroyed by an enormous landslide with the loss of almost 600 people as mentioned by Anand Koul (1913).

The 1751 Guge Earthquake

The earthquake occurred in Tibet roughly 70 km northeast of the source of the 1803 Kumaun earthquake in India. The event is said to be one of the most destructive earthquake in the region, consisting of four shocks that damaged the temples in Daba County on the bank of Sutlej River. Many structures like buildings, offices, and halls were completely destroyed at Daba (31.28 N, 79.96 E) along with a small temple collapsed. Snow slides and avalanches were also reported from the region during the earthquake.

The 1803 Kumaon Earthquake

One of the modern earthquakes occurred in the northwestern Himalaya in the mountainous regions of the Kumaon Himalayas. British pioneers visited the affected regions soon after the earthquake and they documented the damage of the event. The earthquake had disastrous effects in the regions of Tehri Garhwal and Bashhar in the high Himalaya up to the Indo-Gangatic plain (McClelland 1857; Smith 1843).

Temples and houses were ruined at the northernmost extremity of the affected region at the bank of the Bhagirathi River. Considerable loss of life and property incurred in the region with ~ 200–300 livestock lost in the upper reaches of Uttarakhand Himalaya. The Badrinath temple was also affected due to the earthquake and many settlements were completely buried. The destruction was really widespread between the Joshimath and Karnprayag and a Shiva Temple at Panha (Karnaprayag) was ruined. The Rajah's palace was shattered in the capital city of Srinagar at Garhwal, many houses collapsed. Similar damages were also reported from the other regions of Garhwal Himalayas. In Delhi, it is mentioned that Kutb Minar was damaged during this earthquake and rebuilt after that.

The source of the earthquake was believed to be in the districts of Kumaon and Garhwal. No records for this earthquake were found from the Tibetan side.

The 1808 Nepal Earthquake

The following description has been quoted by Pant, 2002 as:

> In the Samvat year 1866 (N.S. 930) on the day of Dasallarii in the third prahara there occurred 21 earthquakes. Houses collapsed and some people and animals perished. Nothing happened to the temple of PaSupatiniitha. The Five Times Venerable Supreme King of Kings Girva~ayuddha Vikrarna Saba was on the royal throne. Since a revolt had spread in the region of Punjab (nowadays Simla district). Gorkha1ese troops were engaged in subduing it and slrengthening the rule. It appears that the earthquake posed no obstacle to the task of that struggle.

From the description, it is quite evident that the earthquake of June 4, 1808 didn't cause much trouble in the Nepalese realm, and even after 21 tremors it didn't cause any significant damage to the valley.

The Wular Lake (1250 BCE) Earthquake

This was one of the most disastrous earthquake in the Kashmir valley and according to the historian Pir Hasan Shah, the event took place during the reign of King Sundersena of the Pandu dynasty in 1250 BCE. The most startling inscription was written by one of the modern authorities in the history of Kashmir by G.D.M. Sufi (1974). He quoted that:

"In the reign of Sundersena, twenty-second in the line (Pandu Dynasty), a great earthquake clove opened the earth in the middle of the capital Samdhimat Nagar, and the whole city was submerged along with its king and inhabitants. The site of the city is now occupied by the Volur (Wular) lake." The king of the northeast side of Kashmir was then elected king of the region.

Given the area of roughly ~ 189 square km the MM intensity of this earthquake should have been XII.

Despite these large to great earthquakes in the western Himalayan region, there are several small earthquakes are tremors were also documented by historians which are later on complied by Iyengar et al. (1999), Ambraseys and Jackson (2003), and Pant (2002).

14.3 Paleoseismic Investigations in the Central Seismic Gap (CSG)

There have been a number of paleoseismic investigations as shown in Fig. 14.3 carried out in the CSG to identify the surface rupture signatures of large to great earthquakes. The section summarizes all the trench studies from Bhatpur in the western periphery of the CSG up to Tribeni in the east (Kumar et al. 2001; Mugnier et al. 2005; Kumar et al. 2006; Yule et al. 2006; Kumhara and Jayangondaperumal 2013; Rajendran et al. 2015; Hossler et al. 2016; Malik et al. 2016; Wesnousky et al. 2017; Jayangondaperumal et al. 2017, 2018; Okumara et al. 2018; Arora et al. 2019; Riesner et al. 2021; Daniels et al. 2023; Kumar et al. 2024 under preparation).

Bhatpur (about 150 km west of Chandigarh)

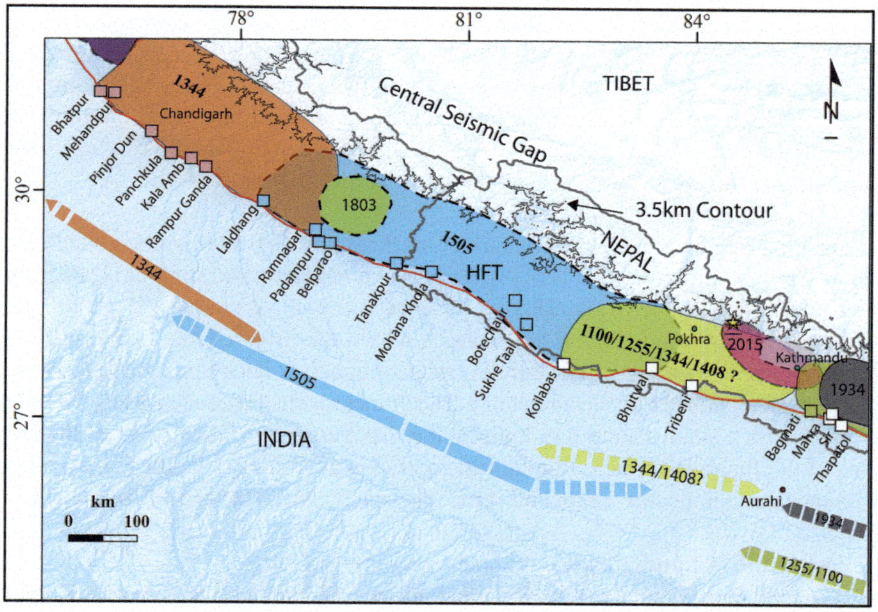

Fig. 14.3 Tectono-geomorphic map of the Central seismic Gap (CSG), showing the mesoseismal zone of 1344 CE, 1505 CE, 1255 CE, 1803 CE, 1934 CE, and 2015 CE earthquakes. The location of the different paleoseismic trenches has been also shown in the map. HFT-Himalayan Frontal Thrust. The trench studies from west to east are (Bhatpur-Kumhara and Jayangondaperumal 2013; Kothyari et al. 2023; Mehandpur-Jayangondaperumal et al. 2017; Pinjor Dun-Arora et al. 2019; Panchkula-Kumar et al. 2006; Kala Amb-Kumar et al. 2001; Rampur Ganda-Kumar et al. 2006; Laldhang-Kumar et al. 2006; Daniels et al. 2023; Ramnagar-Kumar et al. 2006; Padampur-Malik et al. 2016; Belparao-Rajendran et al. 2015; Tanakpur-Kumar et al. 2024 (under preparation); Mohana Khola-Yule et al. 2006; Botechaur-Hossler et al. 2016; Sukhe Taal-Riesner et al. 2021; Koilabas-Mugnier et al. 2005; Bhutwal-Okumara et al. 2018; Tribeni-Wesnousky et al. 2017; Bagmati-Wesnousky et al. 2017; Thapatol-Bollinger et al. 2014; Mahra-Bollinger et al. 2014; Sir-Sapkota et al. 2013)

To gather historical seismic activity data along the HFT in the southern region of the Janauri Anticline, Kumhara and Jayangondaperumal (2013) dug a 22-m-long and 5-m-deep trench. The trench was oriented in a N70°E direction, crossing the fault scarp between the elevated terrace and floodplain near Bhatpur village. The most recent surface movement showed a displacement of at least 9.3 m. Analysis of the deformed sediments suggests that the latest faulting event took place between 1400 and 1460 CE. Recently, Kothyari et al. (2023) investigated the back thrust situated at the northern limb of the Janauri Anticline (JA). Using Optically Stimulated Luminescence (OSL) dating they identified three seismic events at the Back Thrust which are: 535–1269 CE, the penultimate event 1447–1636 CE, and the most recent event 1660–1929 CE.

Mehandpur (east of Bhatpur)

The trench site is located south of Mehandpur village on the northern slope of the Janauri Anticline. At this location, the fault scarp runs in a northwest-southeast direction, perpendicular to the Hum Khad river trend. Jayangondaperumal et al. (2017) dug a trench approximately 16 m long, 8 m deep, and 10 m wide, oriented roughly north–south, crossing the lower scarp of the fluvial terrace. Examination of the trench log reveals that the fault intersects all geological units, and provides an Optically Stimulated Luminescence (OSL) modeled age of 800 $\pm$ 30 years ago. This suggests that the most recent earthquake event occurred after 1200 CE.

Pinjor Dun (near Chandigarh)

The Pinjore Dun is delineated by numerous thrust fault systems, among which the Pinjor Garden Fault, oriented NW–SE, acts as a boundary separating the Dun from the Upper Siwalik succession. This fault has caused both vertical displacement and backward tilting of the alluvial fan surfaces within the Pinjore Dun. It has resulted in irregular scarps facing WSW and SW, ranging in height between 2 to 16 m. Arora et al. (2019) conducted excavations of a trench measuring 4 m in depth and 10 m in length across a south-facing 6 m high fault scarp, situated ~ 100 m west of the left bank of the Jhajra River near Ratpur village. Radiocarbon dating of the faulted units from the trench indicates that the most recent event occurred between 1283 and 1443 CE.

Panchkula (Chandigarh)

A trench approximately 40 m long and about 4 to 5 m deep was dug by Kumar et al. (2006) across the fault where it intersects the Lower River terrace surface close to Chandigarh city. Situated within the Government College campus in Panchkula, the fault trends northwest and is marked by a roughly 2-m-high scarp. Radiocarbon dating of both faulted and unfaulted units exposed in the trench indicates that the displacement observed must have resulted from an event occurring after 1404–1628 CE.

Kala Amb

Kala Amb lies approximately 40 km southeast of the Chandigarh trench site as noted by Kumar et al. (2006). The course of the HFT experiences a significant rightward shift near the town of Kala Amb. This fault step is characterized as a tear fault, evident by a distinct scarp in the alluvial deposits that interrupts and raises terraces along the Markanda River. Kumar et al. (2001) present findings from a trench excavation revealing evidence of two, and potentially three, surface rupture earthquakes, resulting in a cumulative co-seismic slip of 8.6 m on the tear fault over the past 600 years. The two most recent earthquakes occurred after 1294 CE and 1423 CE, respectively, with a possible additional rupture around 260 CE. Minimum displacements during these two recent earthquakes were estimated to be approximately 5.3–5.4 m and 2.4–4.0 m, respectively, with potentially larger displacements observed during the 260 CE event.

Rampur Ganda

The Rampur Ganda study site is located ~ 20 km east of Kala Amb. Here the trace of the HFT is discontinuously expressed as fault scarps in Quaternary alluvium. A ~ 58 m long trench was placed across an east–west trending scarp near the outlet of the Somb River tributary near the small village of Rampur Ganda by Kumar et al. (2006). Dating of detrital charcoal samples from the trench places the most recent fault displacement event after 1222–1380 CE.

Laldhang

A trench was excavated by Kumar et al. (2006) perpendicular to a ~ 10 m high northwest-trending scarp near Laldhang village, revealing the presence of the HFT. The trench, measuring approximately 25 m in length and 4 to 6 m in depth, exposed seven litho-units. Radiocarbon dating from six samples suggests that displacement occurred between 1282 and 1632 CE. Subsequently, Daniels et al. (2023) conducted another trench excavation at the same scarp previously studied by Kumar et al. (2006). Radiocarbon dating of stratigraphic units, analyzed using the OxCal software program, narrowed down the timing of events to 1317–1391 CE and 1447–1572 CE, coinciding with historical earthquakes in 1344 CE and 1505 CE.

Ramnagar

Kumar et al. (2006) conducted a ~ 32-m-long trench across a fault scarp at Ramnagar, adjacent to the Karkat River. The trench exposes five distinct units, distinguished by variations in grain size and facies. Six detrital charcoal samples were collected from the trench exposure. Analysis of the youngest charcoal sample suggests that the displacement likely occurred after 1278–1390 CE.

Padampur

Malik et al. (2016) excavated a 14 m long, 3.5 m wide, and 3 m deep trench, situated along the scarp of the Kotabagh Fault at Dhol Village near the banks of the Dabka River. They collected seven detrital charcoal samples and six Optically Stimulated Luminescence (OSL) samples from various units exposed within the trench. Based on the obtained radiocarbon ages, they deduced the occurrence of at least three earthquakes along the Kotabagh Fault. The first event, referred to as Event I, took place within the range BCE 467– CE 570. Event II occurred between CE 1294 and CE 1587, likely corresponding to CE 1505. Finally, Event III transpired during

CE 1750 and CE 1932, representing the event around CE 1803. However, Rajendran et al. (2016) negated these findings and stated that only a single medieval earthquake ruptured at this site occurring between 1266 and 1636 CE. They further suggested that the 1803 CE earthquake rupture didn't reach the surface and somewhat arrested midway at the decollement flat.

Belparao

Additionally, Rajendran et al. (2015) placed a trench measuring 34 m in length, 3 m in width, and 4.5 m in depth at Belparao, located at the base of a thirteen-meter-high scarp situated between the Dabka and Karkat rivers. The team obtained radiocarbon dates from nine charcoal samples and two Optically Stimulated Luminescence (OSL) dates from sediment samples within the faulted units of the trench. Based on this chronology, they proposed that the initial rupture had likely occurred after 1050 CE, or potentially after 1215 CE, with the final rupture estimated to have taken place in the region between 1259 and 1433 CE.

Mohana Khola

Yule et al. (2006) conducted excavations of two trenches along tectonic scarps measuring 3 and 7 m in height in Mohana Khola, situated in far western Nepal. Radiocarbon dating of charcoal samples from deformed layers yielded ages ranging from ~ 1200 to 2800 years Before Present (BP) for sandy layers, and from ~ 1300 to 1630 CE for a soil layer. Furthermore, analysis of the organic content of the deformed soil, after removing charcoal, provided an age range of CE 1410 to 1470, confirming the displacement associated with the great Himalayan earthquake of 1505 CE.

14.4 Botechaur

Hossler et al. (2016) documented thrusting activity along a reactivated segment of the Main Boundary Thrust known as the Surkhet-Gorahi fault, near Botechaur village. They observed 8 m of displacement in the terraces of the Bheri River, indicating the occurrence of two earthquakes following approximately 1860 $\pm$ 30 years Before Present (BP) and 640 $\pm$ 30 years BP.

Sukhetal

Riesner et al. (2021) documented a normal displacement along the Surkhet-Gorahi segment of the Main Boundary Thrust (MBT) near Sukhetal in Western Nepal. This segment stretches approximately 120 km in length and lies 20–40 km north of the Main Frontal Thrust (MFT). The normal fault dips between 20° and 60°, with a maximum cumulative vertical offset measuring approximately 27 m across the fault. Radiocarbon dating of detrital charcoal samples collected from the hanging wall unit of the fault, colluvial wedge, and overlying strata indicates that the most recent faulting event occurred between 1400 and 1700 CE, which correlates with the significant earthquake of 1505 CE.

Koilabas

Mugnier et al. (2005) analyzed the strath terraces along the Koilabas River, which had been affected by faulting from the Himalayan Frontal Thrust (HFT). He associated approximately 8 m of slip with a fault dip ranging between 45° and 55°. A detrital charcoal sample obtained from the footwall zone of the thrust provided a radiocarbon age of 775 ± 35 years Before Present (BP), corresponding to the time frame between 1190 and 1285 CE.

Bhutwal

Okumara et al. (2018, AGU Abstract) excavated a trench across the HFT 12 km west of Bhutwal at Sorauli (27.70°N 83.36°E). The trench was opened across a small scarp on the tilted surface instead of at the boundary where the groundwater level is too high. There was no fault in the trench, but a fine sand layer was tilted toward the boundary. Preliminary analyses of radiocarbon dates of the tilted layer and the overlapping post-deformation sediments indicate the timing of the last faulting is between 1405 and 1467 CE.

Tribeni

Wesnousky et al. (2017) excavated a ~ 22 m long and ~ 7 m deep trench at Tribeni site at the Nepal Himalayas across a ~ 13 m high scarp. Six different litho-units were identified in the trench and ten detrital charcoal samples were collected to constrain the faulting event in the trench. The modeling of ages using pre and post-date samples in the trench limits the displacement any time between 1221 and 1262 CE.

14.5 Discussion and Conclusion

Paleoseismic studies carried out in the CSG uncover and claim the rupture signatures of some well-known medieval events like the 1100 CE Nepal earthquake, 1255 CE Nepal earthquake, 1344 CE Nepal earthquake, 1505 CE Lo-Mustang, and 1803 CE Garhwal-Kumaun earthquake. These earthquakes are well documented as discussed in the previous sections and are the best possible candidates for the causative seismic events in the CSG, except the 1100 CE earthquake as claimed by Lave et al. (2005) which has no historical documentation. Figure 14.4 shows the OxCal model summarizing all the trench studies carried out in the CSG.

At Bhatpur, Kumhara and Jayangondaperumal, (2013) interpreted the faulting event at this site at any time after 1400 CE. There have been only two earthquakes as per the historical archives which occurred after 1400 CE, the one is located in Nepal and occurred in 1408 CE (Rana, 1936; no properly mentioned source) and the other is 1505 CE Lo-Mustang earthquake (Fig. 14.3). At the Mehandpur paleoseismic site the most recent displacement was declared post-1200 CE, which makes, 1255, 1344, 1408 and 1505 CE earthquakes the most plausible candidates. The Oxcal Fig. 14.4 shows the calibrated ages of pre-dating the most recent event at sites from Bhatpur to Tribeni. As evident from Fig. 14.4 the most probable earthquakes which have ruptured the CSG in medieval times are 1255, 1344, and 1505 CE earthquakes. However, earthquakes like 1344 and 1505 have their sources near Kathmandu (Nepal)

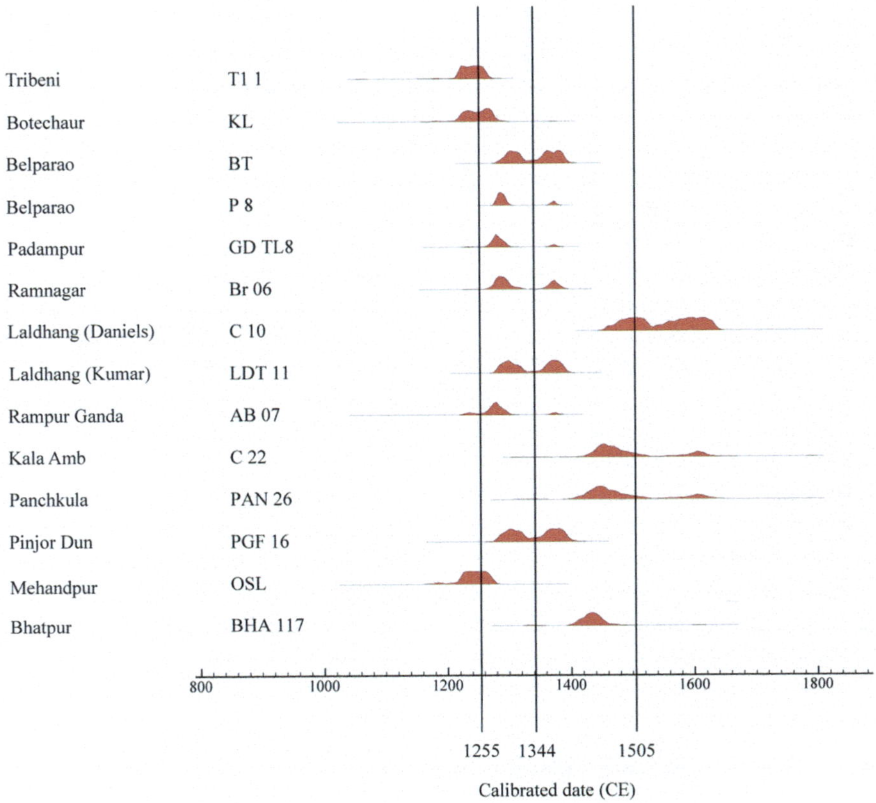

Fig. 14.4 Calibrated calendar ages for all the trenches in the CSG excavated across the HFT from west to east. The ages are calculated and modeled using OxCal version 4.4 (Ramsey et al., 2013) and IntCal13 atmospheric curves (Reimer et al., 2013)

and Lo-Mastang (Tibet), and it is very unlikely that they have reached up to the paleoseismic sites like Bhatpur and Mehandpur covering a distance of > 1000 km. However, Jayangondaperumal et al. (2017) modeled the ages using Oxcal and came up with the results that the western part of the CSG (From Bhatpur to Ramanagar) was ruptured by the 1344 CE earthquake (Fig. 14.4). The 1344 CE rupture has been further extended east up to Laldhang by Daniels et al. (2023). Given this scenario, it's either the same earthquake that devastated Kathmandu or some other contemporaneous pre-historic event that occurred and ruptured the western periphery of the CSG.

The Paleoseismic sites east of Laldhang encompass the deformation event at these sites post-1090 CE. This gives a window for earthquakes in and around Kathmandu to be during 1255, 1344, 1408, 1505, 1751, and 1803 CE to have ruptured the given segment. At Padampur, Malik et al. (2016) corroborated a three events scenario and postulated the penultimate event and most recent event at this site to be 1505 and 1803 CE, respectively. However, the results of Malik et al. (2016) were later on challenged by Rajendran et al. (2016) who emphasized that only a single medieval event occurred between 1266 and 1636 CE that ruptured this site and related it with either 1344 or 1505 CE earthquake. Also, they asserted that the 1803 earthquake had halted halfway across the basal flat, failing to reach the HFT similar to the blind thrust earthquakes such as the 1905 Kangra and the 1833 Nepal. However, Daniels et al. (2023) suggested that the 1803 CE event produced a minor offset in their trench at Laldhang and suggested it might be associated with a likely smaller or distant earthquake.

Studies at Mohana Khola at HFT and further east at MBT at Botechaur and Sukhetal depict the possibility of the 1505 CE earthquake. However, the expression at MBT was preserved in the form of normal faulting and it could be very well related to the post-earthquake seismic relaxation of the 1505 CE earthquake. The displacements at Koilabas and Bhutwal sites are claimed to have occurred because of the aftermaths of the 1255 and 1505 CE earthquakes. Further east at Tribeni, the rupture was initiated post-1221 and could possibly be related to any of the seismic events post-1221 CE.

Depending upon the paleoseismic studies carried out so far in the CSG as shown in Figs. 14.1 and 14.3 the rupture zone of major ruptures like 1344 CE, limited from Bharatpur to Laldhang, 1505 CE, western termination at Laldhang eastern extent unknown, 1255 CE, restricted to western Nepal only (Fig. 14.3). There is a number of historical earthquakes as mentioned in Sect. 2.0 and in Fig. 14.2, whose rupture are still enigmatic. However, the short time span between these earthquakes and the limitations of the radiocarbon dating technique we can't differentiate between the 1223/1255/1344/1505 CE events on the basis of present chronology. Also, the motive of any such paleoseismic investigation is not just to uncover the surface rupture of an event but also to estimate the amount of co-seismic slip it has released which is essential to estimate the future seismic hazard of the area, which is lacking in the CSG. More detailed and closely spaced trenches are warranted in the CSG using more than one chronological technique which will provide a better picture of strain distribution and the probabilistic seismic hazard in the CSG.

References

Ambraseys NN, Jackson D (2003) A note on early earthquakes in northern India and southern Tibet. Curr Sci 84:570–582

Anand Koul P (1913) Geography of the Jammu and Kashmir State. Thacker, Calcutta, p 37

Arora S, Malik JN, Sahoo S (2019) Paleoseismic evidence of a major earthquake event (s) along the hinterland faults: Pinjore Garden Fault (PGF) and Jhajra Fault (JF) in northwest Himalaya, India. Tectonophysics 757:108–122

Bada uni, 'Abd al Qadir (c. 1596) Muntakhab al-tawarikh (eds Ahmed Ali and Lees, W.), i. 319–20, Calcutta, 1864–1869

Bollinger L, Sapkota SN, Tapponnier P, Klinger Y, Rizza M, van Der Woerd J, Bes de Berc S (2014) Estimating the return times of great Himalayan earthquakes in eastern Nepal: evidence from the Patu and Bardibas strands of the Main Frontal Thrust. J Geophys Res Solid Earth 119(9):7123–7163

Daniels RL, Niemi TM, Jayangondaperumal R, Aravind A, Rautela P, Pandey A, Murphy LD (2023) Late Medieval seismicity on the Himalayan Frontal Thrust Fault at Lal Dhang, Uttarakhand. Tectonophysics, India, p 229934

Firishtah, Qasim Hindu-Shah (1610) Tarikh-I Firishtah (ed. Briggs, J.), Lucknow 1865, Transl; Dow, A., The History of Indostan, London, 1803, vol. 2, p. 155

Hasan N (1960) The Encyclopedia of Islam. London 1:252–254

Hossler T, Bollinger L, Sapkota SN, Lavé J, Gupta RM, Kandel TP (2016) Surface ruptures of large Himalayan earthquakes in Western Nepal: evidence along a reactivated strand of the Main Boundary Thrust. Earth Planet Sci Lett 434:187–196

Iyengar RN, Sharma D, Siddiqui JM (1999) Earthquake history of India in medieval times. Indian J. Hist. Sci. 34:181–237

Jackson D (2002) in Tibet, Past and Present (ed. Blezer, H.), Brill's Tibetan Studies Library I, Leiden, pp 147–159

Jayangondaperumal R, Wesnousky SG, Choudhuri BK (2011) Near-surface expression of early to late Holocene displacement along the northeastern Himalayan frontal thrust at Marbang Korong Creek, Arunachal Pradesh, India. Bull Seismol Soc Am 101(6):3060–3064

Jayangondaperumal R, Daniels RL, Niemi TM (2017) A paleoseismic age model for large-magnitude earthquakes on fault segments of the Himalayan Frontal Thrust in the Central Seismic Gap of northern India. Quatern Int 462:130–137

Jayangondaperumal R, Thakur VC, Joevivek V, Rao PS, Gupta AK (2018) Active tectonics of Kumaun and Garhwal Himalaya. Springer, Singapore, p 150

Khattri KN (1987) Great earthquakes, seismicity gaps and potential for earthquakes along the Himalayan plate boundary. Tectonophysics 38:79–92

Khattri KN, Tyagi AK (1983) Seismicity patterns in the Himalayan plate boundary and the identification of areas of high seismic potential. Tectonophysics 96:19–29

Kumahara Y, Jayangondaperumal R (2013) Paleoseismic evidence of a surface rupture along the northwestern Himalayan Frontal Thrust (HFT). Geomorphology 180:47–56

Kumar S, Wesnousky SG, Rockwell TK, Ragona D, Thakur VC, Seitz GG (2001) Earthquake recurrence and rupture dynamics of Himalayan frontal thrust. India Sci 294:2328e2331

Kumar S, Wesnousky SG, Rockwell TK, Briggs RW, Thakur VC, Jayangondaperumal R (2006) Paleoseismic evidence of great surface rupture earthquakes along the Indian Himalaya. J Geophys Res 111:B03304. https://doi.org/10.1029/2004JB003309

Kumar S, Wesnousky SG, Jayangondaperumal R, Nakata T, Kumahara Y, Singh V (2010) Paleoseismological evidence of surface faulting along the northeastern Himalayan front, India: timing, size, and spatial extent of great earthquakes. J Geophy Res Solid Earth, 115(B12)

Lavé J, Yule D, Sapkota S, Basant K, Madden C, Attal M, Pandey R (2005) Evidence for a great medieval earthquake (~ 1100 AD) in the central Himalayas, Nepal. Science 307(5713):1302–1305

Le Roux-Mallouf R, Ferry M, Ritz JF, Berthet T, Cattin R, Drukpa D (2016) First paleoseismic evidence for great surface-rupturing earthquakes in the Bhutan Himalayas. J Geophys Res Solid Earth 121(10):7271–7283

Le Roux-Mallouf R, Ferry M, Cattin R, Ritz JF, Drukpa D, Pelgay P (2020) A 2600-yr-long paleoseismic record for the Himalayan Main Frontal Thrust (Western Bhutan). Solid Earth Discuss 2020:1–36

Malik JN, Sahoo AK, Shah AA, Shinde DP, Juyal N, Singhvi AK (2010) Paleoseismic evidence from trench investigation along Hajipur fault, Himalayan Frontal Thrust, NW Himalaya: implications of the faulting pattern on landscape evolution and seismic hazard. J Struct Geol 32(3):350–361

Malik JN, Naik SP, Sahoo S, Okumura K, Mohanty A (2017) Paleoseismic evidence of the CE 1505 (?) and CE 1803 earthquakes from the foothill zone of the Kumaon Himalaya along the Himalayan Frontal Thrust (HFT), India. Tectonophysics 714:133–145

McClelland, in History of India (ed. Elphinstone, M.) (1857) vol. 3, appendix

Mishra RL, Singh I, Pandey A, Rao PS, Sahoo HK, Jayangondaperumal R (2016) Paleoseismic evidence of a giant medieval earthquake in the eastern Himalaya. Geophys Res Lett 43(11):5707–5715

Mugnier JL, Huyghe P, Gajurel AP, Becel D (2005) Frontal and piggy-back seismic ruptures in the external thrust belt of western Nepal. J Asian Earth Sci 25:707e717

Nakata T, Yagi H, Okumura K, Upreti BN, Rockwell TK, Virid NS, Maemoku H, Kumahara Y (1998) First successful paleoseismic trench study on active faults in the Himalaya. Fall meeting, American Geophysical Union, San Francisco

Nizam al-Din Ahmed, Rabaqat-I Akbari, Bibli. Ind., 1913–1931 [+1594]

Okumura K, Pokhrel P, Kondo H, Furuhashi T, Sapkota SN (2018) The last surface rupture event on the Himalayan frontal thrust in central Nepal near Butwal. In *AGU Fall Meeting Abstracts* (Vol. 2018, pp. T22A-05)

Pandey A, Jayangondaperumal R, Hetényi G, Priyanka RS, Singh I, Srivastava P, Srivastava HB (2021) Establishing primary surface rupture evidence and magnitude of the 1697 CE Sadiya earthquake at the Eastern Himalayan Frontal thrust, India. Sci Rep 11(1):879

Pant MR (2002) A step toward a historical seismicity of Nepal. Adarsa 2:29–60

Priyanka RS, Jayangondaperumal R, Pandey A, Mishra RL, Singh I, Bhushan R, Bhat GR (2017) Primary surface rupture of the 1950 Tibet-Assam great earthquake along the eastern Himalayan front India. Sci Rep 7(1):5433

Priyanka SR (2018) Surface Rupture Investigations of The 1950-Meisoseismal Zone of Assam Earthquake Along Himalayan Foothill Zone, Arunachal Pradesh-Assam Himalaya, PhD thesis, Kumaun University

Rajendran CP, John B, Rajendran K (2015) Medieval pulse of great earthquakes in the central Himalaya: viewing past activities on the frontal thrust. J Geophy Res Solid Earth 120(3):1623–1641

Rajendran CP, John B, Anandasabari K, Sanwal J, Rajendran K, Kumar P, Chopra S (2018) On the paleoseismic evidence of the 1803 earthquake rupture (or lack of it) along the frontal thrust of the Kumaun Himalaya. Tectonophysics 722:227–234

Ramsey CB, Scott EM, van der Plicht J (2013) Calibration for archaeological and environmental terrestrial samples in the time range 26–50 ka cal BP. Radiocarbon 55(4):2021–2027

Rana BS (1936) Mahabhukampa, Bahar Mahal BS1992, Kathmandu., pp 1–250

Reimer PJ, Bard E, Bayliss A, Beck JW, Blackwell PG, Ramsey CB, Van Der Plicht J (2013) IntCal13 and Marine13 radiocarbon age calibration curves 0–50,000 years cal BP. Radiocarbon 55(4):1869–1887

Riesner M, Bollinger L, Hubbard J, Guérin C, Lefèvre M, Vallage A, Sapkota SN (2021) Localized extension in megathrust hanging wall following great earthquakes in western Nepal. Sci Rep 11(1):21521

Sapkota SN, Bollinger L, Klinger Y, Tapponnier P, Gaudemer Y, Tiwari D (2013) Primary surface ruptures of the great Himalayan earthquakes in 1934 and 1255. Nat Geosci 6(1):71–76

Singh I, Pandey A, Mishra RL, Priyanka RS, Brice A, Jayangondaperumal R, Srivastava V (2021) Evidence of the 1950 great Assam earthquake surface break along the Mishmi Thrust at Namche Barwa Himalayan Syntaxis. Geophy Res Lett 48(11):e2020GL090893

Smith RB (1843) J Asiat Soc Bengal 11:1046–1054

Upreti BN, Nakata T, Kumahra Y, Yagi H, Okumura K, Rockwell TK, Virdi NS, Maemoku H (2000) The latest active faulting in southeast Nepal. Proceedings of the Hokudan International symposium and School in active faulting, 17–26, pp 533–536

Upreti BN, Kumahara Y, Nakata T (2007) Paleoseismological study in the Nepal Himalaya–present status. In *Proceedings of the Korea-Nepal Joint symposium on slope stability and landslides* (pp 1–9)

Wesnousky SG, Kumahara Y, Chamlagain D, Pierce IK, Karki A, Gautam D (2017a) Geological observations on large earthquakes along the Himalayan frontal fault near Kathmandu, Nepal. Earth Planet Sci Lett 457:366–375

Wesnousky SG, Kumahara Y, Chamlagain D, Pierce IK, Reedy T, Angster SJ, Giri B (2017b) Large paleoearthquake timing and displacement near Damak in eastern Nepal on the Himalayan Frontal Thrust. Geophys Res Lett 44(16):8219–8226

Wesnousky SG, Kumahara Y, Chamlagain D, Neupane PC (2019) Large himalayan frontal thrust paleoearthquake at Khayarmara in eastern Nepal. J Asian Earth Sci 174:346–351

Yule D, Dawson S, Lave J, Sapkota S, Tiwari D (2006). Possible evidence for surface rupture of the Main Frontal Thrust during the great 1505 Himalayan earthquake, far-western Nepal. In *AGU Fall Meeting Abstracts* (Vol. 2006, pp. S33C-05)

Zhao Y, Grujic D, Baruah S, Drukpa D, Elkadi J, Hetényi G, Welte C (2021) Paleoseismological findings at a new trench indicate the 1714 M8. 1 earthquake ruptured the main frontal thrust over all the Bhutan Himalaya. Front Earth Sci 9:689457

GPSR Compliance

The European Union's (EU) General Product Safety Regulation (GPSR) is a set of rules that requires consumer products to be safe and our obligations to ensure this.

If you have any concerns about our products, you can contact us on ProductSafety@springernature.com

In case Publisher is established outside the EU, the EU authorized representative is:

Springer Nature Customer Service Center GmbH
Europaplatz 3
69115 Heidelberg, Germany

The manufacturer's authorised representative in the EU is Springer Nature Customer Service Centre GmbH, Europaplatz 3, 69115 Heidelberg, Germany. If you have any concerns regarding our products, please contact ProductSafety@springernature.com

Printed and bound by CPI Group (UK) Ltd, Croydon, CR0 4YY

30/09/2026

02225298-0004